Advanced Wireless Communications and Networks

Advanced Wireless Communications and Networks

Edited by **Bernhard Ekman**

New York

Published by NY Research Press,
23 West, 55th Street, Suite 816,
New York, NY 10019, USA
www.nyresearchpress.com

Advanced Wireless Communications and Networks
Edited by Bernhard Ekman

International Standard Book Number: 978-1-63238-026-5 (Hardback)

Contents

Preface

This book has been a concerted effort by a group of academicians, researchers and scientists, who have contributed their research works for the realization of the book. This book has materialized in the wake of emerging advancements and innovations in this field. Therefore, the need of the hour was to compile all the required researches and disseminate the knowledge to a broad spectrum of people comprising of students, researchers and specialists of the field.

This book is a compilation of recent and most popular innovations from the lowest layers to the upper layers of wireless communication networks and consists of "real-time" research developments. The information in this book has been systematically organized in order to make it easily accessible to the readers of all levels. It also preserves the balance between the recent research results and their theoretical support. A huge variety of new techniques in this field are investigated in this book. The authors attempt to present these topics in detail under the following sections wireless communication performance analysis - tools and methods, next generation communication technologies, biological effects of wireless communication, and wireless sensor networks and MANETS. Intelligent and reader-friendly elucidations are provided in this book to serve the readers of all levels, ranging from knowledgeable and practicing communication engineers to beginners or professional researchers.

At the end of the preface, I would like to thank the authors for their brilliant chapters and the publisher for guiding us all-through the making of the book till its final stage. Also, I would like to thank my family for providing the support and encouragement throughout my academic career and research projects.

Editor

Part 1

Wireless Communication Performance Analysis Tools and Methods

Engineering of Communication Systems and Protocols

Pero Latkoski and Borislav Popovski
Faculty of Electrical Engineering and
Information Technologies / Ss Cyril and Methodius University – Skopje
Macedonia

1. Introduction

The complexity of the communication systems and protocols is increasing constantly, while the communication products' time-to-market is becoming shorter. Afterthoughts communication system redesign due to lack of performance is financially and time expensive, and it is unacceptable. This book chapter proposes a method for improving the telecommunication systems, by means of enhancement the performance of the protocols they rely on. The proposed engineering of communication systems is based on a formal method and it provides an early-phase performance evaluation of the underlying communication protocols. The methodology is illustrated through a hands-on case study conducted on an existing wireless communication system.

The development and standardization of new telecommunication and information technologies is a rather complex process which requires a comprehensive framework (Sherif, 2001). The result of such a process is a new agreement that should satisfy all of the involved parties, such as: vendors, providers, and most importantly service users. To create a comprehensive communication standard and consequently a reliable communication system, many strategic and tactical issues need to be considered. The missing question is how to produce a standard that specifies a protocol or a system with high performance. The lack of performance issue might be a major cause for pitfall of entire communication systems. Most of the problems result from poor protocol specifications or from its enormous complexity. Furthermore, the design errors caused by the short and intensive creation period usually remain hidden until the testing and implementation phases of the communication product development. Fixing the problems after product's delivery for communication software and hardware increases the cost of the product by factor of 100 to 1000 compared to the fixing of the problem in the analysis phase.

High performance communication protocols which are untainted of functional errors are crucial in the telecommunications sector where product expectation cycle is denominated in decades instead of years. In order to develop such a protocol, two aspects should be fulfilled: introduction of formal methods during the specification process and integration of the performance-related activities in the early phases of the communication system specification and development. The former one is already taking place as a result of the need for clarity and accuracy in the telecommunication standards, but the last aspect is commonly avoided or even neglected.

The formal methods are always advised for the development process when early functional error detection is needed. Formal Descriptive Techniques (FDTs) provide corrective actions in the more abstract phases by introducing formal syntax and what is more important, precise semantics. In combination with the computer-aided software engineering, FDTs offer a delivery of better communication protocols and systems, sooner. The introduction of the FDTs has brought correctness and reliability into the protocol development, which has been recognized long time ago (Wing, 1990), (Hall, 1990). Today there are many formal languages and tools used in the protocol development process: Specification and Description Language-SDL (SDL, 2011), Simple ProMeLa Interpreter (Spin), Estelle (Estelle, 1989), Language of Temporal Ordering Specifications (LOTOS, 2000), Petri Nets (Petri, 1996), Uppaal (Larsen, 1997), Message Sequence Chart (MCS, 2001) and Unified Modelling Language (Booch, 2000). Among them, SDL has achieved widespread success because of its friendly graphical notation, its standardization by the International Telecommunication Union (ITU-T) as the major specification tool for standards and protocols, and because of its support for other popular notations such as ASN.1 (ASN.1, 1993), MSC and TTCN (TTCN, 2006). The effectiveness of SDL and its ability to develop unambiguous protocols have won it a widespread popularity and have led the standardization institutes, such as ETSI (European Telecommunications Standards Institute) (ETSI), 3GPP (Third Generation Partnership Project) (3GPP) and IEEE (The Institute of Electrical and Electronics Engineers) (IEEE) to include SDL diagrams in their protocols specification. SDL also provides powerful analysis of communication protocols, along with design, comprehensive modelling, protocol prototyping, exhaustive validation and verification, and all that by a user-friendly graphical notation. Along with Message Sequence Chart (MSC) description language, SDL is the most widely used FDT not only in the communication protocol specification area, but also in the industry systems engineering domain. Because of the previously stated advantages, SDL was selected as a protocol description method for the purpose of this chapter's analysis.

The aim of this chapter is to emphasize the importance of conducting an early performance evaluation of the communication protocols and systems, and to suggest an appropriate solution for carrying out such an activity. Performance evaluation activity denotes the actions to evaluate the protocol under development regarding its performance. This process can take place in different phases of the development, and can be based on modelling or measurements. If the designer can control the performance of the product, rather than just manage its functionality, the result will be a much superior creation. This problem is treated in this chapter through a tangible wireless communication protocol example.

The chapter is organized as follows. Section 2 presents the most relevant and most recent work which relates to the target topic of the chapter. In Section 3, the proposed and used methodology is elaborated in details. This methodology is demonstrated in Section 4, where a real engineering problem is provided, involving an IEEE 802.16 wireless communication protocol. Section 5 contains the conclusions of the chapter.

2. Related work

The following section provides an overview of what has been done by other researchers, related to the chapter's topic. Only a part of the most relevant and most recent work has been selected, which is needed for proper introduction of the proposed methodology in Section 3 and for presentation of the example in Section 4.

Engineering of a communication system means to describe, to analyze and to optimize the dynamic, time dependent behavior of the system and its inherent communication protocols. However, as it says in (Mitschele-Thiel, 2001) it is common for a system to be fully designed and functionally tested before an attempt is made to determine its performance characteristics. But it is a necessity to integrate the performance engineering into the design process from the very beginning. In (Mitschele-Thiel, 2001) the author addresses an improvement of the run-time properties by taking into account the characteristics of the applications (communication protocols) and different process scheduling and management strategies. The author concentrates on efficient implementation of behavioural concepts. For the treatment of issues arising from object-oriented concepts the author applies the traditional flattening approach of the language standard. Finally, it is obvious that the book lacks of actual communication system engineering examples, through which the engineering process would have been successfully explained.

The usage of FDT for protocol development has also arisen as a promising way of dealing with the increasing complexity of next generation mobile protocols. In (Showk, 2009) a rudimentary version of the Long Term Evolution (LTE) protocol for the access stratum user plane is modelled using SDL. The LTE radio communication is the upgrade of the current 3G mobile technology with a more complex protocol in order to enable very high data rates. This related work presents a tool which shows easy understanding of the model as well as easy testing of its functionality by simulation in cooperation with Message Sequence Chart. The simulation result presented in (Showk, 2009) shows that the implemented SDL guarantees a good consistency with the target scenarios. The system implementation is mapped to multiple threads and integrated with the operating system to enable execution in multi core hardware platforms. The only obvious drawback of the work is the usability of the created model, as it is only used for functional validation and not for performance evaluation of the analyzed communication protocol.

When developing modern communication systems, the energy consumption is a major concern, especially in the case of wireless networks consisting of battery-powered nodes. In (Gotzhein, 2009) the authors study possibilities of specifying energy aspects in the system a designing phase, with SDL as design language. In particular, they strive for suitable abstractions, by establishing a design view that is largely platform-independent. This objective is achieved by identifying and realizing energy mode signalling and energy scheduling as two complementary approaches to incorporate energy aspects into SDL. A case study illustrates the use of both approaches in a wireless networked control system. These approaches are applied and tested on a hardware platform, but again, the paper does not provide in a sufficient manner any performance metrics of the implemented wireless network.

The security of communication systems is another important aspect which must be considered in the protocol development. In order to study this aspect, (Lopez 2005) have developed a methodology for application of the formal analysis techniques, commonly used in communication protocols, to the analysis of cryptographic protocols. In particular, (Lopez, 2005) have extended the design and analysis phases with security properties. This related work uses a specification notation based on one of the most commonly used standard requirement languages HMSC/MSC, which can be automatically translated into a generic SDL specification. The obtained SDL system can then be used for the analysis of the addressed security properties, by using an observer process scheme. Besides the main goal to provide a notation for describing the formal specification of security systems, (Lopez, 2005) studies the possible attacks to the system, and the possibility of re-using the produced specifications to describe and analyse more complex systems.

The related work (Chen Hui, 2010) analyzes a Networked Control System (NCS), which governs the communication activities and directly affects the communication Quality of Service (QoS). Full or partial reconfiguration of protocol stack offers both optimized communication service and system performance. (Chen Hui, 2010) proposes a formal approach for the design and implementation of reconfiguration protocol stack based on Specification and Description Language for NCS. In Telelogic TAU environment, detail SDL models to support communication and reconfiguration functions of communication link layer, network transmission layer and application layer are discussed respectively. Similarly to the most of the presented related papers, only MSC verification results validate the effectiveness of the reconfiguration concepts of the protocol implementation for NCS.

The methodology which is presented in the following section differs from all previously presented work, as it extends the performance evaluating aspect of the communication systems engineering process. The methodology tries to maintain the functional correctness efforts regarding developing communication entity (similarly to the most of the related work), but at the same time provides the developers with a realistic insight of its performance capabilities.

3. Methodology overview

In order to obtain a performance evaluation-based analysis of telecommunications protocols and systems, the proposed methodology extracts all the necessary information from the available form of the analyzed standard. Taking into consideration that we are talking about an early stage of communication system development, the standard for such particular system is usually available as a draft version which combines the work provided by different working groups. The aim is to build an appropriate model from which the performance of the communication system will be assessed. This kind of model is commonly referred to as performance evaluating model. The communication standard under evaluation generally contains three basic parts: textual, SDL-represented and MSC-represented. Depending on the standard and standardization body, the proportion of these parts can vary. Mainly, the textual part dominates as "in the standardization process, words are still the final product" (Sherif, 1992).

But there is a major setback of the textual part of the standards caused by its inherent ambiguousness which is more deeply related to the natural languages' doublethink. This is the reason why the textual part of standards lacks of scientific foundation and is commonly the reason for miss-communication between standard developers, regarding the communication system requirements and expectations.

On the other hand, the SDL and MSC parts of the standard introduce more rigorous protocol or system specification, brought to a mathematical precision, and these formal parts of the standard are the guarantee for a correct system requirements presentation, as well as for an unambiguous definition of the system behavior. As it was previously said, different standards contain variable amount of formal representation, e.g. IEEE 802.16 (WiMAX, 2010) contains only sequences of formal protocol behavior, IEEE 802.11 (WiFi, 2007) encloses entire Medium Access Control (MAC) Layer presented in SDL, and IEEE 802.15 (Bluetooth, 2005) is completely offered in formal representation.

Using the three basic types of standard representations, the proposed methodology creates a so called behavior model of the analyzed standard, as it is presented in Fig. 1. The behavior model describes the protocol behavior and its abstract data structure by using SDL. In particular, the behavior model evaluates the relationship of single stimuli - response pair applied to the analysed protocol stack. This model is ideal for testing of protocol entity's functional correctness.

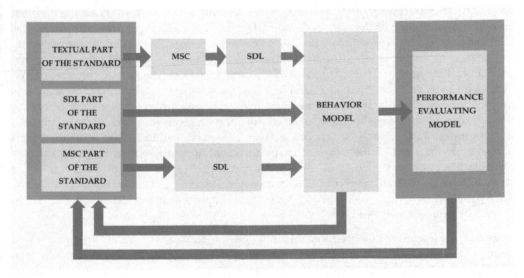

Fig. 1. Transformation of a communication standard into a performance model.

As it can be seen in Fig. 1, the last step of the methodology is the conversion of the behavior model into a so called performance evaluating model. The performance model can emulate a real scenario of communicating devices implementing the described protocol. It is built into a standalone executable that embeds the created behavior model and channelizes its preciseness into an accurate event driven type of simulator. This type of simulator is used for performance evaluation of the communication system or protocol, where every change and tuning of the specification can be easily evaluated. For instance one can measure the achieved data throughput or delay when a group of protocol based devices are communicating between each other.

Both, behavior and performance models provide valuable information regarding the functional and performance issues of the developing communication system, which is then looped-back to the specification process for further improvement of the system.

3.1 Detailed steps in the methodology

The behavior model can be defined as a SDL representation of the requirements, behavior and capabilities of the specified system or protocol. As it was explained earlier, it is created using all three representation parts of standards (text, SDL, and MSC). The SDL part of a standard is easily incorporated in the behavior model. This is the reason why the SDL part of the standard formulates the backbone of this model. Obviously it is the most mathematically rigorous component of the behaviour model. The MSC part of the standard includes all signal exchange sequences occurring among the protocol entities defined by the standard. To incorporate this part into the behavior model, it is necessary first to convert the MSCs into SDL code sequences. Although automated tools for such a conversion exist, the manual step-by-step translation is preferred, as the SDL code produced by the automated tools is not optimized, and also for providing nomenclature and stile consistency of the SDL code. The trickiest part of the behavior model development is to convert all the informal textual representation of the standard into SDL code. This is an unavoidable step, as long as all the missing parts of a complete functional system description are given in a textual form. The SDL code sequences produced from the textual part of the standard act as a glue that connects all the previously created parts of the behavior model. The need to convert text into SDL is also unavoidable because of the fact that for most of the communication standards the SDL and MSC parts are supplementary, while the textual part is mandatory. These are the reason why this step should be taken as the one with greatest importance. Additionally, the communicating signals which are exchanged among entities, along with the signal parameters, are most of the times constructed according to the text of the standard. The practice have shown that it is much easer if the textual represented requirements of the standard are firstly translated into MSC sequences, and after that converted into SDL code. This principle proves to be especially suitable for capturing the complex communication system or protocol behavior.

For the completeness of the previous explanation, this section will present an example of a generic behavior model of an abstract communication standard (Fig. 2). As it can be seen in Figure 2, SDL copes with the protocol complexity by using a hierarchical decomposition and by implying several levels of abstraction. The highest level of abstraction is called the system level. The system level is composed of multiple SDL blocks connected through unidirectional or bidirectional SDL channels. SDL channels are transferring SDL signals, which can carry additional signal parameters. Inside the SDL blocks lays the second level of abstraction represented by groups of processes located in the blocks. The processes use signal routes for transferring the signals among them or to the higher-level channels. Inside the SDL processes, Extended Communicating Finite State Machines (ECFSMs) are used for description of each protocol entity behavior. This is the lowest and the most detailed level of the behavior model. The functionalities of the protocol entity are presented unambiguously by using SDL discrete states, triggers, transitions, tasks, procedures, decisions, manipulation of variables, management of signals, etc. In the same manner, all protocol primitives are described, along with the signal's parameters exchanged among the protocol entities.

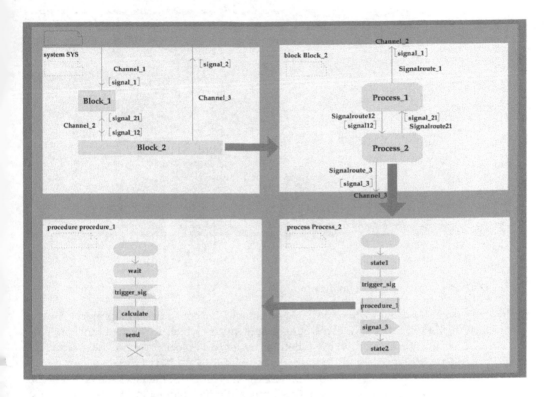

Fig. 2. Insight of a generic behavior model.

In order to assess the real performance of the analyzed system or protocol, it is necessary to build a performance evaluating model. The SDL performance model can emulate real working scenarios of communicating devices. It is a standalone model that embeds the behavior model and translates its preciseness into an accurate event driven type of simulator. The results obtained by the performance model are reliable indicator of the expected performance of the future communication device, which will be built according to the analysed standard.

The process of 'assimilation' and upgrade of a generic behavior model into a generic performance model is depicted in Fig. 3. The presented generic behavior model contains several protocol layers, represented by SDL processes. The following layers are included: Convergence layer, MAC layer, PHY layer, and a vertical Management layer. The behavior model describes all possible interactions between these entities in a single stimuli-response manner.

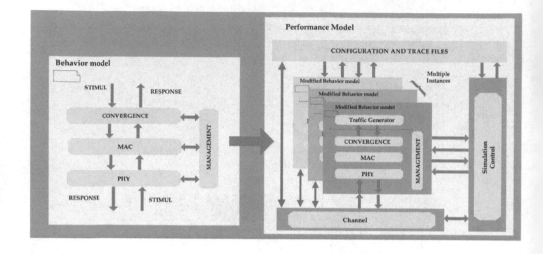

Fig. 3. Building a generic performance model.

Unlike the behavior model where only single stimuli - response pair of a protocol entity is evaluated, the performance model introduces new entities which are needed for completing the communication scenario emulation, both on system and block level. On the system level besides multiple instances of the modified behavior model, it is crucial to introduce a simulation control block and a channel block. The control block manages the generation of block-instances and controls the simulation time, while the channel block ensures a proper emulation of exchanging packets (e.g. radio frequency packets) among the communicating entities. Each block instance is characterized by a unique process identifier (PID), which enables differentiation and addressing of the identical entities.

As it was previously stated, in the foundation of the performance model lays the behavior model. It is necessarily modified (i.e. upgraded), in order to conduct its expected role of real communicating device emulation. Many new processes with an appropriate purpose should be introduced in this upgraded version of the behaviour model: controller of the primitive exchange process, procedures for queuing and prioritization of the signals, processes for segmentation and reassemble of the massages, manipulators of simulation time (timers and clocks), etc. All these processes are introduced according to the textual part of the protocol specification.

After building the performance model using the SDL Graphical Representation (SDL-GR), abstract Data Types (ADTs) are added in order to introduce important functionalities (e.g. reading and writing to file, different kinds of random number generators, etc.).

The analyzer then runs the performance model for detecting all the ambiguities. Next step is the conversion of the built model into a so called Phrasal Representation of SDL (SDL-PR). Using SDL-PR, the code generator produces C++ source code, compiles it and links it with appropriate libraries. The result of these steps is a standalone simulator executable which requires as an input only the configuration files, needed for the description of the desired network scenario.

4. Case study

This section contains an implementation case study of the previously proposed methodology. In particular, the case study extends the findings proposed by (Latkoski, 2010), and provides the needed validation of the analytical and numerical analysis contained in (Latkoski, 2010).

The targeted communication system is based on WiMAX technology, standardized by IEEE 802.16. It belongs to the group of wireless metropolitan area networks (WMANs), which are on the steady track of widespread deployment in many urban environments. This worldwide trend is facilitated by the ever growing demand for "last-kilometre" network connectivity in every part of those urban environments with guaranteed service quality. A significant portion of the WMAN installations are based on the IEEE 802.16 technology, which is mature enough for seamless and low-cost deployment.

The focus of the analysis provided here, is the protocol which is responsible for bandwidth allocation among the WiMAX users. The WiMAX channel access is controlled by one of the several available Medium Access Control (MAC) procedures. (Latkoski, 2010) studies the contention-based bandwidth request procedure based on original analytical model, facilitated by numerical analysis. In (Latkoski, 2010) the key parameters of the contention procedure are optimized in order to minimize the average bandwidth access delay, thus ensuring the highest possible quality of service (QoS) to the WiMAX users.

WiMAX supports several QoS classes: UGS – Unsolicited Grant Service (E1/T1), real-time Polling Service – rtPS (MPEG), non-real-time Polling Service – nrtPS (FTP) and Best Effort – BE (HTTP). Except for the UGS that uses dedicated uplink transmission slots, the remaining service classes use the bandwidth request procedures over the uplink to the base station (BS). Depending on the service class, the access scheme can be either contention-based or based on unicast polling. The vendor-specific implementation can offer two optional non-mandatory procedures: piggybacking and bandwidth stealing procedures. Here, we focus on the IEEE 802.16 contention-based bandwidth request access scheme, which supports the BE class of traffic, generated by most Internet applications (web surfing, FTP, etc.). Additionally we will compare this scheme with the round-robin polling scheme, as well as with the multicast-groping-based principal of bandwidth management. All three access schemes are briefly explained in the following subsections.

4.1 Contention based bandwidth access

The IEEE 802.16 standard supports a mandatory Point-to-Multipoint (PMP) architecture operating in Time Division Duplex (TDD) mode. In such network conditions, the frames are divided to downlink (DL) and uplink (UL) subframes. The BS transmits uplink map (UL-

MAP) messages at the beginning of the DL subframe, in order to schedule the uplink traffic from the subscriber stations (SSs) to the BS. The beginning of UL subframe contains Information Elements (IEs) dedicated for initial ranging and bandwidth request procedures, followed by slots for the actual data transmission. The MAC layer of IEEE 802.16 specifies the rules for the contention-mode bandwidth request procedure. A contention period, as mentioned, is allocated at the beginning of the uplink subframe. It is divided into an integer number of transmission slots and is called an information element. Each transmission slot can be used for a transmission of only one bandwidth request. The SSs use the contention slots to send bandwidth request messages. If a SS's request message transmission is successful, the BS grants contention-free data transmission slot for that particular SS in one of the following frames by placing the SS's Connection ID (CID) in the UL-MAP message.

If more than one SS tries to transmit its request in the same transmission slot, a collision happens. Since it is not practically possible for SSs to sense the UL channel and to detect a collision, the SSs can only know of the success of their bandwidth request transmission if they receive a response in the form of a bandwidth grant from the BS in the subsequent frames. A subscriber station that does not receive a response to its bandwidth request by a certain deadline assumes that either a collision happened or resources are not available at the BS. In either case, since the SS can not determine the cause, it assumes that a collision happened and uses an exponential binary back-off procedure to resolve the collision. In particular, the SS starts a contention-based procedure by setting a so called initial backoff window which is an integer number. Next step is selection of a random number within the window, which determines the number of contention slot for which the SS will defer its next request message transmission. Only the slots for which the SS is eligible to send are counted. When the SS's counter reaches zero, the SS sends its request message. The SS considers the contention transmission as lost if no data grant has been given within the period of time defined by a timer. Then the SS enters in the next stage of the backoff algorithm by doubling the size of the backoff window and selecting another random number. This repeats with each loss of the request massage, until the backoff window size reaches its maximum size.

4.2 Multicast-grouping based bandwidth access

The BS controls the access rights of the SSs for each contention slot. If the BS declares one contention slot as a *broadcast* type of slot, then all SSs have the right to transmit their bandwidth request messages in that particular slot. Contrary to this principle, the BS can mark certain slot as a *multicast* type of slot. In this case only the members of the specified multicast group can access the slot. The BS controls the membership of each SS into multicast groups. For this purpose, it uses a special MAC message called MCA-REQ (Multicast Polling Assignment Request). Each MCA-REQ message contains three basic parameters: the basic CID of the SS, the index of the multicast group, and one of the two possible commands, *join* or *leave*. One SS can belong to several multicast groups.

In order to evaluate the influence of multicast-group implementation over the system's performance, we will calculate the ratio of the number of successful bandwidth request transmissions per frame and the number of active SSs ($maxN_{sud}/n$). The value of this ratio $maxN_{sud}/n = 1$ means that all active SSs are served in one TDD frame. The following figure (Fig. 4) presents the results regarding $maxN_{sud}/n$ for different number of active SSs (n) and different number of transmission slots per frame (N_r). The results are obtained by using the analytical equations provided by (Latkoski, 2010).

Furthermore, Fig. 4 provides additional insight regarding the maximization of the bandwidth procedure success rate. It is obvious that instead of using all N_r slots for the contention of all n subscriber stations, it is better to split the SSs into M groups, and to give each group a portion of N_r/M slots for contention. The colored lines in Fig. 4, give the possibilities for implementation of this idea. For example, instead of using N_r = 16 slots for n = 16 users, it is more efficient to use M = 8 multicast groups, as the N_{suc}/n is higher for (n, N_r) = (2, 2) compared to the case where (n, N_r) = (16, 16).

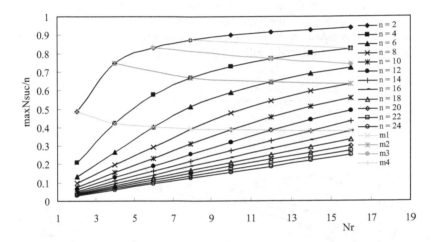

Fig. 4. Normalized success rate for conation-based scheme.

The obvious challenge here is to obtain a precise estimation of the number of active SSs (n) by the serving BS, which controls the contention and multicasting.

4.3 Round robin polling bandwidth access

Instead of using contention based bandwidth distribution among the SSs, the BS has an option to use the round-robin polling based procedure for bandwidth access. In this case, the BS asks each of the registered SSs whether they need bandwidth, starting with the first SS and ending with the last N_{all} SS. Then the circle of polling repeats again from the first SS. Considering that not all N_{all} SSs need bandwidth at a time, but only n of them have such need, we can calculate the efficiency of this method through the performance parameter defined as utilization of the slots. We can compute the utilization of the transmission slots in the case of round-robin polling (RR_{util}) as:

$$RR_{util} = \frac{n}{N_{all}},$$
(1)

while the average bandwidth access delay seen by the SSs (RR_{Td}) is:

$$RR_{Td} = \frac{N_{all}}{N_r} t_{frm} \, , \qquad (2)$$

where the N_r represents the total number of transmitting slots, and t_{frm} is the TDD frame duration. These simple equations reveal that the utilization in the case of round-robin polling scheme does not depend on N_r, while the delay does not depend on n. Consequently, this bandwidth allocation scheme is expected to have higher performance in scenarios where the number of active SSs (SSs which need bandwidth) is close to the number of registered SSs.

The previous conclusion can be proven by making a comparison of the transmission slot utilization in the case of round-robin polling and conation based schemes for different numbers of active users. For this purpose we have used the equations provided by (Latkoski, 2010) for the maximal utilization of the transmitting slots provided by contention scheme. The results presented in the following figure are obtained for different values of N_r and N_{all}.

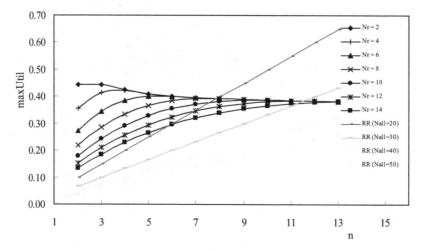

Fig. 5. Comparison of the schemes performance.

The range of values where $n \approx N_{all}$ (right part of the figure) is more appropriate for round-robin polling scheme utilization, compared to the contention-based scheme.

4.4 SDL models

For the purpose of analytical results validation, as well as for testing and improvement of the communication protocol for bandwidth allocation, we have created according to the methodology presented in the previous section, both behavior and performance evaluating models. Actually, we have built several behavior models for different MAC-layer processes involved by the communication protocol, located in both base station and subscriber station. After the functional testing of each protocol entity, the behavior models are implemented into fully operational performance model. The highest level of this model is presented in the following Fig. 6. It contains a behavior model of the BS which contains several processes: Optimizer, Estimator and MsgCreator. The Optimizer determines the optimal values of the following contention parameters: the initial contention window, the number of allowed consequent unsuccessful bandwidth request transmissions, the number of multicast groups, and the number of contention slots per frame. These parameters are sent to the process which constructs the MAC management messages (MsgCreator), as well as to the process Estimator. The purpose of the Estimator is to estimate several network condition related parameters, such as: the number of active users, the probability of collision, the probability of transmission, etc. These parameters are needed for an accurate operation of the Optimizer.

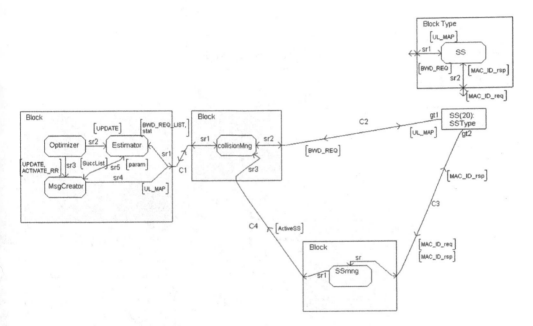

Fig. 6. Case study Performance model.

The performance model also contains blocks for channel emulation (collisionMng) and simulation control block (SSmng) which provides SSs PID management.

Besides the BS, the performance model contains multiple instances of the subscriber station block. All instances of the SS block operate as independent user equipment stations. Through the SSmng block we are able to define and control the number of active stations in the simulation scenario. The BS through the MsgCreator block controls the mode of operation (contention or round-robin, along with the number of multicast groups), according to the Optimizer commands. The Estimator operates dynamically, and feeds the Optimizer with the necessary information regarding the network conditions. The active stations are sending bandwidth request messages and then register the outcome of every attempt (success or failure).

In this communication protocol engineering case study, the Optimizer is the newly proposed entity which operation will be described in details. Actually, the Optimizer performs several steps presented formally in Fig. 7. These steps are:

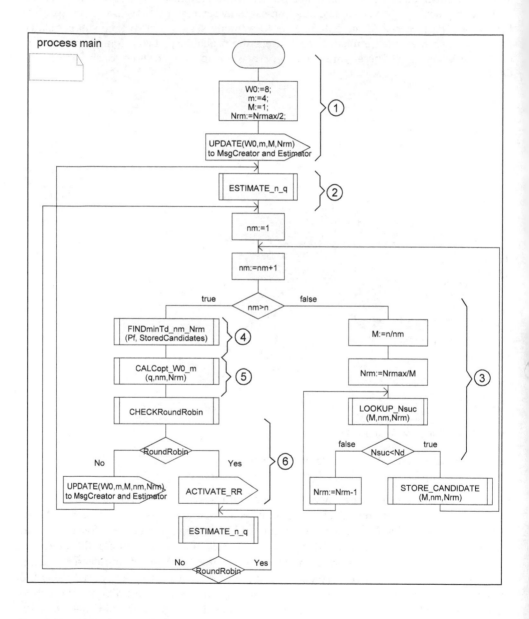

Fig. 7. Functional steps of the Optimizer.

1. The Optimizer initiates the bandwidth request procedure using predefined initial values for the contention parameters, without using multicast ($M = 1$).

2. After a training period of 5 seconds, the Estimator sends to the Optimizer estimated values for the number of active SSs (n) and information regarding their activity dynamics (q), please refer to (Latkoski, 2010).

3. The Optimizer finds the most suitable values for the number of multicast groups and number of SSs per group. For this purpose the Optimizes searches through a LOOKUP table for the possible candidate values of the contention parameters which can provide number of successful bandwidth requests per frame (N_{suc}) such as $N_{suc} < N_d$, where N_d is the number of uplink data slots per frame.

4. From all candidate parameter values, the Optimizer selects those which will provide the lowest value for the average bandwidth access delay.

5. Then the Optimizer calculates the optimal values for the contention window and the number of consecutive unsuccessful attempts.

6. Finally, the Optimizer checks whether the round-robin polling method could provide better performance. After this, it sends its final decision by a command to the MsgCreator.

4.5 Results

The performance model was tested in simulation scenario where the number of active stations (n) is changed with the time, as presented in Fig. 8. The scenario simulates 24 hour network operation. The next two figures (Fig. 9 and Fig. 10) provide the measured performance of the bandwidth request procedure for three different modes of operation: round-robin polling, contention without multicast grouping, and contention with multicast grouping. Fig. 9 presents the transmission slots utilization, while Fig. 10 presents the average bandwidth access delay.

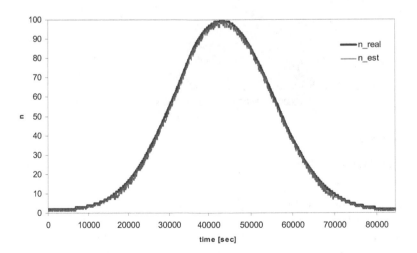

Fig. 8. Number of active SSs during the simulation (actual and estimated).

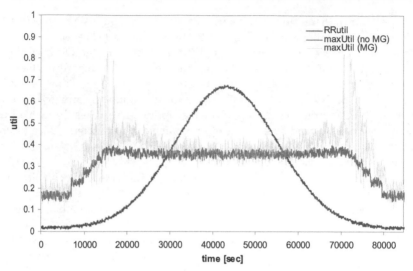

Fig. 9. Measured utilization of the transmission slots.

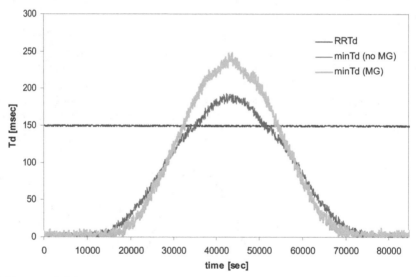

Fig. 10. Measured delay.

During the simulations we have used the following parameter values: $N_r = 10$, $N_{all} = 150$, $q = 1$, $t_{frm} = 10ms$.

From the results we can conclude that the contention based bandwidth request procedure which uses multicast grouping almost always outperforms the case where no multicasting groups are used. This is the case if our comparison criterion is based on the transmission slots utilization. The same conclusion is not entirely valid if the criterion is based on the average bandwidth access delay. The round-robin mode of operation, as expected, outperforms the other modes of operation when the number of active users is close to the number of registered users.

5. Conclusion

The performance of a communication system (or protocol) is the major indicator for the successfulness of the standard that specifies that system. During the development process of new communication technologies, the aim is to increase the overall protocol performance, which automatically means to produce a better standard. The formal-based performance evaluation method described in this chapter, which uses SDL network prototyping, provides the most relevant results of the system performance without the need for its early and prematurely hardware implementation. This is crucial for the production of competitive communication products which will be free of hidden flows during the development process. The aim of this chapter was to emphasize the importance of conducting an early performance evaluation of the communication protocols and systems, and to suggest an appropriate solution for carrying out such an activity.

We have illustrated the proposed framework through an actual case study which targets the WiMAX bandwidth allocation methods. Three schemes were investigated: contention based bandwidth requesting without multicasting, contention based procedure with multicast grouping, and round-robin polling based bandwidth allocation scheme. With the help of the SDL performance model, we have found that the preferred scheme should be selected based on the network working conditions (i.e. number of active subscriber stations) and according to the performance criterion (transmission slots utilization or average bandwidth access delay).

The proposed new protocol etities which are product of the communication protocol engineering process are simple and accurate, and can be easily implemented at the BS side in order to optimize the performance of the WiMAX bandwidth request procedure.

The presented results are only a hint to the possible evaluation outcomes from the network emulations created using the proposed methodology.

6. Acknowledgment

This research is sponsored by the Faculty of Electrical Engineering and Information Technologies - Skopje, Ss. Cyril and Methodius University in Skopje, Macedonia, through the MOBIKS (Modeling and Optimization of Wireless Information-Communications Systems) project. The authors want to express gratitude to all participants involved in this project.

7. References

ASN.1 (1993). ITU-T, Specification of abstract syntax notation one (ASN.1), *ITU- T Reconunendation X.208, Technical Report, Telecommunication Standardization Sector of ITU*, March 1993.

Bluetooth (2005). IEEE Std 802.15.1, Wireless Medium Access Control (MAC) and Physical Layer (PHY) Specifications for Wireless Personal Area Networks (WPANs)

Booch G. et al. (2000). OMG Unified Modeling Language Specification, *Version 1.3 First Edition: March 2000.*

Chen Hui et al., (2010). Formal specification and verification of reconfigurable protocol stack for networked control system, *Proceedings of 2010 International Conference on*

Networking, Sensing and Control (ICNSC), pp. 441 – 446, ISBN: 978-1-4244-6450-0, Chicago, USA, 10-12 April 2010

Estelle (1989). Information Processing Systems - OSI: Estelle, A Formal Description Technique Based on an Extended State Transition Model, *International Standard 9074*, June 1989 ETSI. Available from http://www.etsi.org

Gotzhein R., et al. (2009). Energy-Aware System Design with SDL, *Proceedings of the 14th international SDL conference on Design for motes and mobiles - SDL'09*, pp. 19-33, ISBN:3-642-04553-7 978-3-642-04553-0, September 22-24, Bochum, Germany

Hall A. (1990), Seven Myths of Formal Methods, *IEEE Software*, Vol. 7, No. 5, Sept. 1990, pp. 11-19, September 1990 IEEE. Available from http://www.ieee.org

Latkoski P. et al. (2010). Modeling and optimization of bandwidth request procedure in IEEE 802.16 networks, *Proceedings of the IEEE 21st International Symposium on Personal Indoor and Mobile Radio Communications (PIMRC), 2010*, pp. 1469 – 1474, ISBN: 978-1-4244-8017-3, September 26-30, 2010, Istanbul, Turkey.

Larsen K. G. et al. (1997). UPPAAL in a Nutshell, *International Journal on Software Tools for Technology Transfer (STTT)*, Volume 1, Numbers 1-2, pp.134-152

Lopez J. et al., (2005). Security Protocols Analysis: A SDL-based Approach, *Computer Standards & Interfaces*, Vol. 27, No. 3, pp. 489-499, ISSN: 0920-5489

LOTOS (2000). Information technology Enhancements to LOTOS (E-LOTOS), *SO/IEC JTC1/SC7, International Standard 15437*, July 2000

Mitschele-Thiel A. (2001). *Systems Engineering with SDL: Developing Performance-Critical Communication Systems*, Wiley, ISBN: 978-0-471-49875-9, New York, USA

MSC (2001). Series Z: Languages and General Software Aspects for Telecommunication Systems, Message Sequence Chart, *ITU-T Recommendation Z.120*, Geneva, Switzerland, 2001

Petri C. A. (1996), Nets, Time and Space, *Theoretical Computer Science, Special Volume on Petri Nets*, Vol. 153, No. 1-2, pp. 3-48

SDL (2011). Series Z: Languages and General Software Aspects for Telecommunication Systems, Specification and Description Language, *ITU-T Recommendation Z.100*, Geneva, Switzerland, Latest edition 2011

Sherif M. H. & Sparrell D. K. (1992), Standards and Innovations in Telecommunications, *IEEE Communication Magazine*, Vol. 30, No. 7, July 1992, pp. 22–29, ISSN 0163-6804

Sherif M. H. (2001). A Framework for Standardization in Telecommunications and Information Technology, *IEEE Communications Magazine*, No.4, April 2001, pp. 94-100, ISSN 0163-6804

Showk A., et al. (2009). Modeling LTE protocol for mobile terminals using a formal description technique, *Proceedings of the 14th international SDL conference on Design for motes and mobiles - SDL'09*, pp. 222-238, ISBN:3-642-04553-7 978-3-642-04553-0, September 22-24, Bochum, Germany SPIN. Available from http://netlib.sandia.gov/spin/index.html

TTCN (2006). ITU-T, Recommendation Z.140 Tree and Tabular Combined Notation (TTCN), March 2006.

Wing J. M. (1990), A Specifier's Introduction to Formal Methods, *IEEE Computer*, Vol. 23, No. 9, pp. 8-24, September 1990

WiMAX (2010). IEEE Std 802.16-2004, IEEE Standard for Local and metropolitan area networks, Part 16: Air Interface for Fixed Broadband Wireless Access Systems

WiFi (2007) IEEE Std IEEE 802.11, IEEE Standard for Wireless LAN Medium Access Control and Physical Layer Specification. 3GPP. Available from http://www.3gpp.org

Cell Dwell Time and Channel Holding Time Relationship in Mobile Cellular Networks

Anum L. Enlil Corral-Ruiz[1], Felipe A. Cruz-Pérez[1]
and Genaro Hernández-Valdez[2]
[1]Electrical Engineering Department, CINVESTAV-IPN
[2]Electronics Department, UAM-A
Mexico

1. Introduction

Channel holding time (CHT) is of paramount importance for the analysis and performance evaluation of mobile cellular networks. This time variable allows one to derive other key system parameters such as channel occupancy time, new call blocking probability, and handoff call dropping probability. CHT depends on cellular shape, cell size, user's mobility patterns, used handoff scheme, and traffic flow characteristics. Traffic flow characteristics are associated with unencumbered service time (UST), while the overall effects of cellular shape, users' mobility, and handoff scheme are related to cell dwell time (CDT).

For convenience and analytical/computational tractability, the teletraffic analysis of mobile cellular networks has been commonly performed under the unrealistic assumption that CDT and/or CHT follow the negative exponential distribution (Lin et al., 1994; Hong & Rappaport, 1986). However, a plenty of evidences showed that these assumptions are not longer valid (Wang & Fan, 2007; Christensen et al., 2004, Fang, 2001, 2005; Orlik & Rappaport, 1998; Fang & Chlamtac, 1999; Fang et al., 1999; Alfa & Li, 2002; Rahman & Alfa, 2009; Soong & Barria, 2000; Yeo & Jun, 2002; Pattaramalai, et al., 2007). Recent papers have concluded that in order to capture the overall effects of users' mobility, one needs suitable models for CDT distribution (Lin, 1994; Hong & Rappaport, 1986). In specific, the use of general distributions for modeling this time variable has been highlighted. In this research direction, some authors have used Erlang, gamma, uniform, deterministic, hyper-Erlang, sum of hyper-exponentials, log-normal, Pareto, and Weibull distributions to model the pdf of CDT; see (Wang & Fan, 2007; Fang, 2001, 2005; Orlik & Rappaport, 1998; Fang & Chlamtac, 1999; Fang et al., 1997, 1999; Rahman & Alfa, 2009; Pattaramalai et al., 2007, 2009; Hidata et al., 2002; Thajchayapong & Toguz, 2005; Khan & Zeghlache, 1997; Zeng et al. 2002; Kim & Choi, 2009) and the references therein. Fang in (Fang, 2001)) emphasizes the use of phase-type (PH) distributions for modeling CDT. The reason is twofold. First, PH distributions provide accurate description of the distributions of different time variables in wireless cellular networks, while retaining the underlying Markovian properties of the distribution. Markovian properties are essential in generating tractable queuing models for cellular networks. Second, there have been major advances in fitting PH distributions to real data. Among the PH probability distributions, the use of either Coxian or Hyper-Erlang distributions are of particular interest because their universality property (i.e, they can be used to approximate any non negative distribution arbitrarily close) (Soong & Barria, 2000; Fang, 2001).

Due to the discrepancy and the wide variety of proposed models, it appears mandatory to investigate the implications of the cell dwell time distribution on channel holding time characteristics in mobile wireless networks. This is the topic of research of the present chapter. Let us describe the related work reported in this research direction.

1.1 Previously related work

In (Fang, 2001; Zeng et al. 2002), it is observed that, depending on the variance of CDT, the mean channel holding time for new calls (CHTn) can be greater than the mean channel holding time for handoff calls (CHTh). However, in these works, it is neither explained nor discussed the physical reasons for this observed behavior. This phenomenon (which is addressed in Section 3.1) and the lack of related published numerical results have motivated the present chapter.

Most of the previously published papers that have developed mathematical models for the performance analysis of mobile cellular systems considering general probability distribution for cell dwell time have either only presented numerical results for the Erlang (Wang & Fan, 2007; Fang et al., 1999; Rahman & Alfa, 2009; Kim & Choi, 2009) or Gamma distributions with shape parameter greater than one[1] (Yeo & Jun, 2002; Fang, 2005), and/or only for the CHTh[2] (Fang, 2001; Fang & Chlamtac, 1999), or have not presented numerical results at all (Fang, 2005; Alfa & Li, 2002; Soong & Barria, 2000). Thus, numerical results both for values of the coefficient of variation (CoV) of CDT greater than one and/or for the CHTn have been largely ignored. Exceptions of this are the papers (Orlik & Rappaport, 1998; Fang et al., b, 1997; Pattaramalai, et al., 2009).

On the other hand, probability distribution of CHT has been determined under the assumption of the staged distributions sum of hyper-exponentials, Erlang, and hyper-Erlang for the CDT (Orlik & Rappaport, 1998; Soong & Barria, 2000). However, to the best of the authors' knowledge, probability distribution of CHT in mobile cellular networks with neither hyper-exponential nor Coxian distributed CDT has been previously reported in the literature.

In this Chapter, the statistical relationships among residual cell dwell time (CDTr), CDT, and CHT for new and handoff calls are revisited and discussed. In particular, under the assumption that UST is exponentially distributed and CDT is phase-type distributed, a novel algebraic set of general equations that examine the relationships both between CDT and CDTr and between CDT and channel holding times are obtained. Also, the condition upon which the mean CHTn is greater than the mean CHTh is derived. Additionally, novel mathematical expressions for determining the parameters of the resulting CHT distribution as functions of the parameters of the CDT distribution are derived for hyper-exponentially or Coxian distributed CDT.

---

[1] For the Erlang distribution and for the Gamma distribution with shape parameter greater than one, the coefficient of variation of its associated random variable is smaller than one.

[2] Also referred as handoff call channel occupancy time.

2. System model

A homogeneous multi-cellular system with omni-directional antennas located at the centre of each cell is assumed; that is, the underlying processes and parameters for all cells within the cellular network are the same, so that all cells are statistically identical. As mobile user moves through the coverage area of a cellular network, several variables can be defined: cell dwell time, residual cell dwell time, channel holding time, among others. These time variables are defined in the next section.

2.1 Definition of time interval variables

In this section the different time interval variables involved in the analytical model of a mobile cellular network are defined.

First, the *unencumbered service time* per call x_s (also known as the *requested call holding time* (Alfa and Li, 2002) or *call holding duration* (Rahman & Alfa, 2009)) is the amount of time that the call would remain in progress if it experiences no forced termination. It has been widely accepted in the literature that the unencumbered service time can adequately be modeled by a negative exponentially distributed random variable (RV) (Lin et al., 1994; Hong & Rappaport, 1986). The RV used to represent this time is $\mathbf{X}_s$ and its mean value is $E\{\mathbf{X}_s\} = \frac{1}{\mu}$.

Now, *cell dwell time* or *cell residence time* $x_d(j)$ is defined as the time interval that a mobile station (MS) spends in the *j*-th (for $j = 0, 1, \ldots$) handed off cell irrespective of whether it is engaged in a call (or session) or not. The random variables (RVs) used to represent this time are $\mathbf{X}_d(j)$ (for $j = 0, 1, \ldots$) and are assumed to be independent and identically generally phase-type distributed. For homogeneous cellular systems, this assumption has been widely accepted in the literature (Lin et al., 1994; Hong & Rappaport, 1986; Orlik & Rappaport, 1998; Fang & Chlamtac, 1999; Alfa & Li, 2002; Rahman & Alfa, 2009).

In this Chapter, cell dwell time is modeled as a general phase-type distributed RV with the probability distribution function (pdf) $f_{\mathbf{X}_d}(t)$, the cumulative distribution function (CDF) $F_{\mathbf{X}_d}(t)$, and the mean $E\{\mathbf{X}_d\} = \frac{1}{\eta}$.

The *residual cell dwell time* x_r is defined as the time interval between the instant that a new call is initiated and the instant that the user is handed off to another cell. Notice that residual cell dwell time is only defined for new calls. The RV used to represent this time is $\mathbf{X}_r$. Thus, the probability density function (pdf) of $\mathbf{X}_r$, $f_{\mathbf{X}_r}(t)$, can be calculated in terms of $\mathbf{X}_d$ using the excess life theorem (Lin et al., 1994)

$$f_{\mathbf{X}_r}(t) = \frac{1}{E[\mathbf{X}_d]}\left[1 - F_{\mathbf{X}_d}(t)\right] \tag{1}$$

where $E[\mathbf{X}_d]$ and $F_{\mathbf{X}_d}(t)$ are, respectively, the mean value and   cumulative probability distribution function (CDF) of $\mathbf{X}_d$.

Finally, we define *channel holding time* as the amount of time that a call holds a channel in a particular cell. In this Chapter we distinguish between channel holding times for handed off (CHTh) and channel holding time for new calls (CHTn). CHTh (CHTn) is represented by the random variable $\mathbf{X}_c^{(h)}$ ($\mathbf{X}_c^{(N)}$).

3. Mathematical analysis

3.1 Relationship between X_d and X_r

The relationship between the probability distributions of CDT and CDTr is determined by the residual life theorem. In Table I some particular typically considered CDT distributions and the corresponding CDTr distributions obtained by applying the residual life theorem are shown.

Probability density function of cell dwell time or its Laplace transform.	Probability density function of residual cell dwell time or its Laplace transform.	Parameters of $f_{X_r}(t)$ as a function of the parameters of $f_{X_d}(t)$
Negative Exponential $\eta e^{-\eta t}$	Negative Exponential $\eta e^{-\eta t}$	
Erlang of k order $\dfrac{\eta^k t^{k-1}}{(k-1)!}e^{-\eta t}$	Hyper-Erlang with k stages of $1, 2, \ldots$ and k phases $\displaystyle\sum_{j=1}^{k} P_j^{(N)}\dfrac{\eta^j(t)^{j-1}}{(j-1)!}e^{-\eta t}$	$P_j^{(N)} = \dfrac{1}{k}$
Hyper-exponential of n order $\displaystyle\sum_{i=1}^{n} P_i\lambda_i e^{-\lambda_i t}$	Hyper-exponential of n order $\displaystyle\sum_{i=1}^{n} P_i^{(N)}\lambda_i e^{-\lambda_i t}$	$P_i^{(N)} = \dfrac{P_i\prod_{\substack{j=1\\j\neq i}}^{n}\lambda_i}{\sum_{i=1}^{n} P_i\prod_{\substack{j=1\\j\neq i}}^{n}\lambda_j}$
Hypo-exponential[3] of m order $f_{X_d}^*(S) = \displaystyle\prod_{i=1}^{m}\dfrac{\eta_i}{s+\eta_i}$	Generalized Coxian of m order $f_{X_r}^*(s) = \displaystyle\sum_{i=1}^{m} P_i^{(N)}\prod_{j=i}^{m}\dfrac{\eta_j}{s+\eta_j}$	$P_i^{(N)} = \dfrac{\frac{1}{\eta_i}}{\sum_{j=1}^{m}\frac{1}{\eta_j}}$
Hyper-Erlang of common order (n, m) $\displaystyle\sum_{i=1}^{n} P_i\dfrac{\eta_i^m t^{m-1}}{(m-1)!}e^{-\eta_i t}$	Hyper-Erlang of non common order $\displaystyle\sum_{i=1}^{nm} P_i^{(N)}\dfrac{\left(\eta_{\lfloor\frac{i-1}{m}\rfloor+1}\right)^z t^{z-1}}{(z-1)!}e^{-\eta_{\lfloor\frac{i-1}{m}\rfloor+1}t}$	$P_i^{(N)} = \dfrac{P_i\prod_{\substack{l=1\\l\neq i}}^{n}\eta_l}{\sum_{k=1}^{n} P_k m\prod_{\substack{l=1\\l\neq k}}^{n}\eta_l}$ $z = mod\left(\dfrac{i-1}{m}\right)+1$
Constant $\begin{cases}\delta(t-E\{X_d\}) & ;t=E\{X_d\}\\ 0 & ;otherwise\end{cases}$	Uniform $\dfrac{1}{E\{X_d\}}$; $0\leq t\leq E\{X_d\}$	
Coxian of m order $f_{X_d}^*(s)$ $= \displaystyle\sum_{i=1}^{m} P_i\prod_{j=1}^{i}\dfrac{\eta_j}{(s+\eta_j)}$	Generalized Coxian of m order $f_{X_r}^*(s)$ $= \displaystyle\sum_{j=1}^{\frac{m(m+1)}{2}} P_j^{(N)}\left(\prod_{k=h(j)}^{m}\dfrac{\eta_k}{s+\eta_k}\right)$	$P_j^{(N)}$ $= \dfrac{P_{f(j)}\prod_{\substack{k=1\\k\neq h(j)}}^{m}\eta_k}{\sum_{i=1}^{m}\left[\left(\prod_{\substack{k=1\\k\neq i}}^{m}\eta_k\right)\left(\sum_{l=i}^{m} P_l\right)\right]}$

[3] Also known as Generalized Erlang.

Probability density function of cell dwell time or its Laplace transform.	Probability density function of residual cell dwell time or its Laplace transform.	Parameters of $f_{X_r}(t)$ as a function of the parameters of $f_{X_d}(t)$
Generalized Coxian of m order $f_{X_d}^*(s)$ $$= \sum_{j=1}^{\frac{m(m+1)}{2}} P_j \left(\prod_{k=h(j)} \frac{\eta_k}{s+\eta_k} \right)$$	Gereralized Coxian of m order $f_{X_r}^*(s)$ $$= \sum_{j=1}^{\frac{m(m+1)}{2}} P_j^{(N)} \left(\prod_{k=h(j)} \frac{\eta_k}{s+\eta_k} \right)$$	$P_j^{(N)}$ $$= \frac{\prod_{\substack{k=1 \\ k \neq h(j)}}^m \eta_k \left(\sum_{n=j-h(j)+1}^j P_n \right)}{A}$$ $; A$ $$= \sum_{i=1}^m \left[\prod_{\substack{j=1 \\ j \neq i}}^m \eta_j \left(\sum_{k=i}^m P_{\frac{i^2-i+2}{2}} \right. \right.$$ $$\left. \left. + \sum_{l=1}^{i-1} P_{\frac{i^2-i+2}{2}+l} \right) \right]$$
Gamma $$\frac{x^{k-1}e^{-\frac{x}{\theta}}}{\Gamma(k)\theta^k}$$	$$\frac{1}{k\theta}\left[1 - P\left(k, \frac{x}{\theta}\right)\right]$$	
Weibull $$\frac{k}{\lambda}\left(\frac{x}{\lambda}\right)^{k-1} e^{-\left(\frac{x}{\lambda}\right)^k}$$	$$\frac{1}{\lambda\Gamma\left(1+\frac{1}{k}\right)} e^{-\left(\frac{x}{\lambda}\right)^k}$$	
Pareto $$\frac{\alpha X_m^{\alpha}}{t^{\alpha+1}} \; ; t > X_m$$	$$\begin{cases} \frac{\alpha-1}{\alpha X_m}\left[\left(\frac{X_m}{t}\right)^{\alpha}\right] & ; t > X_m \\ \frac{\alpha-1}{\alpha X_m} & ; 0 \leq t \leq X_m \end{cases}$$	

Table I. Examples of corresponding distribution for X_r given the distribution of X_d.

The functional relationship between the moments of the residual cell dwell time and the cell residual time was obtained in (Kleinrock, 1975) applying the Laplace transform to the residual life theorem. That is,

$$\mathcal{L}\{f_{X_r}(t)\} = \mathcal{L}\left\{\frac{1}{E\{X_d\}}\right\} - \mathcal{L}\left\{\frac{1}{E\{X_d\}} F_{X_d}(t)\right\} \tag{2}$$

This equation can be rewritten as

$$f_{X_r}^*(s) = \left[\frac{1}{s}\right]\frac{1}{E\{X_d\}}[\, 1 - f_{X_d}^*(s)] \tag{3}$$

The n-th moment of the residual cell dwell time in terms of the moments of the cell dwell time can be obtained by deriving n times equation (3) with negative argument and substituting $s=0$. Then (Kleinrock, 1975),

$$E\{(\mathbf{X}_r)^n\} = \frac{E\{(\mathbf{X}_d)^{n+1}\}}{(n+1)E\{\mathbf{X}_d\}} \tag{4}$$

The mean residual cell dwell time as function of the moments of cell dwell time can be obtained as (Kleinrock, 1975)

$$E\{\mathbf{X}_r\} = \frac{E\{\mathbf{X}_d\}}{2} + \frac{VAR(\mathbf{X}_d)}{2E\{\mathbf{X}_d\}} \tag{5}$$

$E\{\mathbf{X}_d\}$ and $VAR(\mathbf{X}_d)$ represent the mean and variance of CDT, respectively. Considering this equation and that $CoV\{\mathbf{X}_d\}$ represents the coefficient of variation of CDT, the condition for which the mean CDTr is greater than the mean CDT ($E\{\mathbf{X}_r\} > E\{\mathbf{X}_d\}$) is given by

$$\frac{E\{\mathbf{X}_d\}}{2} + \frac{VAR\{\mathbf{X}_d\}}{2E\{\mathbf{X}_d\}} > E\{\mathbf{X}_d\}$$

$$CoV\{\mathbf{X}_d\} > 1 \tag{6}$$

In this way, the relationship between mean CDT and mean CDTr only depends on the value of the CoV of CDT. Thus, the mean CDTr is greater than the mean CDT (i.e., $E\{\mathbf{X}_r\} > E\{\mathbf{X}_d\}$) when the CoV of CDT is greater than one. This behavior (i.e., $E\{\mathbf{X}_r\} > E\{\mathbf{X}_d\}$) may seem to be counterintuitive due to the fact that, for a particular realization and by definition, CDTr cannot be greater than CDT[4]. This occurs because in such conditions there is a high variability on the cell dwell times in different cells and it is more probable to start new calls on cells where users spent more time. Then, residual cell dwell times tend to be greater than the mean CDT. This phenomenon that may seem to be counterintuitive is now explained and mathematically formulated in this Chapter.

3.2 Channel holding time distribution for handed off and new calls

Channel holding times for handed off and new calls (denoted by $X_C^{(h)}$ and $X_C^{(N)}$, respectively) are given by the minimum between UST and CDT or CDTr, respectively. The CDF of the CHTh and CHTn are, respectively, given by

$$F_{\mathbf{X}_c^{(h)}}(t) = 1 - \left[1 - F_{\mathbf{X}_s}(t)\right]\left[1 - F_{\mathbf{X}_d}(t)\right] \tag{7}$$

$$F_{\mathbf{X}_c^{(N)}}(t) = 1 - \left[1 - F_{\mathbf{X}_s}(t)\right]\left[1 - F_{\mathbf{X}_r}(t)\right] \tag{8}$$

Due to the fact that the Laplace transform of the pdf of both UST and CDTr are rational functions, the Laplace transform of the pdf of CHTn can be obtained using the Residue Theorem as follows (Wang & Fan, 2007)

$$f_{\mathbf{X}_c^{(N)}}^*(s) = f_{\mathbf{X}_S}^*(s) + s \sum_{p\epsilon\Omega_{X_S}} \frac{Res}{\xi = p + s} \frac{f_{\mathbf{X}_r}^*(\xi)\, f_{\mathbf{X}_S}^*(s-\xi)}{\xi \quad s-\xi} \tag{9}$$

where $p\epsilon\Omega_{X_S}$ is the set of poles of $f_{\mathbf{X}_S}^*(s)$, and $f_{\mathbf{X}}^*(s)$ is the Laplace transform of pdf $f_{\mathbf{X}}(t)$. A similar expression can be obtained for the Laplace transform of the pdf of the channel holding time for handed off calls by replacing residual cell dwell time ($\mathbf{X}_r$) by cell dwell time ($\mathbf{X}_d$).

---

[4] Note that the beginning of CDTr is randomly chosen within the CDT interval.

Under the condition that UST is general phase type (PH) distributed, the authors of (Alfa & Li, 2002) prove that the CDT is PH distributed if and only if the CHTn is PH distributed or the CHTh is PH distributed.

The probability distributions of CHTn and CHTh for different staged probability distributions of CDT assuming that the UST is exponentially distributed are shown in Table II. The first entry of this table is a well known result[5]. In (Soong & Barria, 2000), it was shown that when CDT has Erlang or hyper-Erlang distribution, channel holding times have the uniform Coxian and hyper-uniform Coxian distribution, respectively. Uniform Coxian is a special case of the Coxian distribution where all the phases have the same parameter (Perros & Khoshgoftaar, 1989). The hyper-uniform Coxian distribution is a mixture of uniform Coxian distributions.

pdf of cell dwell time.	pdf of channel holding time for new calls or its Laplace Transform.	pdf of channel holding time for handed off calls or its Laplace Transform.
Exponential (Lin et al., 1994)	Exponential $$(\mu + \eta)e^{-(\mu+\eta)t}$$	Exponential $$(\mu + \eta)e^{-(\mu+\eta)t}$$
Erlang of k-th order (Soong & Barria, 2000)	Uniform Coxian of k-th order $$f^*_{X_c^{(N)}}(s) = \sum_{i=1}^{k} P_i^{O(N)} \prod_{j=1}^{i} \frac{\mu + \eta}{s + \mu + \eta}$$	Uniform Coxian of k-th order $$f^*_{X_c^{(h)}}(s) = \sum_{i=1}^{k} P_i^{O(h)} \prod_{j=1}^{i} \frac{\mu + \eta}{s + \mu + \eta}$$
Hyper-Erlang of common order (n, m) (Soong & Barria, 2000)	Hyper-Uniform Coxian $$f^*_{X_c^{(N)}}(s) = \sum_{i=1}^{k} P_i^{O(N)} \prod_{j=1}^{z} \frac{\mu + \eta_l}{s + \mu + \eta_l}$$ where $$z = mod\left(\frac{i-1}{m}\right) + 1$$ $$l = \left\lfloor \frac{i-1}{m} \right\rfloor + 1$$	Hyper-Uniform Coxian $$f^*_{X_c^{(h)}}(s) = \sum_{i=1}^{k} P_i^{O(h)} \prod_{j=1}^{z} \frac{\mu + \eta_l}{s + \mu + \eta_l}$$ where $$z = mod\left(\frac{i-1}{m}\right) + 1$$ $$l = \left\lfloor \frac{i-1}{m} \right\rfloor + 1$$
Hyper-exponential	Hyper-exponential $$\sum_{i=1}^{n} P_i^{(N)}(\mu + \eta_i)e^{-(\mu+\eta_i)t}$$ where $$P_i^{(N)} = \frac{P_i \prod_{\substack{j=1 \\ j\neq i}}^{n} \eta_i}{\sum_{i=1}^{n} P_i \prod_{\substack{j=1 \\ j\neq i}}^{n} \eta_j}$$	Hyper-exponential $$\sum_{i=1}^{n} P_i(\mu + \eta_i)e^{-(\mu+\eta_i)t}$$

[5] Authors in (Lin *et al.*, 1994) give a condition under which the channel holding time is exponentially distributed, that is, the cell residence time needs to be exponentially distributed.

pdf of cell dwell time.	pdf of channel holding time for new calls or its Laplace Transform.	pdf of channel holding time for handed off calls or its Laplace Transform.
Coxian	Generalized Coxian $f^*_{X_c^{(N)}}(s)$ $$= \sum_{j=1}^{\frac{m(m+1)}{2}} P_j^{O(N)} \left(\prod_{k=h(j)}^{f(j)} \frac{\eta_k}{s+\mu+\eta_k} \right)$$ where $P_j^{O(N)}$ $$= \left[\prod_{i=h(j)}^{f(j)-1} \frac{\eta_i}{\mu+\eta_i} \right] \left[P_j^{(N)} \right.$$ $$\left. + \sum_{k=f(j)+1}^{m} P_{\frac{k^2-k+2}{2}}^{(N)} \left(\frac{\mu}{\mu+\eta_{f(j)}} \right) \right]$$	Coxian $$f^*_{X_c^{(h)}}(s) = \sum_{j=1}^{m} P_j^{O(h)} \prod_{i=1}^{j} \frac{\eta_i}{(s+\mu+\eta_i)}$$ where $$P_j^{O(h)} = \left[\prod_{i=1}^{j-1} \frac{\eta_i}{\mu+\eta_i} \right] \left[P_j \right.$$ $$\left. + \sum_{k=j+1}^{m} P_k \left(\frac{\mu}{\mu+\eta_j} \right) \right]$$
Generalized Coxian (Corral-Ruiz et al., a, 2010)	Generalized Coxian $f^*_{X_c^{(N)}}(s)$ $$= \sum_{j=1}^{\frac{m(m+1)}{2}} P_j^{O(N)} \left(\prod_{k=h(j)}^{f(j)} \frac{\eta_k}{s+\mu+\eta_k} \right)$$ where $P_j^{O(N)}$ $$= \left[\prod_{i=h(j)}^{f(j)-1} \frac{\eta_i}{\mu+\eta_i} \right] \left[P_j^{(N)} \right.$$ $$\left. + \sum_{k=f(j)+1}^{m} P_{\frac{k^2-k+2}{2}}^{(N)} \left(\frac{\mu}{\mu+\eta_{f(j)}} \right) \right]$$	Generalized Coxian $f^*_{X_c^{(h)}}(s)$ $$= \sum_{j=1}^{\frac{m(m+1)}{2}} P_j^{O(h)} \left(\prod_{k=h(j)}^{f(j)} \frac{\eta_k}{s+\mu+\eta_k} \right)$$ where $P_j^{O(h)}$ $$= \left[\prod_{i=h(j)}^{f(j)-1} \frac{\eta_i}{\mu+\eta_i} \right] \left[P_j \right.$$ $$\left. + \sum_{k=f(j)+1}^{m} P_{\frac{k^2-k+2}{2}} \left(\frac{\mu}{\mu+\eta_{f(j)}} \right) \right]$$

Table II. Examples of corresponding distributions for $\mathbf{X}_c^{(N)}$ and $\mathbf{X}_c^{(h)}$.

Next, it is shown that when the UST is exponentially distributed and CDT has hyper-exponential distribution of order n, the distribution of CHTh has also a hyper-exponential distribution of order n. Similarly, when CDT has Coxian distribution of order n, the distribution of CHTn has also a Coxian distribution of order n.

3.2.1 Case 1: Hyper-exponentially distributed cell dwell time

Considering that CDT has a hyper-exponential pdf of order n given by

$$f_{X_d}(t) = \sum_{j=1}^{n} P_j \eta_j e^{-\eta_j t} \tag{10}$$

For exponentially distributed UST and using (4), the CDF of the CHTh can be expressed as follows

$$F_{X_c^{(h)}}(t) = 1 - [e^{-\mu t}]\left[\sum_{i=1}^{n} P_i e^{-\eta_i t}\right]$$

$$F_{X_c^{(h)}}(t) = 1 - \sum_{i=1}^{n} P_i e^{-(\mu+\eta_i)t}$$

(11)

This expression corresponds to a hyper-exponential distribution of order n with phase parameters $\mu + \eta_i$ and probabilities P_i of choosing stage i (for $i = 1, ..., n$).

As the CDTr is hyper-exponentially distributed when CDT has hyper-exponential distribution, the CHTn is also hyper-exponentially distributed. In this case, the probability of choosing stage i (for $i = 1, ..., n$) is given by

$$P_i^{(N)} = \frac{P_i \prod_{\substack{j=1 \\ j \neq i}}^{n} \eta_i}{\sum_{i=1}^{n} P_i \prod_{\substack{j=1 \\ j \neq i}}^{n} \eta_j}$$

(12)

3.2.2 Case 2: Coxian distributed cell dwell time

Considering that cell dwell time has an m-th order Coxian distribution (which diagram of phases is shown in Fig. 1) with Laplace transform of its pdf given by

$$f_{X_d}^*(s) = \sum_{j=1}^{m} P_j \prod_{i=1}^{j} \frac{\eta_i}{(s+\eta_i)}$$

(13)

where

$$P_j = \alpha_j \prod_{i=1}^{j-1}(1 - \alpha_i)$$

(14)

$(1-\alpha_i)$ represents the probability of passing from the i-th phase to the $(i+1)$-th one.

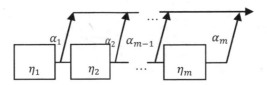

Fig. 1. Diagram of phases of the considered Coxian distribution of order m for modeling cell dwell time.

For exponentially distributed UST and using (9), the Laplace transforms of the pdf of CHTh and CHTn are given by

$$f_{X_c^{(h)}}^*(s) = \frac{\mu}{s+\mu} + \frac{s}{s+\mu}\left[f_{X_d}^*(s + \mu)\right]$$

(15)

$$f_{X_c^{(N)}}^*(s) = \frac{\mu}{s+\mu} + \frac{s}{s+\mu}\left[f_{X_r}^*(s + \mu)\right]$$

(16)

Replacing (13) into (15), it can be written as

$$f^*_{X^{(h)}_c}(s) = \sum_{j=1}^m P_j^{O(h)} \prod_{i=1}^j \frac{\eta_i}{(s+\mu+\eta_i)} \tag{17}$$

where

$$P_j^{O(h)} = \left[\prod_{i=1}^{j-1} \frac{\eta_i}{\mu+\eta_i}\right]\left[P_j + \sum_{k=j+1}^m P_k\left(\frac{\mu}{\mu+\eta_j}\right)\right] \tag{18}$$

for $i = 1, ..., m$. Then, CHTh has also a Coxian distribution of order m but with parameters $(\mu + \eta_i)$, for $i = 1, ..., m$.

On the other hand, the Laplace transform of the residual cell dwell time can be shown to be given by

$$f^*_{X_r}(s) = \sum_{j=1}^{\frac{m(m+1)}{2}} P_j^{(N)} \left(\prod_{k=h(j)}^{f(j)} \frac{\eta_k}{s+\eta_k}\right) \tag{19}$$

where

$$P_j^{(N)} = \frac{P_{f(j)} \prod_{\substack{k=1 \\ k\neq h(j)}}^m \eta_k}{\sum_{i=1}^m \left[\left(\prod_{\substack{k=1 \\ k\neq i}}^m \eta_k\right)\left(\sum_{l=i}^m P_l\right)\right]} \tag{20}$$

$$f(j) = \left\lfloor \frac{1\pm\sqrt{1+8(j-1)}}{2}\right\rfloor \tag{21}$$

$$h(j) = j - \frac{f(j)(f(j)-1)}{2} \tag{22}$$

for $j = 1, ..., m(m+1)/2$. Substituting (19) into (16), Laplace transform of CHTn can be written as

$$f^*_{X^{(N)}_c}(s) = \sum_{j=1}^{\frac{m(m+1)}{2}} P_j^{O(N)} \left(\prod_{k=h(j)}^{f(j)} \frac{\eta_k}{s+\mu+\eta_k}\right) \tag{23}$$

where

$$P_j^{O(N)} = \left[\prod_{i=h(j)}^{f(j)-1} \frac{\eta_i}{\mu+\eta_i}\right]\left[P_j^{(N)} + \sum_{k=f(j)+1}^m P_{\frac{k^2-k+2}{2}}^{(N)}\left(\frac{\mu}{\mu+\eta_{f(j)}}\right)\right] \tag{24}$$

Equation (23) corresponds to the Laplace transform of a generalized Coxian pdf.

The above analytical results show that CHTh (CHTn) has the same probability distribution as CDT (CDTr) but with different parameters of the phases, probabilities of reaching the absorbing state after each phase, and probabilities of choosing each stage. The detailed derivation of the last entry of Tables I and II (i.e., when cell dwell time has generalized Coxian distribution) is addressed in (Corral-Ruiz et al., a, 2010).

3.3 Relationship between $X_c^{(h)}$ and $X_c^{(N)}$

Using (15) and (16) it is straightforward to show that the mean values of CHTn and CHTh are, respectively, given by

$$E\left\{X_c^{(N)}\right\} = \frac{1}{\mu}\left[1 - \frac{\eta}{\mu}\left[1 - f_{X_d}^*(\mu)\right]\right] \tag{25}$$

$$E\left\{X_c^{(h)}\right\} = \frac{1}{\mu}\left[1 - f_{X_d}^*(\mu)\right] \tag{26}$$

At this point, it is important to mention that authors in (Fang, 2001; Zeng et al., 2002) stated that, depending on the variance of CDT, the mean CHTn can be greater than the mean CHTh. However, it was neither explained nor discussed the physical reasons for this observed behavior. This behavior occurs because the residual cell dwell times tend to increase as the variance of cell dwell time increases, as it was explained above.

Using (25) and (26), the condition for which the mean CHTn is greater that the mean CHTh, that is,

$$E\left\{X_c^{(N)}\right\} > E\left\{X_c^{(h)}\right\} \tag{27}$$

can be easily found. This condition is given by

$$f_{X_d}^*(\mu) > \frac{\eta}{\mu + \eta} \tag{28}$$

Thus, the relationship between the mean new and handoff call channel holding times is determined by the mean values of both CDT and UST and by the Laplace transform of the pdf of CDT evaluated at the inverse of the mean UST.

Finally, in a similar way, the squared coefficient of variation for CHTn and CHTh can be shown to be given, respectively, by

$$CoV^2\left(X_c^{(N)}\right) = \frac{-4\eta\, E\{X_c^{(h)}\} + 2\left[1 - \eta\frac{d f_{X_d}^*(\mu)}{d\mu}\right]}{\left[E\{X_c^{(N)}\}\mu\right]^2} - 1 \tag{29}$$

$$CoV^2\left(X_c^{(h)}\right) = \frac{2}{\left(E\{X_c^{(h)}\}\right)^2\mu}\left[\frac{d f_{X_d}^*(\mu)}{d\mu} + E\{X_c^{(h)}\}\right] - 1 \tag{30}$$

It can be shown that the n-th moments for new and handoff call channel holding times are given, respectively, by

$$E\left\{\left(X_c^{(N)}\right)^n\right\} = \frac{1}{\mu}\left[nE\left\{\left(X_c^{(N)}\right)^{n-1}\right\} - \eta E\left\{\left(X_c^{(h)}\right)^n\right\}\right] \tag{31}$$

$$E\left\{\left(X_c^{(h)}\right)^n\right\} = \frac{n}{\mu}\left((-1)^n\frac{d^n\left[f_{X_d}^*(\mu)\right]}{d\mu^n} + E\left\{\left(X_c^{(h)}\right)^{n-1}\right\}\right) \tag{32}$$

4. Numerical results and discussion

In this section, numerical results on how the distribution of cell dwell time (CDT) affects the characteristics of channel holding time (CHT) are presented. We use different distributions to model CDT, say, negative-exponential, constant (deterministic), Pareto with shape parameter α in the range (1, 2] (i.e., when infinite variance is considered), Pareto with $\alpha > 2$

hen finite variance is considered), log-normal, gamma, hyper-Erlang of order (2,2), hyper-exponential of order 2, and Coxian of order 2. Three different mobility scenarios for the numerical evaluation are assumed: $E\{X_d\}=5\cdot E\{X_s\}$ (low mobility), $E\{X_d\}=E\{X_s\}$ (moderate mobility), and $E\{X_d\}=0.2\cdot E\{X_s\}$ (high mobility). In the plots of this section we use $E\{X_s\}=180$ s. In our numerical results, the effect of CoV and skewness of CDT on CHT characteristics is investigated. In the plots presented in this section, "HC" and "NC" stand for channel holding time for handoff calls (CHTh) and channel holding time for new calls (CHTn), respectively.

4.1 Cell dwell time distribution completely characterized by its mean value

Fig. 2 plots the mean value of both CHTn and CHTh versus the mean value of CDT when it is modeled by negative-exponential (EX), constant, and Pareto with $1<\alpha\leq2$ distributions. It is important to remark that all of these distributions are completely characterized by their respective mean values. As expected, Fig. 2 shows that, for the case when CDT is exponentially distributed, mean CHTn is equal to mean CHTh. An interesting observation on the results shown in Fig. 2 is that, irrespective of the mean value of CDT, there exists a significant difference between the mean value of CHTn when CDT is modeled as exponential distributed RV and the corresponding case when it is modeled by a heavy-tailed Pareto distributed RV (this behavior is especially true for the case when α=1.1). Notice, however, that this difference is negligible for the case when α=2 and high mobility scenarios (say, $E\{X_d\}<50$ s) are considered. Similar behaviors are observed if mean CHTh is considered. Consequently, for high mobility scenarios where CDT can be statistical characterized by a Pareto distribution with shape parameter close to 2, the exponential distribution represents a suitable model for the CDT distribution. Fig. 2 also shows that, for

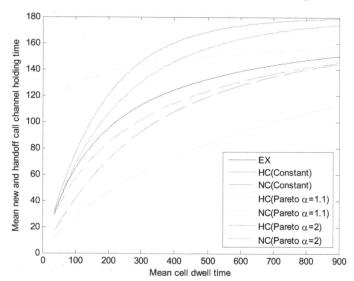

Fig. 2. Mean new and handoff call channel holding time for deterministic, negative exponentially, and Pareto distributed CDT against the mean CDT.

a given value of the mean CDT and considering the case when CDT is Pareto distributed with α=1.1 (α=2), mean CHTn always is greater (lower) than mean CHTh. This behavior can be explained by the combined effect of the following two facts. First, as α comes closer to 1 (2), the probability that CDT takes higher values increases (decreases). This fact contributes to increase (reduce) the mean CHTh. Second, in general, new calls are more probable to start on cells where users spent more time and, as α comes closer to 1, this probability increases. This fact contributes to increase mean CHTn relative to the mean CHTh. Then, the combined effect is dominated by the first (second) fact as α comes closer to 2 (1). This leads us to the behavior explained above and illustrated in Fig.2. It may be interesting to derive the condition upon which the mean CHTn is greater than the mean CHTh when CDT is heavy-tailed Pareto distributed. This represents a topic of our current research.

4.2 Cell dwell time distribution completely characterized by its first two moments

Fig. 3 plots the mean value of both CHTn and CHTh versus the CoV of CDT when it is modeled by Pareto with shape parameter α>2, lognormal, and Gamma distributions; all of them with mean value equal to 180 s. It is important to remark that all of these distributions are completely characterized by their respective first two moments. Fig. 3 shows that both mean CHTn and mean CHTh are highly sensitive to the type of distribution of CDT; this fact is especially true for $CoV>2$. Notice that, for the particular case when $CoV=0$, the mean values of both CHTn and CHTh are identical to the corresponding values for the case when CDT is deterministic with mean value equals 180 s, as expected. Fig. 3 also shows that, for values of CoV of CDT greater than 1 (1.2), mean CHTn is greater that mean CHTh when CDT is Gamma (log-normal) distributed. On the other hand, when CDT is Pareto distributed and irrespective of the value of its CoV, CHTh always is greater that mean CHTn. This behavior is mainly due to the heavy-tailed characteristics of the Pareto distribution.

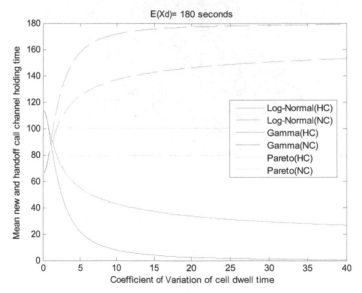

Fig. 3. Mean new and handoff call channel holding time for gamma, log-normal, and Pareto distributed cell dwell time versus CoV of cell dwell time.

4.3 Cell dwell time distribution completely characterized by its first three moments

Figs. 4, 5, and 6 (7, 8, and 9) plot the mean value (CoV) of both CHTn and CHTh versus both the CoV and skewness of CDT when it is modeled by hyper-Erlang (2,2), hyper-exponential of order 2, and Coxian of order 2 distributions, respectively. It is important to remark that all of these distributions are completely characterized by their respective first three moments. Results of (Johnson & Taaffe, 1989; Telek & Heindl, 2003) are used to calculate the parameters of these distributions as function of their first three moments. In Figs. 4 to 9, two different values for the mean CDT are considered: 36 s (high mobility scenario) and 900 s (low mobility scenario). From Figs. 2, 5 and 6 the following interesting observation can be extracted. Notice that, for the case when CDT is modeled by either hyper-exponential or Coxian distributions and irrespective of the mean value of CDT, the particular scenario where skewness and *CoV* of CDT are, respectively, equal to 2 and 1, corresponds to the case when CDT is exponential distributed (in the exponential case mean CHTn and mean CHTh are identical).

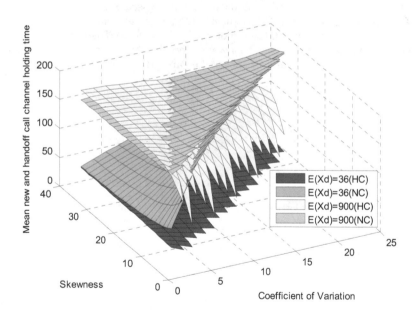

Fig. 4. Mean CHTn and mean CHTh for hyper-Erlang distributed CDT versus CoV and skewness of CDT, with the mean CDT as parameter.

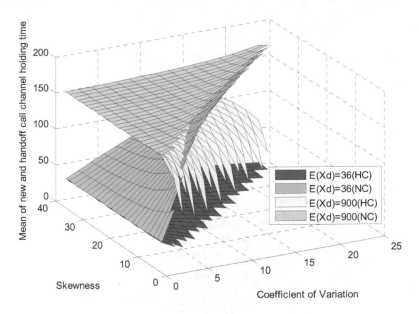

Fig. 5. Mean CHTn and mean CHTh for hyper-exponentially distributed CDT versus CoV and skewness of CDT, with the mean CDT as parameter.

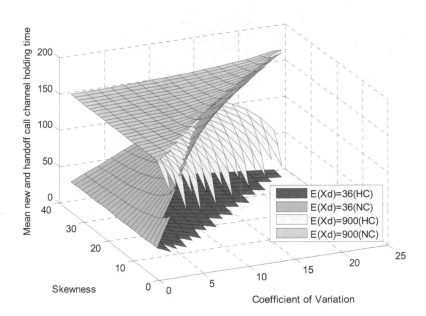

Fig. 6. Mean CHTn and mean CHTh for Coxian distributed cell dwell time versus CoV and skewness of cell dwell time, with the mean CDT as parameter.

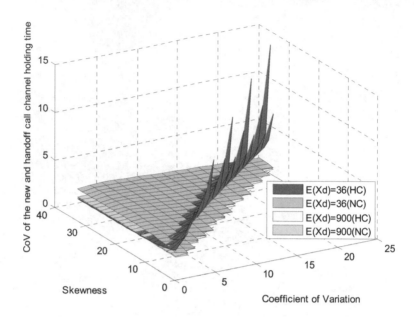

Fig. 7. CoV of CHTn and CHTh for hyper-Erlang distributed CDT versus CoV and skewness of CDT, with the mean CDT as parameter.

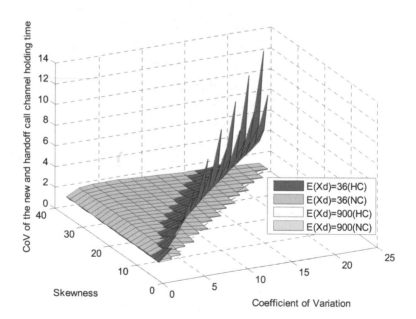

Fig. 8. CoV of CHTn and CHTh for hyper-exponential distributed CDT versus CoV and skewness of CDT, with the mean CDT as parameter.

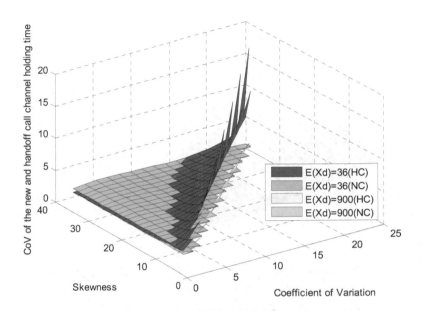

Fig. 9. CoV of CHTn and CHTh for Coxian distributed CDT versus CoV and skewness of cell dwell time, with the mean cell dwell time as parameter.

On the other hand, Fig. 4 shows that the case when hyper-Erlang distribution with skewness equals 2 and *CoV* equals 1 is used to model CDT does not strictly correspond to the exponential distribution; however, the exponential model represents a suitable approximation for CDT in this particular case. From Figs. 4 to 9, it is observed that the qualitative behavior of mean and *CoV* of both CHTn and CHTh is very similar for all the phase-type distributions under study. The small quantitative difference among them is due to moments higher than the third one. Analyzing the impact of moments of CDT higher than the third one on channel holding time characteristics represents a topic of our current research.

From Fig. 10 is observed that the difference among the mean values of CHTn and CHTh is strongly sensitive to the CoV of the CDT, while it is practically insensitive to the skewness of the CDT. This difference is higher for the case when the CDT is modeled as hyper-exponential distributed RV compared with the case when it is modeled as hyper-Erlang distributed RV. Also, it is observed that this difference remains almost constant for the entire range of values of the CoV of the CDT.

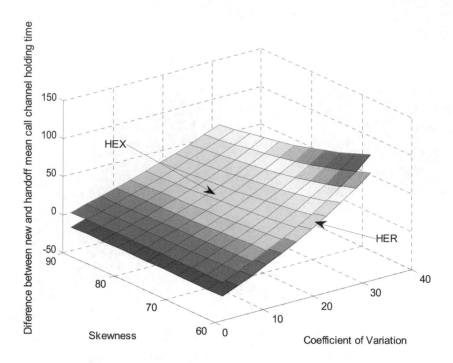

Fig. 10. Difference among the mean values of new and handoff call channel holding times for hyper-Erlang and hyper-exponential distributed cell dwell time versus CoV and skewness of cell dwell time, for the moderate-mobility scenario.

Finally, in Fig. 11 the mean channel holding time for new and handoff calls considering the gamma, hyper-Erlang (2,2), hyper-exponential of order 2, and Coxian of order 2 distributions for the cell dwell time are shown for different values of the coefficient of variation. The numerical results shown in Fig. 11 are obtained by equaling the first three moments of the different distributions to those of the gamma distribution. From Fig. 11, it is observed that for the hyper-exponential and Coxian distributions practically the same results are obtained for the mean channel holding time for both new and handoff calls. The differences among the other distributions are due to the fact that they differ on the higher order moments. To show this, the forth standardized moment (i.e., excess kurtosis) of the different distributions is shown in Fig. 12 for different values of the coefficient of variation, equaling the first three moments of the different distributions to those of the gamma distribution. From Fig. 12, it is observed that the hyper-exponential and Coxian distributions practically have the same value of excess kurtosis but this differs for that of the gamma and hyper-Erlang distributions. The gamma distribution shows the more different value of the excess kurtosis and, therefore, for this distribution the more different values of the mean channel holding times in Fig. 11 are obtained. Then, it could be necessary to capture more than three moments, even though the lower order moments dominate in importance. Similar conclusion was drawn in (Gross & Juttijudata, 1997).

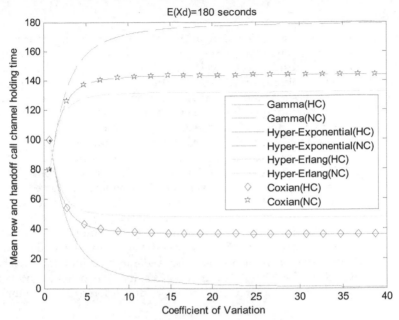

Fig. 11. Mean new and handoff call channel holding time for gamma, hyper-exponential (2), hyper-Erlang (2,2) and Coxian (2) distributed cell dwell time versus CoV of cell dwell time.

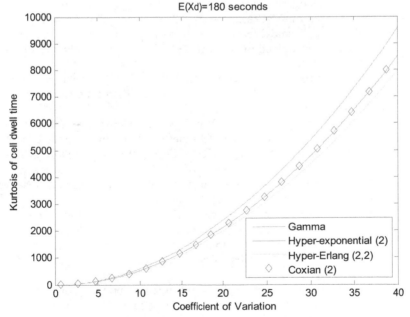

Fig. 12. Kurtosis of cell dwell time for gamma, hyper-exponential (2), Coxian (2) and hyper-Erlang (2,2) distributed cell dwell time versus CoV of cell dwell time.

5. Conclusions

In this Chapter, under the assumption that unencumbered service time is exponentially distributed, a set of novel general-algebraic equations that examines the relationships between cell dwell time and residual cell dwell time as well as between cell dwell time and new and handoff channel holding times was derived. This work includes relevant new analytical results and insights into the dependence of channel holding time characteristics on the cell dwell time probability distribution. For instance, we found that when cell dwell time is Coxian or hyper-exponentially distributed, channel holding times are also Coxian or hyper-exponentially distributed, respectively. Also, our analytical results showed that the mean and coefficient of variation of the new and handoff call channel times depend on Laplace transform and first derivative of the Laplace transform of the probability density function of cell dwell time evaluated at the inverse of the mean unencumbered service time as well as on the mean of both cell dwell time and unencumbered service time. Additionally, we derive the condition upon which the mean new call channel holding time is greater than the mean handoff call channel holding time. Similarly, the condition upon which the mean residual cell dwell time is greater than the mean cell dwell time was also derived. To the best authors' knowledge, this phenomenon that may seem to be counterintuitive has been explained and mathematically formulated in this Chapter. We believe that the study presented here is important for planning, designing, dimensioning, and optimizing of mobile cellular networks.

6. References

Alfa A.S. and Li W., "A homogeneous PCS network with Markov call arrival process and phase type cell dwell time," Wirel. Net., vol. 8, no. 6, pp. 597-605, 2002.

Christensen T.K., Nielsen B.F., and Iversen V.B., "Phase-type models of channel-holding times in cellular communication systems," IEEE Trans. Veh. Technol., vol. 53, no. 3, pp. 725-733, May 2004.

Corral-Ruiz A.L.E., Cruz-Pérez F.A., and Hernández-Valdez G., a, "Channel holding time in mobile cellular networks with generalized Coxian distributed cell dwell time," IEEE PIMRC'2010, Istanbul, Turkey, Sep. 2010.

Corral-Ruiz A.L.E., Rico-Páez Andrés, Cruz-Pérez F.A., and Hernández-Valdez G., b, "On the Functional Relationship between Channel Holding Time and Cell Dwell Time in Mobile Cellular Networks," IEEE GLOBECOM'2010, Miami, Florida, USA, Dec. 2010.

Fang Y., "Hyper-Erlang distribution model and its application in wireless mobile networks," Wirel. Networks, vol. 7, no. 3, pp. 211-219, May. 2001.

Fang Y., a, "Performance evaluation of wireless cellular networks under more realistic assumptions," Wirel. Commun. Mob. Comp., vol. 5, no. 8, pp. 867-885, Dec. 2005.

Fang Y., b, "Modeling and performance analysis for wireless mobile networks: a new analytical approach," IEEE Trans. Networking, vol. 13,no. 5, pp. 989-1002, Oct. 2005.

Fang Y. and Chlamtac I., a, "Teletraffic analysis and mobility modeling of PCS networks," IEEE Trans. Commun., vol 47, no. 7, pp. 1062-1072, July 1999.

Fang Y., Chlamtac I., and Lin Y.-B., b, "Channel occupancy times and handoff rate for mobile computing and PCS networks," IEEE Trans. Computers, vol 47, no. 6, pp. 679-692, 1999.

Fang Y., Chlamtac I., and Lin Y.-B, a, "Modeling PCS networks under general call holding time and cell residence time distributions," IEEE/ACM Trans. Networking, vol. 5, no. 6, pp. 893-906, Dec. 1997.

Fang Y., Chlamtac I., and Lin Y.-B., b, "Call performance for a PCS network," IEEE J. Select. Areas Commun., vol. 15, no. 8, pp. 1568-1581, Oct. 1997.

Gross D. and Juttijudata M., "Sensitivity of output performance measures to input distributions in queueing," in Proc. Winter Simulation Conference (WSC'97), Atlanta, GA, Dec. 1997.

Hidaka H., Saitoh K., Shinagawa N., and Kobayashi T., "Self similarity in cell dwell time caused by terminal motion and its effects on teletraffic of cellular communication networks," IEICE Trans. Fund., vol. E85-A, no. 7, pp. 1445-1453, 2002.

Hong D. and Rappaport S. S., "Traffic model and performance analysis for cellular mobile radio telephone systems with prioritized and nonprioritized handoff procedures," IEEE Trans. Veh. Technol., vol. 35, no. 3, pp. 77–92, Aug. 1986.

Johnson M.A. and Taaffe M.R.. "Matching moments to phase distributions: mixture of Erlang distributions of common order," Stochastic Models, vol. 5, no. 4, pp. 711-743, 1989.

Khan F. and Zeghlache D., "Effect of cell residence time distribution on the performance of cellular mobile networks," Proc. IEEE VTC'97, Phoenix, AZ, May 1997, pp. 949-953.

Kim K. and Choi H., "A mobility model and performance analysis in wireless cellular network with general distribution and multi-cell model," Wirel. Pers. Commun., published on line: 10 March 2009.

Kleinrock L. Queueing Systems. John Wiley and Sons: New York, NY, 1975.

Lin Y.-B., Mohan S. and Noerpel A., "Queuing priority channel assignment strategies for PCS and handoff initial access," IEEE Trans. Veh. Technol., vol. 43. no. 3, pp. 704-712, Aug. 1994.

Orlik P.V. and Rappaport S.S., a, "A model for teletraffic performance and channel holding time characterization in wireless cellular communication with general session and dwell time distributions," IEEE J. Select. Areas Commun., vol. 16, no. 5, pp. 788-803, June 1998.

Orlik P.V. and Rappaport S.S., b, "Traffic performance and mobility modeling of cellular communications with mixed platforms and highly variable mobility," Proc. of the IEEE, vol. 86, no. 7, pp. 1464-1479, July 1998.

Pattaramalai S., Aalo V.A., and Efthymoglou G.P., "Evaluation of call performance in cellular networks with generalized cell dwell time and call-holding time distributions in the presence of channel fading," IEEE Trans. Veh. Technol., vol. 58, no. 6, pp. 3002-3013, July 2009.

Pattaramalai S., Aalo V.A., and Efthymoglou G.P., "Call completion probability with Weibull distributed call holding time and cell dwell time," in Proc. IEEE Globecom'2007, Washington, DC, Nov. 2007, pp. 2634-2638.

Perros H. and Khoshgoftaar T., "Approximating general distributions by a uniform Coxian distribution," in Proc. 20th Annual Pittsburgh Conf. on Modeling and Simulation, Instrument Society of America, 1989, 325-333.

Rahman M.M. and Alfa A.S., "Computationally efficient method for analyzing guard channel schemes," Telecomm. Systems, vol. 41, pp. 1-11, published on line: 22 April 2009.

Sinclair B., "Coxian Distributions," Connexions module: m10854. Available on line: http://cnx.org/content/m10854/latest/

Soong B.H. and Barria J.A., "A Coxian model for channel holding time distribution for teletraffic mobility modeling," IEEE Commun. Letters, vol. 4, no. 12, pp. 402-404, Dec. 2000.

Telek M. and Heindl A., "Matching moments for acyclic discrete and continuous phase-type distributions of second order," International Journal of Simulation, vol. 3, no. 3-4, pp. 47-57, 2003.

Thajchayapong S. and Tonguz O. K., "Performance implications of Pareto-distributed cell residual time in distributed admission control scheme (DACS)," in Proc. IEEE WCNC'05, New Orleans, LA, Mar. 2005, pp 2387-2392.

Wang X. and Fan P., "Channel holding time in wireless cellular communications with general distributed session time and dwell time," IEEE Commun. Letters, vol. 11, no. 2, Feb. 2007.

Yeo K. and Jun C.-H., "Teletraffic analysis of cellular communication systems with general mobility based on hyper-Erlang characterization," Computer & Industrial Engineering, vol. 42, pp. 507-520, 2002.

Zeng H., Fang Y., and Chlamtac I., "Call blocking performance study for PCS Networks under more realistic mobility assumptions," Telecomm. Systems, vol. 19, no. 2, pp. 125-146, 2002.

3

Generalized Approach to Signal Processing in Wireless Communications: The Main Aspects and some Examples

Vyacheslav Tuzlukov
Kyungpook National University
South Korea

1. Introduction

The additive and multiplicative noise exists forever in any wireless communication system. Quality and integrity of any wireless communication systems are defined and limited by statistical characteristics of the noise and interference, which are caused by an electromagnetic field of the environment. The main characteristics of any wireless communication system are deteriorated as a result of the effect of the additive and multiplicative noise. The effect of addition of noise and interference to the signal generates an appearance of false information in the case of the additive noise. For this reason, the parameters of the received signal, which is an additive mixture of the signal, noise, and interference, differ from the parameters of the transmitted signal. Stochastic distortions of parameters in the transmitted signal, attributable to unforeseen changes in instantaneous values of the signal phase and amplitude as a function of time, can be considered as multiplicative noise. Under stimulus of the multiplicative noise, false information is a consequence of changed parameters of transmitted signals, for example, the parameters of transmitted signals are corrupted by the noise and interference. Thus, the impact of the additive noise and interference may be lowered by an increase in the signal-to-noise ratio (SNR). However, in the case of the multiplicative noise and interference, an increase in SNR does not produce any positive effects.

The main functional characteristics of any wireless communication systems are defined by an application area and are often specific for distinctive types of these systems. In the majority of cases, the main performance of any wireless communication systems are defined by some initial characteristics describing a quality of signal processing in the presence of noise: the precision of signal parameter measurement, the definition of resolution intervals of the signal parameters, and the probability of error.

The main idea is to use the generalized approach to signal processing (GASP) in noise in wireless communication systems (Tuzlukov, 1998; Tuzlukov, 1998; Tuzlukov, 2001; Tuzlukov, 2002; Tuzlukov, 2005; Tuzlukov, 2012). The GASP is based on a seemingly abstract idea: the introduction of an additional noise source that does not carry any information about the signal and signal parameters in order to improve the qualitative performance of wireless communication systems. In other words, we compare statistical data defining the statistical parameters of the probability distribution densities (pdfs) of the observed input stochastic samples from two independent frequency-time regions – a "yes" signal is possible in the first region and it is known a priori that a "no" signal is obtained in the second region. The proposed GASP allows us to formulate a decision-making rule based on the determination of *the jointly sufficient statistics of the mean and variance* of the likelihood function (or functional). Classical and modern signal processing theories allow us to define *only the mean* of the likelihood function (or functional). Additional information about the statistical characteristics of the likelihood function (or functional) leads us to better quality signal detection and definition of signal parameters in compared with the optimal signal processing algorithms of classical or modern theories.

Thus, for any wireless communication systems, we have to consider two problems – analysis and synthesis. The first problem (analysis) – the problem to study a stimulus of the additive and multiplicative noise on the main principles and performance under the use of GASP – is an analysis of impact of the additive and multiplicative noise on the main characteristics of wireless communication systems, the receivers in which are constructed on the basis of GASP. This problem is very important in practice. This analysis allows us to define limitations on the use of wireless communication systems and to quantify the additive and multiplicative noise impact relative to other sources of interference present in these systems. If we are able to conclude that the presence of the additive and multiplicative noise is the main factor or one of the main factors limiting the performance of any wireless communication systems, then the second problem – the definition of structure and main parameters and characteristics of the generalized detector or receiver (GD or GR) under a dual stimulus of the additive and multiplicative noise – the problem of synthesis – arises.

GASP allows us to extend the well-known boundaries of the potential noise immunity set by classical and modern signal processing theories. Employment of wireless communication systems, the receivers of which are constructed on the basis of GASP, allows us to obtain high detection of signals and high accuracy of signal parameter definition with noise components present compared with that systems, the receivers of which are constructed on the basis of classical and modern signal processing theories. The optimal and asymptotic optimal signal processing algorithms of classical and modern theories, for signals with amplitude-frequency-phase structure characteristics that can be known and unknown a priori, are constituents of the signal processing algorithms that are designed on the basis of GASP.

2. GASP: Brief description

GASP is based on the assumption that the frequency-time region Z of the noise exists where a signal may be present; for example, there is an observed stochastic sample from this region, relative to which it is necessary to make the decision a "yes" signal (the hypothesis H_1) or a "no" signal (the hypothesis H_0). We now proceed to modify the initial premises of the classical and modern signal processing theories. Let us suppose there are two independent frequency-time regions Z and Z^* belonging to the space A. Noise from these regions obeys the same pdf with the same statistical parameters (for simplicity of considerations). Generally, these parameters are differed. A "yes" signal is possible in the noise region Z as before. *It is known a priori that a "no" signal is obtained in the noise region Z^*.* It is necessary to make the decision a "yes" signal (the hypothesis H_1) or a "no" signal (the hypothesis H_0) in the observed stochastic sample from the region Z, by comparing statistical parameters of pdf of this sample with those of the sample from the reference region Z^*. Thus, there is a need to accumulate and compare statistical data defining the statistical parameters of pdf of the observed input stochastic samples from two independent frequency-time regions Z and Z^*. If statistical parameters for two samples are equal or agree with each other within the limits of a given before accuracy, then the decision of a "no" signal in the observed input stochastic process $X_1,...,X_N$ is made – the hypothesis H_0. If the statistical parameters of pdf of the observed input stochastic sample from the region Z differ from those of the reference sample from the region Z^* by a value that exceeds the prescribed error limit, then the decision of a "yes" signal in the region Z is made – the hypothesis H_1.

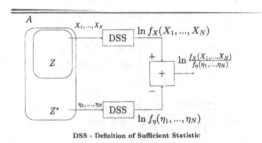

DSS - Definition of Sufficient Statistic

It is known *a priori* that a "no" signal obtains in
the noise region Z^*.

Fig. 1. Definition of sufficient statistics under GASP.

The simple model of GD in form of block diagram is represented in Fig.2. In this model, we use the following notations: MSG is the model signal generator (the local oscillator), the AF is the additional filter (the linear system) and the PF is the preliminary filter (the linear system) A detailed discussion of the AF and PF can be found in (Tuzlukov, 2001 and Tuzlukov, 2002).

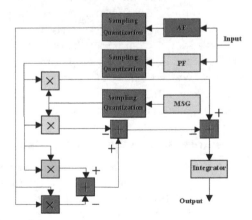

Fig. 2. Principal flowchart of GD.

Consider briefly the main statements regarding the AF and PF. There are two linear systems at the GD front end that can be presented, for example, as bandpass filters, namely, the PF with the impulse response $h_{PF}(\tau)$ and the AF with the impulse response $h_{AF}(\tau)$. For simplicity of analysis, we think that these filters have the same amplitude-frequency responses and bandwidths. Moreover, a resonant frequency of the AF is detuned relative to a resonant frequency of PF on such a value that signal cannot pass through the AF (on a value that is higher the signal bandwidth). Thus, the signal and noise can be appeared at the PF output and *the only noise* is appeared at the AF output. It is well known, if a value of detuning between the AF and PF resonant frequencies is more than $4 \div 5\Delta f_a$, where Δf_a is the signal bandwidth, the processes forming at the AF and PF outputs can be considered as independent and uncorrelated processes (in practice, the coefficient of correlation is not more than 0.05). In the case of signal absence in the input process, the statistical parameters at the AF and PF outputs will be the same, because the same noise is coming in at the AF and PF inputs, and we may think that the AF and PF do not change the statistical parameters of input process, since they are the linear GD front end systems.

By this reason, the AF can be considered as a generator of reference sample with *a priori information a "no" signal is obtained in the additional reference noise* forming at the AF output. There is a need to make some comments regarding the noise forming at the PF and AF outputs. If the Gaussian noise $n(t)$ comes in at the AF and PF inputs (the GD linear system front end), the noise forming at the AF and PF outputs is Gaussian, too, because the AF and PF are the linear systems and, in a general case, take the following form:

$$n_{PF}(t) = \int_{-\infty}^{\infty} h_{PF}(\tau)n(t-\tau)d\tau \quad \text{and} \quad n_{AF}(t) = \int_{-\infty}^{\infty} h_{AF}(\tau)n(t-\tau)d\tau . \tag{1}$$

If, for sake of simplicity, the additive white Gaussian noise (AWGN) with zero mean and two-sided power spectral density $2N_0$ is coming in at the AF and PF inputs (the GD linear system front end), then the noise forming at the AF and PF outputs is Gaussian with zero mean and variance given by $\sigma_n^2 = \frac{2N_0\omega_0^2}{8\Delta_F}$ (Tuzlukov, 2002) where in the case if AF (or PF) is the RLC oscillatory circuit, the AF (or PF) bandwidth Δ_F and resonance frequency ω_0 are defined in the following manner $\Delta_F = \pi\beta$, $\omega_0 = \frac{1}{\sqrt{LC}}, \beta = \frac{R}{2L}$ .The main functioning condition of GD is an equality over the whole range of parameters between the model signal $u^*(t)$ at the GD MSG output and the transmitted signal $u(t)$ forming at the GD input liner system (the PF) output, i.e. $u(t) = u^*(t)$. How we can satisfy this condition in practice is discussed in detail in (Tuzlukov, 2002; Tuzlukov, 2012). More detailed discussion about a choice of PF and AF and their impulse responses is given in (Tuzlukov, 1998).

3. Diversity problems in wireless communication systems with fading

In the design of wireless communication systems, two main disturbance factors are to be properly accounted for, i.e. fading and additive noise. As to the former, it is usually taken into account by modeling the propagation channel as a linear-time-varying filter with random impulse response (Bello, 1963 & Proakis, 2007). Indeed, such a model is general enough to encompass the most relevant instances of fading usually encountered in practice, i.e.

frequency- and/or time-selective fading, and flat-flat fading. As to the additive noise, such a disturbance has been classically modeled as a possibly correlated Gaussian random process.

However, the number of studies in the past few decades has shown, through both theoretical considerations and experimental results, that Gaussian random processes, even though they represent a faithful model for the thermal noise, are largely inadequate to model the effect of real-life noise processes, such as atmospheric and man-made noise (Kassam, 1988 & Webster, 1993) arising, for example, in outdoor mobile communication systems. It has also been shown that non-Gaussian disturbances are commonly encountered in indoor environments, for example, offices, hospitals, and factories (Blankenship & Rappaport, 1993), as well as in underwater communications applications (Middleton, 1999). These disturbances have an impulsive nature, i.e. they are characterized by a significant probability of observing large interference levels.

Since conventional receivers exhibit dramatic performance degradations in the presence of non-Gaussian impulsive noise, a great attention has been directed toward the development of non-Gaussian noise models and the design of optimized detection structures that are able to operate in such hostile environments. Among the most popular non-Gaussian noise models considered thus far, we cite the alpha-stable model (Tsihrintzis & Nikias, 1995), the Middleton Class-A and Class-B noise (Middleton, 1999), the Gaussian-mixture model (Garth & Poor, 1992) which, in turn, is a truncated version, at the first order, of the Middleton Class-A noise, and the compound Gaussian model (Conte et al., 1995). In particular, in the recent past, the latter model, subsuming, as special cases, many marginal probability density functions (pdfs) that have been found appropriate for modeling the impulsive noise, like, for instance, the Middleton Class-A noise, the Gaussian-mixture noise (Conte, 1995), and the symmetric alpha-stable noise (Kuruoglu, E. et al., 1998). They can be deemed as the product of a Gaussian, possibly complex random process times a real non-negative one.

Physically, the former component, which is usually referred to as speckle, accounts for the conditional validity of the central limit theorem, whereas the latter, the so-called texture process, rules the gross characteristics of the noise source. A very interesting property of compound-Gaussian processes is that, when observed on time intervals whose duration is significantly shorter than the average decorrelation time of the texture component, they reduce to spherically invariant random processes (SIRPs) (Yao, 1973), which have been widely adopted to model the impulsive noise in wireless communications (Gini, F et al., 1998), multiple access interference in direct-sequence spread spectrum cellular networks (Sousa, 1990), and clutter echoes in radar applications (Sangston & Gerlach, 1994).

We consider the problem of detecting one of M signals transmitted upon a zero-mean fading dispersive channel and embedded in SIRP noise by GD based on the GASP in noise. The similar problem has been previously addressed. In (Conte, 1995), the optimum receiver for flat-flat Rayleigh fading channels has been derived, whereas in (Buzzi et al., 1999), the case of Rayleigh-distributed, dispersive fading has been considered. It has been shown therein that the receiver structure consists of an estimator of the short-term conditional, i.e. given the texture component, noise power and of a bank of M estimators-correlators keyed to the estimated value of the noise power. Since such a structure is not realizable, a suboptimum detection structure has been introduced and analyzed in (Buzzi et al., 1997).

We design the GD extending conditions of (Buzzi et al., 1997) and (Buzzi et al., 1999) to the case that a diversity technique is employed. It is well known that the adoption of diversity techniques is effective in mitigating the negative effects of the fading, and since conventional diversity techniques can incur heavy performance loss in the presence of impulsive disturbance (Kassam & Poor, 1985), it is of interest to envisage the GD for optimized diversity reception in non-Gaussian noise. We show that the optimum GD is independent of the joint pdf of the texture components on each diversity branch. We also derive a suboptimum GD, which is amenable to a practice. We focus on the relevant case of binary frequency-shift-keying (BFSK) signaling and provide the error probability of both the optimum GD and the suboptimum GD. We assess the channel diversity order impact and noise spikiness on the performance.

3.1 Problem statement

The problem is to derive the GD aimed at detecting one out of M signals propagating through single-input multiple-output channel affected by dispersive fading and introducing the additive non-Gaussian noise. In other words, we have to deal with the following M-ary hypothesis test:

$$
H_i \Rightarrow \begin{cases} x_1(t) = s_{1,i}(t) + n_1(t) \\ \dots\dots\dots\dots\dots\dots\dots \\ x_P(t) = s_{P,i}(t) + n_P(t) \end{cases} \quad i = 1,\dots,M \quad t \in [0,T], \tag{2}
$$

where P is the channel diversity order and $[0,T]$ is the observation interval; the waveforms $\{x_p(t)\}_{p=1}^P$ are the complex envelopes of the P distinct channel outputs; $\{s_{p,i}(t)\}_{p=1}^P, i = 1,\dots$ $,M$ represent the baseband equivalents of the useful signal received on the P diversity branches under the ith hypothesis. Since the channel is affected by dispersive fading, we may assume (Proakis, 2007) that these waveforms are related to the corresponding transmitted signals $u_i(t)$

$$
s_{p,i}(t) = \int_{-\infty}^{\infty} h_p(t,\tau) u_i(t - \tau) d\tau , \quad t \in [0,T] \tag{3}
$$

where $h_p(t,\tau), p = 1,\dots,P$ is the random impulse response of the channel pth diversity branch and is modeled as a Gaussian random process with respect to the variable t. In keeping with the uncorrelated-scattering model, we assume that the random processes $h_p(t,\tau), p = 1,\dots,P$ are all statistically independent; as a consequence, the waveforms $\{s_{p,i}(t)\}_{p=1}^P$ are themselves independent complex Gaussian random processes that we assume to be zero-mean and with the covariance function

$$
Cov(t,\tau) = E[s_{p,i}(t)s_{p,i}^*(\tau)] , \quad i = 1,\dots,M \quad t,\tau \in [0,T] \tag{4}
$$

independent of p (the channel correlation properties are identical of each branch) and upper bounded by a finite positive constant. This last assumption poses constraint on the average receive energy in the i-th hypothesis $E_i = \int_0^T Cov_i(t,t)dt < \infty$. We also assume in keeping with the model (Van Trees, 2001) that $E[s_{p,i}(t)s_{p,i}(\tau)] = 0$. This is not a true limitation in most practical instances, and it is necessarily satisfied if the channel is wide sense stationary. Finally, as to the additive non-Gaussian disturbances $\{n_p(t)\}_{p=1}^{P}$, we resort to the widely adopted compound model, i.e. we deem the waveform $n_p(t)$ as the product of two independent processes:

$$n_p(t) = v_p(t)g_p(t) , \qquad p = 1,\ldots,P \tag{5}$$

where $v_p(t)$ is a real non-negative random process with marginal pdf $f_{v_p}(\cdot)$ and $g_p(t)$ is a zero-mean complex Gaussian process. If the average decorrelation time of $v_p(t)$ is much larger than the observation interval $[0,T]$, then the disturbance process degenerates into SIPR (Yao, 1973)

$$n_p(t) = v_p g_p(t) , \qquad p = 1,\ldots,P . \tag{6}$$

From now on, we assume that such a condition is fulfilled, and we refer to (Conte, 1995) for further details on the noise model, as well as for a list of all of the marginal pdfs that are compatible with (5). Additionally, we assume $E[v_p^2] = 1$ and that the correlation function of the random process $g_p(t)$ is either known or has been perfectly estimated based on (5). While previous papers had assumed that the noise realization $n_1(t),\ldots,n_P(t)$ were statistically independent, in this paper, this hypothesis is relaxed. To be more definite, we assume that the Gaussian components $g_1(t),\ldots,g_P(t)$ are uncorrelated (independent), whereas the random variables $v_1,\ldots,v_P$ are arbitrary correlated. We thus denote by $f_{v_1,\ldots,v_P}(v_1,\ldots,v_P)$ their joint pdf. It is worth pointing out that the above model subsumes the special case that the random variables $v_1,\ldots,v_P$ are either statistically independent or fully correlated, i.e. $v_1 = \cdots = v_P$. Additionally, it permits modeling a much wider class of situations that may occur in practice. For instance, if one assumes that the P diversity observations are due to a temporal diversity, it is apparent that if the temporal distance between consecutive observations is comparable with the average decorrelation time of the process $v(t)$, then the random variables $v_1,\ldots,v_P$ can be assumed to be neither independent nor fully correlated. Such a model also turns out to be useful in clutter modeling in that if the diversity observations are due to the returns from neighboring cells, the corresponding texture components may be correlated (Barnard & Weiner, 1996). For sake of simplicity, consider the white noise case, i.e. $n_p(t)$ possesses an impulsive covariance $\forall p$

$$Cov_n(t,\tau) = 2N_0 E[v_p^2]\delta(t-\tau) = 2N_0\delta(t-\tau) , \tag{7}$$

where $2N_0$ is the power spectral density (PSD) of the Gaussian component of the noise processes $g_1(t),\ldots,g_P(t)$. Notice that this last assumption does not imply any loss of generality should the noise possess a non-impulsive correlation Then, due to the closure of SIRP with respect to linear transformations, the classification problem could be reduced to the above form by simply preprocessing the observables through a linear whitening filter. In such a situation, the $s_{p,i}(t)$ represent the useful signals at the output of the cascade of the channel and of the whitening filter. Due to the linearity of such systems, they are still Gaussian processes with known covariance functions. Finally, we highlight here that the assumption that the useful signals and noise covariance functions (3) and (6) are independent of the index p has been made to simplify notation.

3.2 Synthesis and design

3.2.1 Optimum GD structure design

Given the M-ary hypothesis test (1), the synthesis of the optimum GD structure in the sense of attaining the minimum probability of error P_E requires evaluating the likelihood functionals under any hypothesis and adopting a maximum likelihood decision-making rule. Formally, we have

$$\hat{H} = H_i \Rightarrow \Lambda[\mathbf{x}(t); H_i] > \max_{k \neq i} \Lambda[\mathbf{x}(t); H_k] \tag{8}$$

with $\mathbf{x}(t) = [x_1(t), \ldots, x_P(t)]^T$. The above functionals are usually evaluated through a limiting procedure. We evaluate the likelihood $f_{\mathbf{x}_Q|H_i}(\mathbf{x}_Q)$ of the Q-dimensional random vector $\mathbf{x} = [x_1, \ldots, x_Q]^T$ whose entries are the projections of the received signal along the first Q elements of suitable basis B_i. Therefore, the likelihood functional corresponding to H_i is

$$\Lambda[\mathbf{x}(t); H_i] = \lim_{Q \to \infty} \frac{f_{\mathbf{x}_Q|H_i}(\mathbf{x}_Q)}{f_{\mathbf{n}_{AF_Q}}(\mathbf{n}_{AF_Q})}, \tag{9}$$

where $f_{\mathbf{n}_{AF_Q}}(\mathbf{n}_{AF_Q})$ is the likelihood corresponding to the reference sample with *a priori* information a "no" signal is obtained in the additional reference noise forming at the AF output, i.e. no useful signal is observed at the P channel outputs. In order to evaluate the limit (9), we resort to a different basis for each hypothesis. We choose for the i-th hypothesis the Karhunen-Loeve basis B_i determined by the covariance function of the useful received signal under the hypothesis H_i. Projecting the waveform received on the p-th diversity branch along the first N axes of the i-th basis yields the following N-dimensional vector:

$$\mathbf{x}_{N,p}^i = \mathbf{s}_{N,p}^i + v_p \mathbf{g}_{N,p}^i, \qquad p = 1, \ldots, P \tag{10}$$

where $\mathbf{s}_{N,p}^i$ and $\mathbf{g}_{N,p}^i$ are the corresponding projections of the waveforms $s_{p,i}(t)$ and $g_p(t)$. Since B_i is the Karhunen-Loeve basis for the random processes $s_{1,i}(t), \ldots, s_{P,i}(t)$, the entries of $\mathbf{s}_{N,p}^i$ are a sequence of uncorrelated complex Gaussian random variables with the variances $(\sigma_{s_{1,i}}^2, \ldots, \sigma_{s_{N,i}}^2)$ which are the first N eigenvalues of the covariance function $Cov_i(t, u)$, whereas the entries of $\mathbf{g}_{N,p}$ are a sequence of uncorrelated Gaussian variables with variance $2N_0$. Here we adopt the common approach of assuming that any complete orthonormal system is an orthonormal basis for white processes (Conte, 1995 and Poor, 1988). Upon defining the following NP-dimensional vector

$$\mathbf{x}_N^i = [\mathbf{x}_{N,1}^{iT}, \mathbf{x}_{N,2}^{iT}, \ldots, \mathbf{x}_{N,P}^{iT}]^T \tag{11}$$

the likelihood functional taking into consideration subsection 3.1 and (Tuzlukov, 2001) can be written in the following form

$$\Lambda[\mathbf{x}_N^i;H_i] = \frac{f_{\mathbf{x}_N^i|H_i}(\mathbf{x}_N^i)}{f_{\mathbf{n}_{AF_N}^i|H_0}(\mathbf{n}_{AF_N}^i)} = \frac{\int \prod_{p=1}^{P} \prod_{j=1}^{N} \frac{1}{\sigma_{s_{j,i}}^2 + 4\sigma_n^4 y_p^2} \exp\left[-\frac{|x_{j,p}^i|^2}{\sigma_{s_{j,i}}^2 + 4\sigma_n^4 y_p^2}\right] f_{\mathbf{v}}(\mathbf{y})d\mathbf{y}}{\int \prod_{p=1}^{P} \frac{1}{(4\sigma_n^4 y_p^2)^N} \exp\left[-\frac{|n_{AF_{j,p}}^i|^2}{4\sigma_n^4 y_p^2}\right] f_{\mathbf{v}}(\mathbf{y})d\mathbf{y}}, \tag{12}$$

where $x_{j,p}^i$ is the j-th entry of the vector $\mathbf{x}_{N,p}^i$, the integrals in (12) are over the set $[0,\infty)^P$, $\mathbf{v} = [v_1,...,v_P]$, $\mathbf{y} = [y_1,...,y_P]$, and $d\mathbf{y} = \prod_{i=1}^{P} dy_i$. The convergence in measure of (12) for increasing N to the likelihood functional $\Lambda[\mathbf{x}(t);H_i]$ is ensured by the Grenander theorem (Poor, 1988). In order to evaluate the above functional, we introduce the substitution

$$y_p = \frac{\|\mathbf{x}_{N,p}^i\|}{\sqrt{4\sigma_n^4 z_p}}, \qquad p = 1,2,...,P \tag{13}$$

where $\|\ \|$ denotes the Euclidean norm. Applying the same limiting procedure as in (Buzzi, 1999), we come up with the following asymptotical expression:

$$\Lambda[\mathbf{x}(t);H_i] = \lim_{N\to\infty} \prod_{p=1}^{P} \Lambda_{gN}^p\left[\mathbf{x}_{N,p}^i, \frac{\|\mathbf{x}_{N,p}^i\|^2}{4\sigma_n^4 N};H_i\right], \tag{14}$$

where

$$\Lambda_{gN}^p\left(\mathbf{x}_{N,p}^i, y_p^2;H_i\right) = \exp\left\{\sum_{j=1}^{N}\left[\frac{\sigma_{s_{j,i}}^2 |x_{j,p}^i|^2}{4\sigma_n^4 y_p^2(\sigma_{s_{j,i}}^2 + 4\sigma_n^4 y_p^2)} - \ln\left(1 + \frac{\sigma_{s_{j,i}}^2}{4\sigma_n^4 y_p^2}\right)\right]\right\} \tag{15}$$

represents the ratio between the conditional likelihoods for H_i and H_0 based on the observation of the signal received on the p-th channel output only. Equation (14) also requires evaluating

$$Z_p = \lim_{N\to\infty} \frac{\|\mathbf{x}_{N,p}^i\|^2}{N}, \tag{16}$$

that, following in (Buzzi, 1999), can be shown to converge in the mean square sense to the random variable $4\sigma_n^4 v_p^2$ for any of the Karhunen-Loeve basis $B_i, i = 1,...,M$. Due to the fact that the considered noise is white, this result also holds for the large signal-to-noise ratios even though, in this case, a large number of summands is to be considered in order to achieve a given target estimation accuracy. Notice also that $4\sigma_n^4 v_p^2$ can be interpreted as a short-term noise power spectral density (PSD), namely, the PSD that would be measured on sufficiently short time intervals on the p-th channel output. Thus, the classification problem under study admits the sufficient statistics

$$\ln \Lambda[\mathbf{x}(t);H_i] = \sum_{p=1}^{P}\sum_{j=1}^{\infty}\left[\frac{\sigma_{s_{j,i}}^2 \, |x_{j,p}^i|^2}{Z_p(\sigma_{s_{j,i}}^2 + Z_p)} - \ln\left(1 + \frac{\sigma_{s_{j,i}}^2}{Z_p}\right)\right]. \tag{17}$$

The above equations demonstrate that the optimum GD structure for the problem given in (1) is completely canonical in that for any $f_{v_1,\ldots,v_p}(v_1,\ldots,v_p)$ and, for any noise model in the class of compound-Gaussian processes and for any correlation of the random variables v_1, $\ldots,v_p$, the likelihood functional is one and the same. Equation (17) can be interpreted as a bank of P estimator-GDs (Van Trees, 2003) plus a bias term depending on the eigenvalues of the signal correlation under the hypothesis H_i. The optimum test based on GASP can be written in the following form:

$$\hat{H} = H_i \Rightarrow \sum_{p=1}^{P}\frac{1}{Z_p}\left\{\int_0^T[2x_p(t)\hat{s}_{p,i}^*(t) - x_p(t)x_p(t-\tau)]dt + \int_0^T n_{AF_p}^2(t)dt\right\} - b_{p,i}$$

$$> \sum_{p=1}^{P}\frac{1}{Z_p}\left\{\int_0^T[2x_p(t)\hat{s}_{p,k}^*(t) - x_p(t)x_p(t-\tau)]dt + \int_0^T n_{AF_p}^2(t)dt\right\} - b_{p,k}, \qquad \forall k \neq i \tag{18}$$

where $\hat{s}_{p,i}(t)$ is the linear minimum mean square estimator of $s_{p,i}(t)$ embedded in white noise with PSD Z_p, namely,

$$\hat{s}_{p,i}(t) = \int_0^T h_{p,i}(t,u)x_p(u)du, \tag{19}$$

where $h_{p,i}(t,u)$ is the solution to the Wiener-Hopf equation

$$\int_0^T Cov_i(t,z)h_{p,i}(z,\tau)dz + Z_p h_{p,i}(t,\tau) = Cov_i(t,\tau). \tag{20}$$

As to the bias terms $b_{p,i}$, they are given by

$$b_{p,i} = \sum_{j=1}^{\infty}\ln\left[1 + \frac{\sigma_{s_{j,i}}^2}{Z_p}\right] \qquad i=1,\ldots,M \quad \text{and} \quad p=1,\ldots,P. \tag{21}$$

The block diagram of the corresponding GD is outlined in Fig.3. The received signals $x_1(t)$, $\ldots,x_P(t)$ are fed to P estimators of the noise short-term PSD, which are subsequently used for synthesizing the bank of MP minimum mean square error filters $h_{p,i}(\cdot,\cdot), \forall p=1,\ldots,P, \forall i = 1,\ldots,M$ to implement the test (18). The newly proposed GD structure is a generalization, to the case of multiple observations, of that proposed in (Buzzi, 1999), to which it reduces to $P=1$.

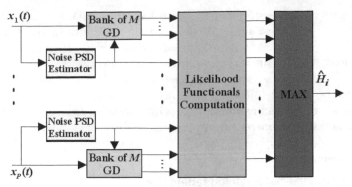

Fig. 3. Flowchart of optimum GD in compound Gaussian noise.

3.2.2 Suboptimal GD: Low energy coherence approach

Practical implementation of the decision rule (18) requires an estimation of the short-term noise PSDs on each diversity branch and evaluation of the test statistic. This problem requires a real-time design of MP estimator-GDs that are keyed to the estimated values of the short-term PSDs. This would require a formidable computational effort, which seems to prevent any practical implementation of the new receiving structure. Accordingly, we develop an alternative suboptimal GD structure with lower complexity. Assume that the signals $\{s_{p,i}(t): \forall p = 1,\ldots,P, \forall i = 1,\ldots,M\}$ possess a low degree of coherence, namely, that their energy content is spread over a large number of orthogonal directions. Since

$$\bar{E}_i = \sum_{j=1}^{+\infty} \sigma_{s_{j,i}}^2 , \tag{22}$$

The low degree of coherence assumption implies that the covariance functions $Cov_i(t,\tau)$ have a large number of nonzero eigenvalues and do not have any dominant eigenvalue. Under these circumstances, it is plausible to assume that the following low energy coherence condition is met:

$$\sigma_{s_{j,i}}^2 << 2N_0 \qquad i = 1,\ldots,M \qquad j = 1,2,\ldots. \tag{23}$$

If this is the case, we can approximate the log-likelihood functional (17) with its first-order McLaurin series expansion with starting point $\sigma_{s_{j,i}}^2 / Z_p = 0$. Following the same steps as in (Buzzi, 1997), we obtain the following suboptimal within the limits GASP decision-making rule:

$$\hat{H} = H_i \Rightarrow \sum_{p=1}^{P} \frac{1}{Z_p^2} \left\{ \int_0^T \int_0^T x_p(t) x_p^*(\tau) Cov_i(t,\tau) dt d\tau + \int_0^T n_{AF_p}^2(t) dt \right\} - \frac{\bar{E}_i}{Z_p}$$

$$> \sum_{p=1}^{P} \frac{1}{Z_p^2} \left\{ \int_0^T \int_0^T x_p(t) x_p^*(\tau) Cov_k(t,\tau) dt d\tau + \int_0^T n_{AF_p}^2(t) dt \right\} - \frac{\bar{E}_k}{Z_p} \qquad \forall k \neq i . \tag{24}$$

The new GD again requires estimating the short-term noise PSDs $Z_1, \ldots, Z_p$. Unlike the optimum GD (18), in the suboptimum GD (24), the MP minimum mean square error filters $h_{p,i}(\cdot, \cdot), \forall p = 1, \ldots, P, \forall i = 1, \ldots, M$ whose impulse responses depend on $Z_1, \ldots, Z_p$ through (20) are now replaced with M filters whose impulse response $Cov_i(t, \tau)$ is independent of the short-term noise PSDs realizations, which now affect the decision-making rule as mere proportionality factors. The only difficulty for practical implementation of such a GD scheme is the short-term noise PSD estimation through (16). However, as already mentioned, such a drawback can be easily circumvented by retaining only a limited number of summands.

3.3 Special cases

3.3.1 Channels with flat-flat Rayleigh fading

Let us consider the situation where the fading is slow and non-selective so that the signal observed on the p-th channel output under the hypothesis H_i takes the form

$$s_{p,i}(t) = A_p \exp\{j\theta_p\} u_i(t), \tag{25}$$

where $A_p \exp\{j\theta_p\}$ is a complex zero-mean Gaussian random variable. The signal covariance function takes a form:

$$Cov_i(t, \tau) = \overline{E}_i u_i(t) u_i^*(\tau), \tag{26}$$

where the assumption has been made that $u_i(t)$ possesses unity norm. Notice that this equation represents the Mercer expansion of the covariance in a basis whose first unit vector is parallel to $u_i(t)$. It should be noted that since the Mercer expansion of the useful signal covariance functions contains just one term, the low energy coherence condition is, in this case, equivalent to a low SNR condition. It thus follows that the low energy coherence GD can be now interpreted as a locally optimum GD, thus implying that for large $SNRs$, its performance is expectedly much poorer than that of the optimum GD. The corresponding eigenvalues are

$$\sigma_{s_{1,i}}^2 = \overline{E}_i, \quad \sigma_{s_{k,i}}^2 = 0, \quad \forall k \neq 1. \tag{27}$$

Accordingly, the minimum mean square error filters to be substituted in (18) have the following impulse responses:

$$h_{p,i}(t, \tau) = \frac{\overline{E}_i}{\overline{E}_i + Z_p} u_i(t) u_i^*(\tau), \tag{28}$$

where the bias term is simply $b_{p,i} = \ln\left\{1 + \frac{\overline{E}_i}{Z_p}\right\}$. We explicitly notice here that such a bias term turns out to depend on the estimated PSD Z_p. Substituting into (18), we find the optimum test

$$\hat{H} = H_i \Rightarrow \sum_{p=1}^{P} \frac{\overline{E}_i}{Z_p(\overline{E}_i + Z_p)} \left| \int_0^T [2x_p(t) u_i^*(t) - x_p(t) x_p(t - \tau)] dt + \int_0^T n_{AF_p}^2(t) dt \right|^2 - b_{p,i}$$

$$> \max_{k \neq i} \sum_{p=1}^{P} \frac{\bar{E}_k}{Z_p(\bar{E}_k + Z_p)} \left| \int_0^T [2x_p(t)u_k^*(t) - x_p(t)x_p(t-\tau)]dt + \int_0^T n_{AF_p}^2(t)dt \right|^2 - b_{p,k} , \qquad (29)$$

whereas its low energy coherence suboptimal approximation can be written in the following form:

$$\hat{H} = H_i \Rightarrow \sum_{p=1}^{P} \frac{1}{Z_p^2} \left| \int_0^T [2x_p(t)u_i^*(t) - x_p(t)x_p(t-\tau)]dt + \int_0^T n_{AF_p}^2(t)dt \right|^2 - \frac{\bar{E}_i}{Z_p}$$

$$> \max_{k \neq i} \sum_{p=1}^{P} \frac{1}{Z_p^2} \left| \int_0^T [2x_p(t)u_i^*(t) - x_p(t)x_p(t-\tau)]dt + \int_0^T n_{AF_p}^2(t)dt \right|^2 - \frac{\bar{E}_i}{Z_p} . \qquad (30)$$

It is worth pointing out that both GDs are akin to the "square-law combiner" (Tuzlukov, 2001) GD that is well known to be the optimum GD in GASP (Tuzlukov, 2005 and Tuzlukov, 2012) viewpoint for array signal detection in Rayleigh flat-flat fading channels and Gaussian noise. The relevant difference is due to the presence of short-term noise PSDs $Z_1, \dots, Z_P$ which weigh the contribution from each diversity branch. In the special case of equienergy signals, the bias terms in the above decision-making rules end up irrelevant, and the optimum GD test (29) reduces to a generalization of the usual incoherent GD, with the exception that the decision statistic depends on the short-term noise PSD realizations.

3.3.2 Channels with slow frequency-selective Rayleigh fading

Now, assume that the channel random impulse response can be written in the following form:

$$\chi_p(t,\tau) = \chi_p(\tau) = \sum_{k=0}^{L-1} A_{p,k} \exp\{j\theta_{p,k}\}\delta(\tau - kW^{-1}) , \qquad (31)$$

where $A_{p,k} \exp\{j\theta_{p,k}\}$ is a set of zero-mean, independent complex Gaussian random variables, and L is the number of paths. Equation (31) represents the well known taped delay line channel model, which is widely encountered in wireless mobile communications. It is readily shown that in such a case, the received useful signal, upon transmission of $u_i(t)$, has the following covariance function:

$$Cov_i(t,\tau) = \sum_{k=0}^{L-1} \overline{A_k^2} u_i(t-kW^{-1})u_i^*(\tau - kW^{-1}) , \qquad i = 1,\dots,M \qquad (32)$$

where $\overline{A_k^2}$ is the statistical expectation (assumed independent of p) of the random variables $A_{p,k}^2$. These correlations admit L nonzero eigenvalues, and a procedure for evaluating their eigenvalues and eigenfunctions can be found in (Matthews, 1992). In the special case that the L paths are resolvable, i.e. $T \leq W^{-1}$, the optimum GD (18) assumes the following simplified form:

$$\hat{H} = H_i \Rightarrow \sum_{p=1}^{P}\sum_{j=0}^{L-1}\left\{\frac{\overline{A_j^2}\left|\int_0^T x_p(t)u_i^*(t-jW^{-1})dt + \int_0^T n_{AF_p}^2(t)dt\right|^2}{Z_p(Z_p + E_i\overline{A_j^2})} - \ln\left\{1 + \frac{E_i\overline{A_j^2}}{Z_p}\right\}\right\}$$

$$> \max_{k \neq i}\sum_{p=1}^{P}\sum_{j=0}^{L-1}\left\{\frac{\overline{A_j^2}\left|\int_0^T x_p(t)u_k^*(t-jW^{-1})dt + \int_0^T n_{AF_p}^2(t)dt\right|^2}{Z_p(Z_p + E_k\overline{A_j^2})} - \ln\left\{1 + \frac{E_k\overline{A_j^2}}{Z_p}\right\}\right\},$$

(33)

where $\overline{E}_i$ is the energy of the signal $u_i(t)$. The low energy coherence suboptimal GD (24) is instead written as

$$\hat{H} = H_i \Rightarrow \sum_{p=1}^{P}\sum_{j=0}^{L-1}\left\{\frac{\overline{A_j^2}}{Z_p^2}\left|\int_0^T x_p(t)u_i^*(t-jW^{-1})dt + \int_0^T n_{AF_p}^2(t)dt\right|^2 - \frac{E_i\overline{A_j^2}}{Z_p}\right\}$$

$$> \max_{k \neq i}\sum_{p=1}^{P}\sum_{j=0}^{L-1}\left\{\frac{\overline{A_j^2}}{Z_p^2}\left|\int_0^T x_p(t)u_k^*(t-jW^{-1})dt + \int_0^T n_{AF_p}^2(t)dt\right|^2 - \frac{E_k\overline{A_j^2}}{Z_p}\right\}.$$

(34)

Optimality of (33) obviously holds for one-short detection, namely, neglecting the intersymbol interference induced by the channel band limitedness.

3.4 Performance assessment

In this section, we focus on the performance of the proposed GD structures. A general formula to evaluate the probability of error P_E of any receiver in the presence of spherically invariant disturbance takes the following form:

$$P_E = \int P_E(e\,|\,\mathbf{v})f_\mathbf{v}(\mathbf{v})d\mathbf{v},$$

(35)

where $P_E(e\,|\,\mathbf{v})$ is the receiver probability of error in the presence of Gaussian noise with PSD on the p-th diversity branch $2N_0v_p^2$. The problem to evaluate P_E reduces to that of first analyzing the Gaussian case and then carrying out the integration (35). In order to give an insight into the GD performance, we consider a BFSK signaling scheme, i.e. the baseband equivalents of the two transmitted waveforms are related as

$$u_2(t) = u_1(t)\exp\{j2\pi\Delta ft\},$$

(36)

where $\Delta f = T^{-1}$ denotes the frequency shift. Even for this simple case study, working out an analytical expression for the probability of error of both the optimum GD and of its low energy coherence approximation is usually unwieldy even for the case of Gaussian noise. With regard to the optimum GD structure, upper and lower bounds for the performance may be established via Chernoff-bounding techniques. Generalizing to the case of multiple observations, the procedure in (Van Trees, 2003), the conditional probability of error given $v_1,\ldots,v_P$ can be bounded as

$$\frac{\exp\{2\mu(0.5\,|\,\mathbf{v})\}}{2\left(1+\sqrt{0.25\pi\ddot{\mu}(0.5\,|\,\mathbf{v})}\right)} \leq P_E(e\,|\,\mathbf{v}) \leq \frac{\exp\{2\mu(0.5\,|\,\mathbf{v})\}}{2\left(1+\sqrt{0.25\pi\ddot{\mu}(0.5\,|\,\mathbf{v})}\right)}, \tag{37}$$

where $\mu(\cdot\,|\,\mathbf{v})$ is the following conditional semi-invariant moment generating the function

$$\mu(x\,|\,\mathbf{v}) = \lim_{N\to\infty} \ln E\left\{\exp\left[x\sum_{p=1}^{P}\ln\Lambda_{gN}^{P}\left(x_{N,p}^{i};v_p^2:H_1\right)\right]\,\Big|\,H_0,\mathbf{v}\right\}$$

$$= \sum_{j=1}^{\infty}\sum_{p=1}^{P}\left\{(1-x)\ln\left[1+\frac{\sigma_{s_j}^2}{4\sigma_n^4 v_p^2}\right] - \ln\left[1+\frac{\sigma_{s_j}^2(1-x)}{4\sigma_n^4 v_p^2}\right]\right\} \tag{38}$$

with $\{\sigma_{s_j}^2\}_{j=1}^{\infty}$ being the set of common eigenvalues. Substituting this relationship into (37) and averaging with respect to $v_1,\ldots,v_P$ yields the unconditional bounds on the probability of error for the optimum GD (18).

3.5 Simulation results

To proceed further in the GD performance there is a need to assign both the marginal pdf, as well as the channel spectral characteristics. We assume hereafter the generalized Laplace noise, i.e. the marginal pdf of the p-th noise texture component takes the following form:

$$f_{v_p}(x) = \frac{2v^v}{\Gamma(v)}x^{2v-1}\exp\{-vx^2\}, \qquad x>0 \tag{39}$$

where v is a shape parameter, ruling the distribution behavior. In particular, the limiting case $v\to\infty$ implies $f_{v_p}(x) = \delta(x-1)$ and, eventually, Gaussian noise, where increasingly lower values of v account for increasingly spikier noise distribution. Regarding the channel, we consider the case of the frequency-selective, slowly fading channel, i.e. the channel random impulse response is expressed by (31), implying that the useful signal correlation is that given in (32). For simplicity, we also assume that the paths are resolvable. In the following plots the P_E is evaluated a) through a semianalytic procedure, i.e. by numerically averaging the Chernoff bound (37) with respect to the realizations of the $v_1,\ldots,v_P$, and b) by resorting to a Monte Carlo counting procedure. In this later case, the noise samples have been generated by multiplying standard, i.e. with zero-mean and unit-variance, complex Gaussian random variates times the random realizations of $v_1,\ldots,v_P$.

The Chernoff bound for the optimum GD versus the averaged received radio-frequency energy contrast that is defined as $\gamma_0 = P\sum_{j=1}^{L}\sigma_{s_j}^2(4\sigma_n^4)^{-0.5}$ at $P=2$ and for two values of the noise shape parameter v is shown in Fig.4. The noise texture components have been assumed to be independent. Inspecting the curves, we see that the Chernoff bound provides a very reliable estimate of the actual P_E, as the upper and lower bound very tightly follow each other. As expected, the results demonstrate that in the low P_E region, the spikier the noise, i.e. the lower ν, the worse the GD performance. Conversely, the opposite behavior is observed for small values of γ_0. This fact might appear, at a first look, surprising. It may be analytically justified in light of the local validity of Jensen's inequality (Van Trees, 2003) and is basically the same phenomenon that makes digital modulation schemes operating in Gaussian noise to achieve, for low values of γ_0, superior performance in Rayleigh flat-flat fading channels than in no-fading channels. Notice, this phenomenon is in accordance with that observed in (Conte 1995). In order to validate the Chernoff bound, we also show, on the same plots, some points obtained by Monte Carlo simulations. These points obviously lie between the corresponding upper and lower probability of error bounds. Additionally, we compare the GD Chernoff bound with that for the conventional optimum receiver (Buzzi et al., 2001). A superiority of GD structure is evident.

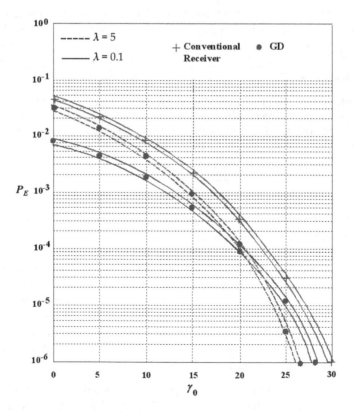

Fig. 4. Chernoff bounds for P_E of the optimum GD.

In Fig. 5, the effect of the channel diversity order is investigated. Indeed, the optimum GD performance versus γ_0 is represented for several values of P and with $v = 1$. The $Z_1, \ldots, Z_P$ have been assumed exponentially correlated with correlation coefficient $\rho = 0.2$. A procedure for generating these exponentially correlated random variables for integer and semi-integer values of v is reported in (Lombardo et al., 1999). As expected, as P increases, the GD performance ameliorates, thus confirming that diversity represents a suitable means to restore performance in severely hostile scenarios. Also, we compare the GD performance with that for the conventional optimum receiver (Buzzi et al., 2001) and we see that the GD keeps superiority in this case, too.

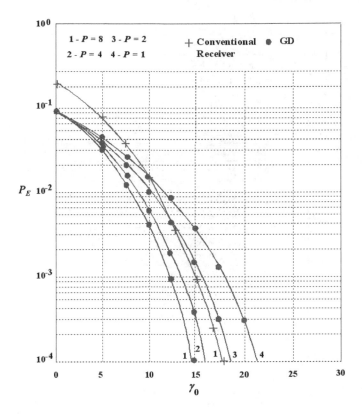

Fig. 5. P_E at several values of P.

The optimum GD performance versus γ_0 for the generalized Laplace noise at $v = 1, P = 4$ and for several values of the correlation coefficient ρ is demonstrated in Fig. 6. It is seen that the probability of error improves for vanishingly small ρ. For small ρ, the GD takes much advantage of the diversity observations. For high values of ρ, the realizations $Z_1, ..., Z_P$ are very similar and much less advantage can be gained through the adoption of a diversity strategy. Such GD performance improvement is akin to that observed in signal diversity detection in the presence of flat-flat fading and Gaussian noise. We see that the GD outperforms the conventional optimum receiver (Buzzi et al., 2001) by the probability of error.

In Fig. 7, we compare the optimum GD performance versus that of the low energy coherence GD. We assumed $\rho = 0.2$ and $P = 4$. It is seen that the performance loss incurred by the low energy coherence GD with respect to the optimum GD is kept within a fraction of 1 dB at $P_E = 10^{-4}$. Simulation results that are not presented in the paper show that the crucial factor ruling the GD performance is the noise shape parameter, whereas the particular noise distribution has a rather limited effect on the probability of error.

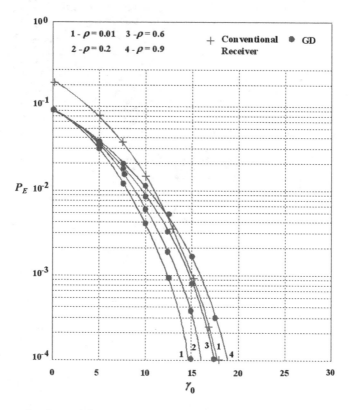

Fig. 6. P_E at several values of the correlation coefficient.

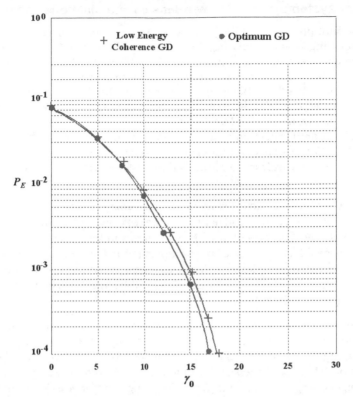

Fig. 7. P_E for the optimum and low energy coherence GDs.

3.6 Discussion

We have considered the problem of diversity detection of one out of M signals transmitted over a fading dispersive channel in the presence of non-Gaussian noise. We have modeled the additive noise on each channel diversity branch through a spherically invariant random process, and the optimum GD has been shown to be independent of the actual joint pdf of the noise texture components present on the channel diversity outputs. The optimum GD is similar to the optimum GD for Gaussian noise, where the only difference is that the noise PSD $2N_0$ is substituted with a perfect estimate of the short-term PSD realizations of the impulsive additive noise. We also derived a suboptimum GD matched with GASP based on the low energy coherence hypothesis. At the performance analysis stage, we focused on frequency-selective slowly fading channels and on a BFSK signaling scheme and evaluated the GD performance through both a semianalytic bounding technique and computer simulations. Numerical results have shown that the GD performance is affected by the average received energy contrast, by the channel diversity order, and by the noise shape parameter, whereas it is only marginally affected by the actual noise distribution. Additionally, it is seen that in impulsive environments, diversity represents a suitable strategy to improve GD performance.

4. MIMO radar systems applied to wireless communications based on GASP

Multiple-input multiple-output (MIMO) wireless communication systems have received a great attention owing to the following viewpoints: a) MIMO wireless communication systems have been deemed as efficient spatial multiplexers and b) MIMO wireless communication systems have been deemed as a suitable strategy to ensure high-rate communications on wireless channels (Foschini, 1996). Space-time coding has been largely investigated as a viable means to achieve spatial diversity, and thus to contrast the effect of fading (Tarokh, et al., 1998 and Hochwald, et al., 2000). We apply GASP to the design and implementation of MIMO wireless communication systems used space-time coding technique. Theoretical principles of MIMO wireless communication systems were discussed and the potential advantages of MIMO wireless communication systems are thoroughly considered in (Fishler, et al., 2006).

MIMO architecture is able to provide independent diversity paths, thus yielding remarkable performance improvements over conventional wireless communication systems in the medium-high range of detection probability. As was shown in (Fishler, et al., 2006), the MIMO mode can be conceived as a means of bootstrapping to obtain greater coherent gain. Some practical issues concerning implementation (equipment specifications, dynamic range, phase noise, system stability, isolation and spurs) of MIMO wireless communication systems are discussed in (Skolnik, 2008).

MIMO wireless communication systems can be represented by m transmit antennas, spaced several wavelengths apart, and n receive antennas, not necessarily collocated, and possibly forwarding, through a wired link, the received echoes to a fusion center, whose task is to make the final decision about the signal in the input waveform. If the spacing between the transmit antennas is large enough and so is the spacing between the receive antennas, a rich scattering environment is generated, and each receive antenna processes l statistically independent copies of incoming signal. The concept of rich scattering environment is borrowed from communication theory, and models a situation where the MIMO architecture yields interchannel interference, eventually resulting into a number of independent random channels. Unlike a conventional wireless communication array system, which attempts to maximize the coherent processing, MIMO wireless communication system resorts to the fading diversity in order to improve the detection performance. Indeed, it is well known that, in conventional wireless communication array system, multiple access interference (MAI) of the order of 10 dB may arise. This effect leads to severe degradations of the detection performance, due to the high signal correlation at the array elements. This drawback might be partially circumvented under the use of MIMO wireless communication system, which exploits the channel diversity and fading. Otherwise, uncorrelated signals at the array elements are available. Based on mentioned above statements, it was shown in (Fishler, et al., 2006) that in the case of additive white Gaussian noise (AWGN), transmitting orthogonal waveforms result into increasingly constrained fluctuation of interference.

Our approach is based on implementation of GASP and employment of some key results from communication theory, and in particular, the well-known concept that, upon suitably space-time encoding the transmitted waveforms, a maximum diversity order given by $m \times n$ can be achieved. Importing these results in a wireless communication system scenario poses a number of problems, which forms the object of the present study, and in particular: a) the issue of waveform design, which exploits the available knowledge as to space-time codes; b) the issue of designing a suitable detection structure based on GASP, also in the light of the fact that the disturbance can no longer be considered as AWGN, due to the presence of interferences; and c) at the performance assessment level, the issue of evaluating the maximum diversity order that can be achieved and the space-time coding ensuring it under different types of interferences. The first and third tasks are merged in the unified problem of determining the space-time coding achieving maximum diversity order in signal detection, for constrained BER, and for given interference covariance. As to the second task, the decision-making criterion exploiting by GASP is employed.

Unlike (Fishler, et al., 2006), no assumption is made on either the signal model or the disturbance covariance. Thus, a family of detection structures is derived, depending upon the number of transmitting and receiving antennas and the disturbance covariance. A side result, which paves the way to further investigations on the feasibility of fully adaptive MIMO wireless communication systems is that the decision statistic, under the null hypothesis of no signal, is an ancillary statistic, in the sense that it depends on the actual interference covariance matrix, but its probability density function (pdf) is functionally independent of such a matrix. Therefore, threshold setting is feasible with no prior knowledge as to the interference power spectrum. As to the detection performance, a general integral form of the probability of detection P_D is provided, holding independent of the signal fluctuation model. The formula is not analytically manageable, nor does it appear to admit general approximate expressions, that allow us to give an insightful look in the wireless communication system behavior. We thus restrict our atention to the case of Rayleigh-distributed attenuation, and use discussed in (De Maio & Lops, 2007) an information-theoretic approach to code construction, which, surprisingly enough, leads to the same solution found through the optimization of the Chernoff bound.

4.1 System model

We consider MIMO radar system composed of m fixed transmitters and n fixed receivers and assume that the antennas as the two ends of the wireless communication system are sufficiently spaced such that a possible incoming message and/or interference provides uncorrelated reflection coefficients between each transmit/receive pair of sensors. Denote by $s_i(t)$ the baseband equivalent of the coherent pulse train transmitted by the i-th antenna, for example,

$$s_i(t) = \sum_{j=1}^{N} a_{i,j} p[t - (j-1)T_p], \qquad i = 1,\ldots,m \qquad (40)$$

where $p(t)$ is the signature of each transmitted pulse, which we assume, without loss of generality, with unit energy and duration τ_p; T_p is the pulse repetition time;

$$\mathbf{a}_i = [a_{i,1},\ldots,a_{i,N}]^T \qquad (41)$$

is an N-dimensional column vector whose entries are complex numbers which modulate both in amplitude and in phase the N pulses of the train, where $(\cdot)^T$ denotes transpose. In the sequel, we refer to $\mathbf{a}_i$ as the code word of the i-th antenna. The baseband equivalent of the signal received by the i-th sensor, from a target with two-way time delay τ, can be presented in the following form

$$x_i(t) = \sum_{l=1}^{m} a_{i,l} \sum_{j=1}^{N} a_{l,j} p[t - \tau - (j-1)T_p] + n_i(t) , \qquad i = 1, \ldots, n , \qquad (42)$$

where $a_{i,l}$, $i = 1, \ldots, n$ and $l = 1, \ldots, m$, are complex numbers accounting for both the target backscattering and the channel propagation effects between the l-th transmitter and the i-th receiver; $n_i(t)$, $i = 1, \ldots, n$, are zero-mean, spatially uncorrelated, complex Gaussian random processes accounting for both the external and the internal disturbance. For simplicity, we assume a zero-Doppler target, but all the derivations can be easily extended to account for a possible known Doppler shift. We explicitly point out that the validity of the above model requires the narrowband assumption

$$\frac{d_{\max}^m + d_{\max}^n}{c} << \frac{1}{B} \qquad (43)$$

where B is the bandwidth of the transmitted pulse, $d_{\max}^m$ and $d_{\max}^n$ denote the maximum spacing between two sensors at the transmitter and the receiver end, respectively. The signal $x_i(t)$, at each of the receive elements, is matched filtered to the pulse $p(t)$ by preliminary filter of the GD and the filter output is sampled at the time instants $\tau + (k-1)T_p$, $k = 1, \ldots, N$. Thus, denote by $x_i(k)$ the k-th sample, i.e.,

$$x_i(k) = \sum_{l=1}^{m} a_{i,l} a_{l,k} + n_i(k) , \qquad (44)$$

where $n_i(k)$ is the filtered noise sample. Define the N-dimensional column vectors

$$\mathbf{x}_i = [x_i(1), \ldots, x_i(N)]^T \qquad (45)$$

and rewrite them as

$$\mathbf{x}_i = \mathbf{A}\boldsymbol{\alpha}_i + \boldsymbol{\xi}_{PF_i} , \qquad i = 1, \ldots, n \qquad (46)$$

where

$$\boldsymbol{\xi}_{PF_i} = [\xi_{PF_i}(1), \ldots, \xi_{PF_i}(N)]^T , \qquad (47)$$

$$\boldsymbol{\alpha}_i = [a_{i,1}, \ldots, a_{i,m}]^T , \qquad (48)$$

and the $(N \times m)$-dimensional matrix $\mathbf{A}$, defined in the following form

$$\mathbf{A} = [\mathbf{a}_1, \ldots, \mathbf{a}_m] \tag{49}$$

has the code words as columns. This last matrix is referred to as the code matrix. We assume that $\mathbf{A}$ is full rank matrix. It is worth underlining that the model given by (46) applies also to the case that space-time coding is performed according to (De Maio & Lops, 2007), namely, by dividing a single pulse in N sub-pulses. The code matrix $\mathbf{A}$ thus defines m different code words of length N, which can be received by a single receive antenna, thus defining the multiple-input single-output (MISO) structure, as well as by a set of n receive antennas, as in the present study.

4.2 GD design for MIMO radar systems applied to wireless communications

The problem of detecting a target return signal with a MIMO radar system can be formulated in terms of the following binary hypothesis test

$$\begin{cases} H_0 \Rightarrow \mathbf{x}_i = \xi_{PF_i}, & i = 1, \ldots, n \\ H_1 \Rightarrow \mathbf{x}_i = \mathbf{A}\alpha_i + \xi_{PF_i}, & i = 1, \ldots, n \end{cases} \tag{50}$$

where ξ_{PF_i}, $i = 1, \ldots, n$, are statistically independent and identically distributed (i.i.d.) zero-mean complex Gaussian vectors with covariance matrix

$$E[\xi_{PF_i} \xi_{PF_i}^*] = E[\xi_{AF_i} \xi_{AF_i}^*] = \mathbf{M}. \tag{51}$$

Here $E[\cdot]$ denotes the statistical expectation and $(*)$ denotes conjugate transpose. The covariance matrix (51) is assumed positive definite and known. According to the Neyman-Pearson criterion, the optimum solution to the hypotheses testing problem (50) must be the likelihood ratio test. However, for the case at hand, it cannot be implemented since total ignorance of the parameters α_i is assumed. One possible way to circumvent this drawback is to resort to the generalized likelihood ratio test (GLRT) (Van Trees, 2003), which is tantamount to replacing the unknown parameters with their maximum likelihood (ML) estimates under each hypothesis. Applying GASP to the GLRT, we obtain the following decision rule

$$\frac{\max_{\alpha_1, \ldots, \alpha_n} f(\mathbf{x}_1, \ldots, \mathbf{x}_n \mid H_1, \mathbf{M}, \alpha_1, \ldots, \alpha_n)}{f(\xi_{AF_1}, \ldots, \xi_{AF_n} \mid H_0, \mathbf{M})} \underset{<H_0}{\overset{>H_1}{\gtrless}} K_g, \tag{52}$$

where $f(\mathbf{x}_1, \ldots, \mathbf{x}_n \mid H_1, \mathbf{M}, \alpha_1, \ldots, \alpha_n)$ is the probability density function (pdf) of the data under the hypothesis H_1 and $f(\xi_{AF_1}, \ldots, \xi_{AF_n} \mid H_0, \mathbf{M})$ is pdf of the data under the hypothesis H_0, respectively, K_g is a suitable modification of the original threshold. Previous assumptions imply that the aforementioned pdfs can be written in the following form:

$$f(\xi_{AF_1}, \ldots, \xi_{AF_n} \mid H_0, \mathbf{M}) = \frac{1}{\pi^{Nn} \det^n(\mathbf{M})} \exp\left[-\sum_{i=1}^{n} \xi_{AF_i}^* \mathbf{M}^{-1} \xi_{AF_i} \right] \tag{53}$$

at the hypothesis H_0 and

$$f(\mathbf{x}_1,\ldots,\mathbf{x}_n \mid H_1,\mathbf{M},\boldsymbol{\alpha}_1,\ldots,\boldsymbol{\alpha}_n) = \frac{1}{\pi^{Nn}\det^n(\mathbf{M})}\exp\left[-\sum_{i=1}^{n}(\mathbf{x}_i - \mathbf{A}\boldsymbol{\alpha}_i)^*\mathbf{M}^{-1}(\mathbf{x}_i - \mathbf{A}\boldsymbol{\alpha}_i)\right] \quad (54)$$

under the hypothesis H_1, where $\det(\cdot)$ denotes the determinant of a square matrix. Substituting (16) and (17) in (15), we can recast the GLRT based on GASP, after some mathematical transformations, in the following form

$$\sum_{i=1}^{n}\mathbf{\S}_{AF_i}^*\mathbf{M}^{-1}\mathbf{\S}_{AF_i} - \sum_{i=1}^{n}\min_{\boldsymbol{\alpha}_i}(\mathbf{x}_i - \mathbf{A}\boldsymbol{\alpha}_i)^*\mathbf{M}^{-1}(\mathbf{x}_i - \mathbf{A}\boldsymbol{\alpha}_i)\mathop{\gtrless}_{H_0}^{H_1} K_g. \quad (55)$$

In order to solve the n minimization problems in (55) we have to distinguish between two different cases.

Case 1: $N > m$. In this case, the quadratic forms in (55) achieve the minimum at

$$\hat{\boldsymbol{\alpha}}_i = (\mathbf{A}^*\mathbf{M}^{-1}\mathbf{A})^{-1}\mathbf{A}^*\mathbf{M}^{-1}\mathbf{x}_i, \quad i = 1,\ldots,n \quad (56)$$

and, as a consequence, the GLRT based on GASP at the main condition of GD functioning, i.e., equality in whole range of parameters between the transmitted information signal and refe-rence signal (signal model) in the receiver part, becomes

$$2\sum_{i=1}^{n}\mathbf{x}_i^*\mathbf{M}^{-1}\mathbf{A}(\mathbf{A}^*\mathbf{M}^{-1}\mathbf{A})^{-1}\mathbf{A}^*\mathbf{M}^{-1}\mathbf{x}_i - \sum_{i=1}^{n}\mathbf{x}_i^*\mathbf{M}^{-1}\mathbf{A}\mathbf{A}^*\mathbf{M}^{-1}\mathbf{x}_i + \sum_{i=1}^{n}\mathbf{\S}_{AF_i}^*\mathbf{M}^{-1}\mathbf{M}^{-1}\mathbf{\S}_{AF_i}\mathop{\gtrless}_{H_0}^{H_1} K_g. \quad (57)$$

Case 2: $N \leq m$. In this case, the minimum of the quadratic forms in (55) is zero, since each linear system

$$\mathbf{A}\hat{\boldsymbol{\alpha}}_i = \mathbf{x}_i, \quad i = 1,\ldots,n \quad (58)$$

is determined. As a consequence the GLRT based on GASP at the main condition of GD functioning, i.e., equality in whole range of parameters between the transmitted information signal and reference signal (signal model) in the receiver part, becomes

$$\sum_{i=1}^{n}\mathbf{\S}_{AF_i}^*\mathbf{M}^{-1}\mathbf{M}^{-1}\mathbf{\S}_{AF_i} - \sum_{i=1}^{n}\mathbf{x}_i^*\mathbf{M}^{-1}\mathbf{A}\mathbf{A}^*\mathbf{M}^{-1}\mathbf{x}_i\mathop{\gtrless}_{H_0}^{H_1} K_g. \quad (59)$$

4.3 Performance analysis

In order to define possible design criteria for the space-time coding, it is useful to establish a direct relationship between the probability of detection P_D and the transmitted waveform, which is thus the main goal of the present section. Under the hypothesis H_0, the left hand side of the GLRT based on GASP can be written in the following form

$$\sum_{i=1}^{n}\mathbf{\S}_{AF_i}^*\mathbf{M}^{-1}\mathbf{\S}_{AF_i} - \sum_{i=1}^{n}\mathbf{\S}_{PF_i}^*\mathbf{M}^{-1}\mathbf{\S}_{PF_i} \quad (60)$$

and, represents the GD background noise. It follows from (Tuzlukov 2005) that the decision statistic is defined by the modified second-order Bessel function of an imaginary argument or, as it is also called, McDonald's function with $m \times n$ degrees of freedom. Thus, the decision statistic is independent of dimensionality N of the column vector given by (41) whose entries are complex numbers, which modulate both in amplitude and in phase the N pulses of the train. Consequently, the probability of false alarm P_{FA} can be evaluated in the following form

$$P_{FA} = \exp(-K_g) \sum_{k=0}^{n} \frac{(K_g)^k}{k!}. \tag{61}$$

This last expression allows us to note the following observations: a) the decision statistic is ancillary, in the sense that it depends on the actual interference covariance matrix, but its pdf is functionally independent of such a matrix; and b) the threshold setting is feasible with no prior knowledge as to the interference power spectrum, namely, the GLRT based on GASP ensures the constant false alarm (CFAR) property.

Under the hypothesis H_1, given $\boldsymbol{\alpha}_i$, the vectors $\mathbf{x}_i$, $i = 1, \ldots, n$, are statistically independent complex Gaussian vectors with the mean value $\mathbf{M}^{-1} \mathbf{A} \boldsymbol{\alpha}_i$ and identity covariance matrix. It follows that, given $\boldsymbol{\alpha}_i$, the GLRT based on GASP is no the central distributed modified second-order Bessel function of an imaginary argument, with the no centrality parameter $\sum_{i=1}^{n} \boldsymbol{\alpha}_i^* \mathbf{A}^* \mathbf{M}^{-1} \mathbf{A} \boldsymbol{\alpha}_i$ and degrees of freedom $m \times n$. Consequently, the conditional probability of detection P_D based on statements in (Van Trees, 2003) and discussion in (Tuzlukov, 2005) can be represented in the following form

$$P_D = Q_{m \times n} \left(\sqrt{2q}, \sqrt{2K_g} \right), \tag{62}$$

where

$$q = \sum_{i=1}^{n} \boldsymbol{\alpha}_i^* \mathbf{A}^* \mathbf{M}^{-1} \mathbf{A} \boldsymbol{\alpha}_i \tag{63}$$

and $Q_k(\cdot, \cdot)$ denotes the generalized Marcum Q function of order k. An alternative expression for the conditional probability of detection P_D, in terms of an infinite series, can be also written in the following form:

$$P_D = \sum_{k=0}^{\infty} \frac{\exp(-q)q^k}{k!} \left[1 - \Gamma_{inc}(K_g, k + m \times n) \right], \tag{64}$$

where

$$\Gamma_{inc}(p,r) = \frac{1}{\Gamma(r)} \int_0^w \exp(-z) z^{r-1} dz \tag{65}$$

is the incomplete Gamma function. Finally, the unconditional probability of detection P_D can be obtained averaging the last expression over the pdf of α_i, $i = 1,...,n$.

4.4 Code design by information-theoretic approach

In principle, the basic criterion for code design should be the maximization of the probability of detection P_D given by (62) over the set of admissible code matrices, i.e.,

$$\arg \max_{\mathbf{A}} E\left[Q_{m \times n}\left(\sqrt{2q}, \sqrt{2K_g}\right)\right] = \arg \max_{\mathbf{A}} E\left[Q_{m \times n}\left(\sqrt{2\sum_{i=1}^{n} \alpha_i^* \mathbf{A}^* \mathbf{M}^{-1} \mathbf{A} \alpha_i}, \sqrt{2K_g}\right)\right], \quad (66)$$

where $\arg \max_{\mathbf{A}}(\cdot)$ denotes the value of $\mathbf{A}$, which maximizes the argument and the statistical average is over α_i, $i = 1,...,n$. Unfortunately, the above maximization problem does not appear to admit a closed-form solution, valid independent of the fading law, whereby we prefer here to resort to the information-theoretic criterion supposed in (De Maio & Lops, 2007). Another way is based on the optimization of the Chernoff bound over the code matrix $\mathbf{A}$. As was shown in (De Maio & Lops, 2007), these ways lead to the same solution, which subsumes some well-known space-time coding, such as Alamouti code and, more generally, the class of space-time coding from orthogonal design (Alamouti, 1998) and (Tarokh et al., 1999), which have been shown to be optimum in the framework of communication theory. In subsequent derivations, we assume that α_i, $i = 1,...,n$, are independent and identically distributed (i.i.d.) zero-mean complex Gaussian vectors with scalar covariance matrix, i.e.,

$$E\left[\alpha_i \alpha_i^*\right] = \sigma_a^2 \mathbf{I}, \quad (67)$$

where σ_a^2 is a real factor accounting for the backscattered useful power, and $\mathbf{I}$ denotes the identity matrix.

Roughly speaking, the GLRT strategy overcomes the prior uncertainty as to the target fluctuations by ML estimating the complex target amplitude, and plugging the estimated value into the conditional likelihood in place of the true value. Also, it is well known that, under general consistency conditions, the GLRT converges towards the said conditional likelihood, thus achieving a performance closer and closer to the perfect measurement bound, i.e., the performance of an optimum test operating in the presence of known target parameters. Diversity, on the other hand, can be interpreted as a means to transform an amplitude fluctuation in an increasingly constrained one. It is well known, for example that, upon suitable receiver design, exponentially distributed square target amplitude may be transformed into a central chi-square fluctuation with d degrees of freedom through a diversity of order d in any domain. More generally, a central chi-square random variable with $2m$ degrees of freedom may be transformed into a central chi-square with $2m \times d$ degrees of freedom. In this framework, a reasonable design criterion for the space-time coding is the maximization of the mutual information between the signals received from the various diversity branches and the fading amplitudes experienced thereupon. Thus, denoting by $I(\alpha, \mathbf{X})$ the mutual information (Cover & Thomas, 1991) between the random matrices

$$\alpha = [\alpha_1, \ldots, \alpha_n] \tag{68}$$

and

$$X = [x_1, \ldots, x_k] = A\alpha + \Xi \tag{69}$$

the quantity to be maximized is

$$I(\alpha, X) = H(X) - H(X \mid \alpha), \tag{70}$$

where

$$\Xi = [\xi_1, \ldots, \xi_n], \tag{71}$$

$H(X)$ denotes the entropy of the random matrix Ξ, and $H(X \mid \alpha)$ is the conditional entropy of X given α (Cover & Thomas, 1991). Exploiting the statistical independence between α and X, we can write (70) in the following form

$$I(\alpha, X) = H(X) - H(X \mid \alpha) = H(X) - H(\Xi), \tag{72}$$

where $H(\Xi)$ is the entropy of the random matrix Ξ. Assuming that the columns of α are i.i.d. zero-mean complex Gaussian vectors with covariance matrix $\sigma_a^2 I$, we can write $H(X)$ and $H(\Xi)$, respectively, in the following form:

$$H(X) = x \lg[(\pi e)^N \det(M + \sigma_a^2 AA^*)] \tag{73}$$

and

$$H(\Xi) = x \lg[(\pi e)^N \det(M)]. \tag{74}$$

As design criterion we adopt the maximization of the minimum probability of detection P_D, which can be determined as the lower Chernoff bound, under an equality constraint for the average signal-to-clutter power ratio (SCR) given by

$$SCR = \frac{1}{Nmn} E\left[\sum_{i=1}^{n} \alpha_i^* A^* M^{-1} A \alpha_i\right] = \frac{\sigma_a^2}{Nm} \operatorname{tr}(A^* M^{-1} A) = \frac{\sigma_a^2}{Nm} \sum_{j=1}^{m} \lambda_j, \tag{75}$$

where $\operatorname{tr}(\cdot)$ denotes the trace of a square matrix and λ_j are the elements or corresponding ordered (in decreasing order) eigenvalues of the diagonal matrix Λ defined by the eigenvalue decomposition $V^* \Lambda V$ of the matrix $M^{-1} AA^* M^{-1}$, where V is an $N \times N$ unitary matrix. The considered design criterion relies on the maximization of the mutual information (70) under equality constraint (75) for SCR. This is tantamount to solving the following constrained minimization problem since $H(\Xi)$ does not exhibit any functional dependence on A.

$$\min_{\lambda_1, \ldots, \lambda_m} \prod_{j=1}^{m} \left[\frac{1}{1 + \gamma(\lambda_j \sigma_a^2 + 1)}\right]^n \quad \text{and} \quad \frac{\sigma_a^2}{Nm} \sum_{j=1}^{m} \lambda_j = \mu \tag{76}$$

which, taking the logarithm, is equivalent

$$\max_{\lambda_1,\dots,\lambda_m} \sum_{j=1}^{m} \lg[1+\gamma(\sigma_a^2\lambda_j +1)] \quad \text{and} \quad \sum_{j=1}^{m}\lambda_j = \frac{\mu m N}{\sigma_a^2}, \tag{77}$$

where γ is the variable defining the upper Chernoff bound (Benedetto & Biglieri, 1999).

Since $\lg[1+\gamma(\sigma_a^2 y +1)]$ is a concave function of y, we can apply Jensen's inequality (Cover & Thomas, 1991) to obtain

$$\sum_{j=1}^{m} \lg[1+\gamma(\sigma_a^2\lambda_j +1)] \le m\lg\left[1+\gamma\left(\frac{1}{m}\sum_{j=1}^{m}\lambda_j\sigma_a^2 +1\right)\right]. \tag{78}$$

Moreover, forcing in the right hand side of (78), the constraint of (77), we obtain

$$\sum_{j=1}^{m} \lg[1+\gamma(\sigma_a^2\lambda_j +1)] \le m\lg[1+\gamma(\mu N +1)]. \tag{79}$$

The equality in (79) is achieved if

$$\lambda_k = \frac{\mu N}{\sigma_a^2}, \quad k=1,\dots,m \tag{80}$$

implying that an optimum code must comply with the condition

$$\mathbf{M}^{-1}\mathbf{A}\mathbf{A}^*\mathbf{M}^{-1} = \begin{cases} \dfrac{\mu N}{\sigma_a^2}[2\mathbf{A}(\mathbf{A}^*\mathbf{M}^{-1}\mathbf{A})^{-1}\mathbf{A}^* - \mathbf{A}\mathbf{A}^*] & \text{Case 1} \\[2mm] \dfrac{\mu N}{\sigma_a^2}\mathbf{I} & \text{Case 2 .} \end{cases} \tag{81}$$

In particular, if the additive disturbance is white, i.e., $\mathbf{M} = \sigma_n^2\mathbf{I}$, the above equation reduces to

$$\mathbf{A}\mathbf{A}^* = \begin{cases} \dfrac{4\sigma_n^4\mu N}{\sigma_a^2}(\mathbf{A}^*\mathbf{M}^{-1}\mathbf{A})^{-1} & \text{Case 1} \\[2mm] \dfrac{4\sigma_n^4\mu N}{\sigma_a^2}\mathbf{I} & \text{Case 2 .} \end{cases} \tag{82}$$

The last equation subsumes, as a relevant case, the set of orthogonal space-time codes. Indeed, assuming $N = n = m$, the condition (82) yields, for the optimum code matrix,

$$\mathbf{A}\mathbf{A}^* = \frac{4\sigma_n^4\mu N}{\sigma_a^2}\mathbf{I}, \tag{83}$$

i.e., the code matrix $\mathbf{A}$ should be proportional to any unitary $N \times N$ matrix. Thus, any orthonormal basis of F^N can be exploited to construct an optimum code under the Case 2 and white Gaussian noise. If, instead, we restrict our attention to code matrices built upon Galois Fields (GF), there might be limitations to the existing number of optimal codes. Deffering to (Tarokh, 1999) and to the Urwitz-Radon condition exploited therein, we just remind here that, under the constraint of binary codes, unitary matrices exist only for limited values of N: for 2×2 coding, we find the normalized Alamouti code (Alamouti, 1998), which is an ortho-normal basis, with elements in GF (2), for F^2.

Make some comments. First notice, that under the white Gaussian noise, both performance measures considered above are invariant under unitary transformations of the code matrix, while at the correlated clutter they are invariant with respect to right multiplication of **A** by a unitary matrix. Probably, these degrees of freedom might be exploited for further optimization in different radar functions. Moreover, (70) represents the optimum solution for the case that no constraint is forced upon the code alphabet; indeed, the code matrices turn out in general to be built upon the completely complex field. If, instead, the code alphabet is co-nstrained to be finite, then the optimum solution (70) may be no longer achievable for arbitrary clutter covariance. In fact, while for the special case of white clutter and binary alphabet the results of (Tarokh, 1999) may be directly applied for given values of m and n, for arbitrary clutter covariance and (or) transmit/receive antennas number, a code matrix constructed on GF (q) and fulfilling the conditions (70) is no longer ensured to exist. In these situations, which however form the object of current investigations, a brute-force approach could consist of selecting the optimum code through an exhaustive search aimed at solving (66), which would obviously entail a computational burden $O\left(q^{mN}\right)$ floating point operations. Herein we use the usual Landau notation. $O(n)$.; hence, an algorithm is $O(n)$ if its implementation requires a number of floating point operations proportional to n (Golub & Van Loan, 1996). Fortunately, the exhaustive search has to be performed off line. The drawback is that the code matrix would inevitably depend on the target fluctuation law; moreover, if one would account for possible nonstationarities of the received clutter, a computationally acceptable code updating procedure should be envisaged so, as to optimally track the channel and clutter variations.

4.5 Simulation

The present section is aimed at illustrating the validity of the proposed encoding and detection schemes under diverse scenarios. In particular, we first assume uncorrelated disturbance, whereby orthogonal space-time codes are optimal. In this scenario, simulations have been run, and the results have been compared to the Chernoff bounds of the conventional GLRT receiver discussed in (De Maio & Lops, 2007) and to the GD performance achievable through a single-input single-output (SISO) radar system. Next, the effect of the disturbance correlation is considered, and the impact of an optimal code choice is studied under different values of transmit/receiver antenna numbers. In all cases, the behavior of the mutual information between the observations and the target replicas can be also represented, showing that such a measure is itself a useful tool for system design and assessment, but this analysis is outside of a scope of the present chapter.

Figure 8 represents the white Gaussian disturbance and assesses the performance of the GLRT GD. To elicit the advantage of waveform optimization, we consider both the optimum coded wireless communication system and the uncoded one, corresponding to pulses with equal amplitudes and phases. The probability of detection P_D is plotted versus SCR assuming $P_{FA} = 10^{-4}$ and $N = m = n = 2$. This simulation setup implies that the Alamouti code is optimum in the sense specified by (82). For comparison purposes, we also plot the performance of the uncoded SISO GD. We presented the performance of the conventional GLRT to underline a superiority of GD employment.

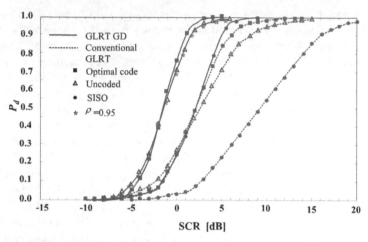

Fig. 8. P_D versus SCR; white Gaussian disturbance and disturbance with exponentially shaped covariance matrix ($\rho = 0.95$); $P_{FA} = 10^{-4}; N = m = n = 2$.

The curves highlight that the optimum coded wireless communication system employing the GD and exploiting the Alamouti code, achieves a significant performance gain with respect to both the uncoded and the SISO radar systems. Precisely, for $P_D = 0.9$, the performance gain that can be read as the horizontal displacement of the curves corresponding to the analyzed wireless communication systems, is about 1 dB with reference to the uncoded GLRT GD wireless communication system and 5 dB with respect to the SISO GD. Superiority of employment GD with respect to the conventional GLRT wireless communication systems achieves 6 dB for the optimum coded wireless communication system, 8 dB for the uncoded wireless communication systems, and 12 dB for SISO wireless communication systems. It is worth pointing out that the uncoded wireless communication system performs slightly better the coded one for low detection probabilities. This is a general trend in detection theory, which predicts that less and less constrained fluctuations are detrimental in the high SCR region, while being beneficial in the low SCR region. On the other hand, the code optimization results in a more constrained fluctuation, which, for low SCRs, leads to slight performance degradation as compared with uncoded systems. The effect of disturbance correlation is elicited in Fig.8 too, where the analysis is produced assuming an overall disturbance with exponentially shaped covariance matrix, whose one-lag correlation coefficient ρ is set to 0.95. In this case, the Alamouti code is no longer optimum. The plots show that the performance gain of the optimum coded GLRT GD wireless communication system over both the uncoded and the SISO GD detector is almost equal to that resulting when the disturbance is white. On the other hand, setting $N = m = n = 2$ in (81), shows that, under correlated disturbance, the optimum code matrix is proportional to **M**: namely, an optimal code tends to restore the "white disturbance condition." This also explains why the conventional Alamouti code follows rather closely the performance of the uncoded GLRT GD wireless communication system.

The effect of number n of receive antennas on the performance is analyzed in Fig.9, where P_D is plotted versus SCR for $N = m = 8$, exponentially shaped clutter covariance matrix with $\rho = 0.95$, and several values of n. The curves highlight that the higher n, namely the higher the diversity order, the better the performance. Specifically, the performance gap between the case $n = 8$ and the case of a MISO GLRT GD radar system (i.e., $n = 1$) is about 2.5 dB, while, in the case of the conventional GLRT radar systems, is about 7 dB for $P_D = 0.9$. A great superiority between the radar systems employing GLRT GD and conventional GLRT is evident and estimated at the level of 6 dB at $n = 8$ and 10 dB in the case of a MISO (i.e., $n = 1$) for $P_D = 0.9$. Notice that this performance trend is also in accordance with the expression of the mutual information that exhibits a linear, monotonically increasing, dependence on n. The same qualitative, but not quantitative, performance can be presented under study of the number m of available transmit antennas on the GLRT GD wireless communication system performance.

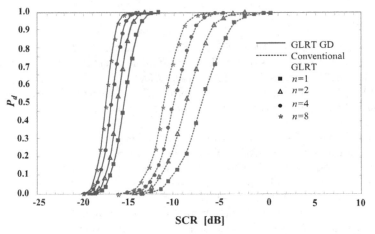

Fig. 9. P_D of optimum coded system versus SCR; disturbance with exponentially shaped covariance matrix ($\rho = 0.95$); and several values of m; $P_{FA} = 10^{-4}$; $N = m = 8$.

4.6 Discussion

We have addressed the synthesis and the analysis of MIMO radar systems employing the GD and exploiting space-time coding. To this end, after a short description of the MIMO radar signal model applied to wireless communications, we have devised the GLRT GD under the assumption of the additive white Gaussian disturbance. Remarkably, the decision statistic is ancillary and, consequently, CFAR property is ensured, namely, the detection threshold can be set independent of the disturbance spectral properties. We have also assessed the performance of the GLRT GD providing closed-form expressions for both P_D and P_{FA}. Lacking a manageable expression for P_D under arbitrary target fluctuation models, we restricted our attention to the case of Rayleigh distributed amplitude fluctuation. The performance assessment that has been undertaken under several instances of number of receive and transmit antennas, and of clutter covariance, has confirmed that MIMO GD radar systems with a suitable space-time coding achieve significant performance gains over SIMO, MISO, SISO, or conventional SISO radar systems employing the conventional GLRT detector. Also, these MIMO GD radar systems outperform the listed above systems employing the conventional GD. Future research might concern the extension of the proposed framework to the case of an unknown clutter covariance matrix, in order to come up with a fully adaptive detection system. Moreover, another degree of freedom, represented by the shapes of the transmitted pulses could be exploited to further optimize the performance. More generally, the impact of space-time coding in MIMO CD radar systems to estimate the target parameters is undoubtedly a topic of primary concern. Finally, the design of GD and space-time coding strategies might be of interest under the very common situation of non-Gaussian radar clutter.

5. Acknowledgment

This research was supported by Kyungpook National University research Grant, 2011.

6. References

Barnard, T. & Weiner, D. (1996). Non-Gaussian Clutter Modelling with Generalized Spherically Invariant Random Vectors. *IEEE Transactions on Signal Processing*, Vol. SP-44, No. 10, pp. 2384-2390, ISNN 1053-587X.

Bello, P. (1963), Characterization of Randomly Time-Invariant Linear Channels. *IEEE Transactions on Communications Systems*, Vol. CS-11, No. 12, pp. 360-393, ISSN 0096-1965.

Benedetto, S. & Biglieri, E. (1999). *Principle of Digital Transmission with Wireless Applications*, Plenum Press, ISBN 0-3064-5753-9, New York, USA.

Blankenship, T. & Rappaport, T. (1998). Characteristics of Impulsive Noise in the 450-MHz band in hospitals and clinics. *IEEE Transactions on Antennas and Propagation*, Vol. 46, No. 2, pp.194-203, ISNN 0018-926X.

Buzzi, S. et all. (1999). Optimum Detection over Rayleigh-Fading, Dispersive Channels with Non-Gaussian Noise. *IEEE Transactions on Communications*, Vol. COM-35, No. 7, pp. 926-934, ISSN 0096-1965.

Buzzi, S. et all. (1997). Signal Detection over Rayleigh-Fading Channels with Non-Gaussian Noise. In *Proc. Inst. Elect. Eng., Commun.*, Vol. 144, No. 6, pp. 381-386.

Buzzi, S. et all. (2001). Optimum Diversity Detection over Fading Disperive Channels with Non-Gaussian Noise. *IEEE Transactions on Communications*, Vol. COM-49, No. 4, pp. 767-775, ISSN 0096-1965.

Gini, F. et al. (1998). The Modified Cramer-Rao Bound in Vector Parameter Estimation. *IEEE Transactions on Communications*, Vol. COM-46, No. 1, pp. 52-60, ISSN 0096-1965.

Conte, E. et al. (1995). Canonical Detection in Spherically Invariant Noise. *IEEE Transactions on Communications*, Vol. COM-43, No. 2-4, pp. 347-353, ISSN 0096-1965.

Conte, E. et al. (1995). Optimum Detection of Fading Signals in Impulsive Noise. *IEEE Transactions on Communications*, Vol. COM-43, No. 2-4, pp. 869-876, ISSN 0096-1965.

Cover, T. & Thomas, J. (1991). *Elements of Information Theory*, Wiley & Sons, Inc., ISBN 0-4710-6259-6, New York, USA.

De Maio, A. & Lops, M. (2007). Design Principles of MIMO Radar Detectors. *IEEE Transactions on Aerospace and Electronic Systems*, Vol. AES-43, No. 3, pp. 886-898, ISSN 0018-9251.

Fishler, E. et al. (2006). Spatial Diversity in Radars – Models and Detection Performance. *IEEE Transactions on Signal Processing*, Vol. SP-54, No. 3, pp. 823-838, ISNN 1053-587X.

Foschini, G. (1996). Layered Space-Time Architecture for Wireless Communication in a Fading Environment Using Multi-Element Antennas. *BLTJ*, Vol. 1, No. 2, pp. 41-59, ISSN 1538-7305.

Golub, G. & Van Loan, C. (1996). *Matrix Computations*, 3rd Ed. The John Hopkins Press, ISBN 0-8018-5414-8, Baltimore, MD, USA.

Kassam, S. (1988), *Signal Detection in Non-Gaussian Noise*, Springer-Verlag, ISBN 0-3879-6680-3, New York, USA.

Kassam, S. & Poor, H. (1985). Robust Technique for Signal Processing: A Survey. *Proceedings IEEE*, Vol. 73, No. pp. 433-481, ISNN 0018-9219.

Kuruoglu, E. et al. (1998). Near Optimal Detection of Signals in Impulsive Noise modelled with a Symmetric α-Stable Distribution. *IEEE Communications Letters*, Vol. 2, No. 10, pp. 282-284.

Lombardo, P et all. (1999). MRC Performance for Binary Signals in Nakagami Fading with General Branch Correlation. *IEEE Transactions on Communications*, Vol. COM-47, No. 1, pp. 44-52, ISSN 0096-1965.

Matthews, J. (1992). Eigenvalues and Troposcatter Multipath Analysis, *IEEE Journal of Selected Areas in Communications*, Vol. 10, No. 4, pp. 497-505, ISNN 0733-8716.

Middleton, D. (1999). New Physical-Statistical Methods and Models for Clutter and Reverberation: The KA-Distribution and Related Probability Structures. *IEEE Journal of Oceanic Engineering*, Vol. 24, No. 7, pp. 261-284, ISSN 0364-9059.

Poor, H. (1988). *An Introduction to Signal Detection and Estimation*. Springer-Verlag, ISBN 0-3879-4173-8, New York, USA.

Proakis, J. (2007), *Digital Communications*, 5th Ed. McGraw-Hill, ISBN 0-0729-5716-6, New York, USA.

Sangston, K. & Gerlach, K. (1994). Coherent Detection of Radar Targets in a Non-Gaussian Background. *IEEE Transactions on Aerospace and Electronic Systems*, Vol. AES-30, No. 4, pp. 330-334, ISSN 0018-9251.

Skolnik, M. (2008). *Radar Handbook*, 3rd Ed. McGraw-Hill, ISBN 978-0-07-148547-0, New York, USA

Sousa, E. (1990). Interference Modelling in a Direct-Sequence Spread-Spectrum Packet Radio Network. *IEEE Transactions on Communications*, Vol. COM-38, No. 9, pp. 1475-1482, ISSN 0096-1965.

Tuzlukov, V. (1998). A New Approach to Signal Detection Theory. *Digital Signal Processing*, Vol. 8, No. 3, pp. 166-184, ISSN 1051-2204

Tuzlukov, V. (1998), *Signal Processing in Noise: A New Methodology*, IEC, ISBN 985-6453-16-X, Minsk, Belarus

Tuzlukov, V. (2001), *Signal Detection Theory*, Springer-Verlag, ISBN 0-8176-4152-1, New York, USA

Tuzlukov, V. (2002), *Signal Processing Noise*, CRC Press, Taylor & Francis Group, ISBN 0-8493-1025—3, Boca Raton, USA

Tuzlukov, V. (2005), *Signal and Image Processing in Navigational Systems*, CRC Press, Taylor & Francis Group, ISBN 0-8493-1598-0, Boca Raton, USA

Tuzlukov, V. (2012), *Signal Processing in Radar Systems*, CRC Press, Taylor & Francis Group, ISBN 0-8493-....-.., Boca Raton, USA (in press).

Van Trees, H. (2003), *Detection, Estimation, and Modulation Theory*. 2nd Ed. Wiley & Sons, Inc., ISBN 978-0-471-44967-6, New York, USA.

Webster, R. (1993), Ambient Noise Statistics. *IEEE Transactions on Signal Processing*, Vol. SP-41, No. 6, pp. 2249-2253, ISNN 1053-587X.

Yao, K. (1973). A Representation Theorem and Its Applications to Spherically Invariant Random Processes. *IEEE Transactions on Information Theory*, Vol. IT-19, No. 7-8, pp. 151-155, ISNN 0018-9448.

Part 2

Next Generation Wireless Communication Technologies

4

Introduction to the Retransmission Scheme Under Cooperative Diversity in Wireless Networks

Yao-Liang Chung[1] and Zsehong Tsai[2]
Graduate Institute of Communication Engineering,
National Taiwan UniversityTaipei
Taiwan, R.O.C.

1. Introduction

As implied by the word "Cooperative Diversity (CD)," mobile users in a multi-user environment can share their antennas in a manner that creates a virtual Multiple-Input Multiple-Output (MIMO) system, which can be conceptually viewed as a multichannel transmission environment in the network layer, to achieve individual or common purposes of those users. By employing CD for transmissions, the quality and reliability of users' data in wireless networks can thus be improved, mainly owning to the reason that the effect of wireless channel fading can be reduced. In this chapter, we aim to introduce existing representative retransmission schemes under various environments and further present a novel packet retransmission scheme for Quality-of-Service (QoS)-constrained applications in a general CD environment.

Transmit diversity of MIMO systems is an important technique which can bring significant gain to wireless systems with multiple transmit antennas. This technique is clearly advantageous to be employed on a cellular base station; however, it may not be practical for other scenarios. To be more specific, due to size, cost, or hardware limitations, small handsets/cellular phones may not be able to support certain types of multiple transmit antennas. For example, the size of an antenna must be several times the wavelength of the carrier frequency. Therefore, the use of multiple antennas is not an attractive way to achieve the transmit/receiving diversity in small handsets/cellular phones. To overcome such a naturally fundamental problem, CD is in nature an effective strategy to allow a single-antenna mobile device to achieve the benefit of MIMO systems with the help of cooperative mobile devices.

CD, which is a form of spatial diversity, is through cooperating users' (usually called partners) relaying signals to the destination. This technique is achieved without the use of additional antennas of any user. That is to say, the antennas of the sender and partners together form a multiple-transmit antenna situation. Basically, the relay mechanism can be decode-and-forward or amplify-and-forward. Moreover, CD is an emerging and powerful technique that can mitigate fading and improve robustness to interference in wireless environments. Thus, CD becomes a promising candidate for emulating MIMO systems.

Recently, many research groups have turned their attention to the CD-related topics. Individual aspects of these problems have been considered, for example, in various papers [1-4]. In [1], Mahinthan et al. proposed a Quadrature Signaling (QS) mechanism in the CD system for transmissions. CD transmissions considering issues related to power allocation algorithms were explored by Mahinthan et al. in [2-3]. In [4], Chen et al. exploited that the use of space-time block coding in the multi-user CD to improve the performance of the transmission in wireless local area networks. Other abounding literature survey and investigation regarding the issue related to CD including principles and applications can be referred to Ray Liu et al. in [5] and Fitzek et al. in [6], respectively. Recently, a simple method to evaluate the performance of complex networks under CD using sampling property of a delta function was proposed by Jang in [7].

However, because of fundamental physical characteristics of wireless channels, data packets often cannot be delivered to the destination successfully. As a result, the design focusing on the efficient retransmission scheme under such a CD environment still plays a highly crucial role. Due to the evolution of the communication technology, most packet retransmission schemes under CD in literatures were based on the rich results from those retransmission schemes on point-to-point transmissions. Thus, we will first provide an overview of retransmission schemes on point-to-point transmissions, and then, investigate the issue on the retransmission scheme under the CD environment.

While there have been many papers exploring various retransmission schemes in the CD environment, there were no elaborations on the issue considering the time constraint for delay-sensitive services. Consequently, in such a CD environment their throughput formulas did not reflect the effective throughput (goodput) that must satisfy the typical delay constraints of streaming-type or real-time multimedia flows. Motivated by the above point, we therefore pay our attention to design a novel fast packet retransmission scheme to be employed in a general CD environment for delay-sensitive flows as a case study.

The rest of this chapter is organized as follows. A survey of various retransmission schemes is included in Section 2. Next, Section 3 proposes a novel fast packet retransmission scheme in a general CD environment for delay-sensitive applications as a case study. Section 4 makes a summary of this chapter and suggests the future work of interest. Finally, the list of references is provided in the end.

2. A survey of various retransmission schemes

The traditional retransmission scheme designed to combat the loss of transmission data for single-radio single-channel environments was first introduced by Lin et al. in [8]. Thanks to the advances of multi-band radio technologies, for many broadband wireless systems and short range communication networks, there may be many communication channels available to use. Consequently, it is natural to arrange the link layer packets to be transmitted over multiple channels to boost bandwidth. Issues regarding how to design and analyze the multi-channel transmission schemes have recently become an important research direction. A vast amount of research groups have thus started to pay attention to related topics. Individual aspects of these problems have been separately considered in many related papers [9-18]. Most literatures in this area can be conceptually categorized into 2 research directions; one is the single-radio multi-channel transmission discussed in [9-11], and the other direction is the multi-radio multi-channel transmission discussed in [12-18]. Meanwhile, for the single-radio single-channel cases, the performance results of the Automatic-Repeat-reQuest (ARQ) retransmission schemes based on Markov analysis were available in [19-22], and optimized in both the power and the packet drop probability aspects respectively were recently studied in [23]. These studies [9-23] together provide a basis for the analysis and comparison of the multi-channel transmission. Furthermore, various kinds of ARQ and Hybrid ARQ schemes designed to be employed for CD environments were explored in [24-27]. The design approaches of these related research works are elaborated as follows.

In [9], a protocol was proposed to enable hosts to utilize multiple channels by switching channels dynamically, and their simulation showed that the effective throughput was improved, especially when the network was highly congested. Centralized and distributed algorithms to perform efficient channel assignments in component-based approach were proposed and implemented in [10]. A joint multi-channel and multi-path control protocol was proposed in [11], where it combined multi-channel link layer with multi-path routing, and simulations showed that the scheme can improve the throughput significantly.

The uses of switchable interfaces and multi-channel routing were proposed in [12], and simulation results showed that the throughput in a wireless ad hoc network can be improved. In [13], a hybrid channel assignment scheme was modeled into an integer linear programming formulation, and an approximation method for simplification was also studied. In the ad hoc network related areas, [14] maximized the network throughput subject to fairness requirements using the proposed wireless network coding schemes for a variety of multi-radio multi-channel environments with different routing strategies.

Several works focusing on the ARQ schemes in multi-radio multi-channel transmissions have been discussed in [15-18]. Formulas of the link layer throughput for the Stop-and-Wait (SW), the Go-Back-N (GBN) or the Selective-Repeat (SR) schemes were derived under different assumptions of channel characteristics. The multi-channel SR ARQ scheme in [15] assumed equal transmission rate and allowed the transmitter dynamically to assign the retransmission link packet to the channel using the link packet error probability as the selection criterion, and retransmissions can continue till its success. In [16], both throughput and delay performances of the multi-channel SW, GBN, and SR ARQ schemes were investigated and validated via simulations, while all channels were assumed statistically independent and identical. Throughput analysis of the multi-channel SW, GBN, and SR ARQ schemes were generalized in [17], where the generalization took the form of packet-to-channel assignment rules, and radio channels can be with different transmission rates and different link packet error probabilities. Two fast ARQ/HARQ packet retransmission schemes have been proposed to transport delay-sensitive flows in a multi-radio multi-channel environment in [18], where they can incorporate various retransmission policies, which are adjusted by the channel signal-to noise ratio (SNR) and the APDU size.

Closed form equations of the service data unit delay under the SR ARQ scheme were successfully derived and validated via simulation in [19]. An exact Markov model proposed to evaluate the delay statistics of the link packet for the SR ARQ scheme was available in [20]. The queueing models using dynamic link adaptation for the GBN and the SR ARQ schemes were formulated in [21], where the exact queue length and the delay statistics were obtained. For HARQ schemes, a Markov model was presented to analyze the SR truncated type II HARQ scheme employing Reed Solomon linear erasure block codes in [22], where the link packet throughput, error probability, and delay performance were analyzed. In [23], a suboptimal root-finding solution was developed to solve the exhaustive search for the optimization problem formulated based on the incremental-redundancy HARQ scheme.

The delay performance of several truncated ARQ and HARQ schemes in a CD environment under the assumption of Poisson arriving packets were evaluated in detail by Boujemâa in [24]. An analytical model to quantify end-to-end performances for a CD ARQ scheme in a cluster-based multi-hop wireless network was proposed by Le et al. in [25]. Markov models developed to evaluate the CD system were also investigated by Mahinthan et al. in [26] and by Issariyakul et al. in [27], respectively.

While papers [24-27] have widely explored various ARQ/HARQ schemes in the CD environments, the issues regarding time constraints for delay-sensitive flows were not addressed and elaborated on. Therefore, their throughput formulas did not reflect the effective throughput that must satisfy the typical delay constraints of streaming-type or real-time multimedia flows in such an environment.

Due to the aforementioned reasons, we herein propose a novel fast packet retransmission scheme, where a new approach of retransmission strategy is designed and appropriately combines the encoding/decoding mechanism presented in [18], in such a CD environment for delay-sensitive flows as a case study. In the proposed scheme, there are 2 retransmission policies that can be employed adaptively according to both the channel quality and the Application layer Protocol Data Unit (APDU) size. The retransmission is designed to be allowed only one time. Here, APDU flows in the sender are further assumed to always have a link packet ready for transmission. As a result, it is not much meaningful to analyze the packet delay involving the queueing analysis. In this paper, we only focus on the complete throughput analysis to gain the main insight of optimizing the number of channels for retransmission between the 2 proposed retransmission policies under such the CD environment. All of the derived formulas are then verified via simulations. The effective throughput of our proposed scheme is shown better than that of other CD retransmission schemes (such as [26]) and non-CD retransmission schemes.

3. Case study: On the effective throughput gain of cooperative diversity with a fast retransmission scheme for delay-sensitive flows [33-34]

3.1 System description

3.1.1 Cooperative diversity system

A general CD system model composing of a sender, a partner, and a receiver is considered, as shown in Fig. 1, where two cooperative users (i.e., sender and partner) transmit their information to the same destination (i.e., receiver). It is assumed that each user' device in this system only has one radio transceiver. Additionally, Orthogonal Frequency Division Multiplexing (OFDM) is employed as the underlying transmission technique.

In the present system, channels among sender, partner and receiver are modeled as non-identical but independent Nakagami-m slow-fading channels corrupted by additive white Gaussian noise. The fading channels and the noise are assumed to be independent of each other.

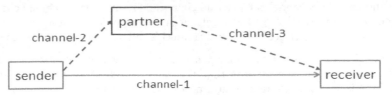

Fig. 1. System model for two cooperative users (sender and partner) transmission.

Generally speaking, for the transmission of flows of the sender, application layer flows are composed of APDUs. We assume that an APDU consists of s link packets. Each link packet will be encoded (described in detail in Section 3.1.3) in sequence for transmission. Here, APDU flows in the sender are assumed to always have a link packet ready for transmission.

For the convenience of the following analysis, channels that between sender and destination, between sender and partner, and between partner and destination are denoted as channel-j, j=1,2,3, respectively.

Last, but not least, we herein choose to employ only 1 partner for study since the significant improvement of the overall system performance with CD is usually owing to the contribution of the best partner [3].

3.1.2 Principles of fast retransmission strategy

The design philosophy of the retransmission strategy is to improve the application layer throughput while the effective control of the transmission delay is also assured. We assume that the underlying coding scheme is HARQ and that if a packet is retransmitted, then only its complementary packet is sent.

The packet retransmission strategy can be described via the following 4 principles:

- A link packet will be duplicated a copy in the sender buffer before its first transmission. When the sender begins to transmit this link packet, it will be broadcasted to the receiver and the partner.
- When an original link packet is transmitted to the receiver, an acknowledgment (ACK) or Negative ACK (NACK) packet (assumed error-free) will be sent to both the sender and the partner.
- There are 2 retransmission policies designed for the retransmission, indexed as *policy_k*: for k=0, 1. If a NACK is received, a complementary link packet will be retransmitted only one time via the partner (indexed as *policy_0*) or via both the sender and the partner (indexed as *policy_1*).
- The best retransmission policy is selected based upon both the APDU length and the expected long term link packet error probability among sender, partner and receiver, using the average application throughput as the performance objective. Under different APDU sizes and different link packet error probabilities, the corresponding best retransmission policy can be different.

Since an OFDM system is assumed, we assume each channel only uses a subset of OFDM subchannels. For the above retransmission strategy, note that both channel-1 and channel-2 are with orthogonal subchannel set 1, while channel-3 is with orthogonal subchannel set 2. The intersection of subchannel set 1 and set 2 is arranged to be an empty set; therefore, for the receiver, the signals from channel-1 and channel-3 will not interfere with each other.

Typical retransmission operations under *policy_0* and *policy_1* are illustrated in Fig. 2 and Fig. 3, respectively. Also, we assume that the delay threshold of the considered delay-sensitive APDU is set equal to the maximum of maximum delays for 2 policies. Notice that in this model the terminology *delay* only indicates the air-transmission delay component for the APDU under the proposed retransmission principle.

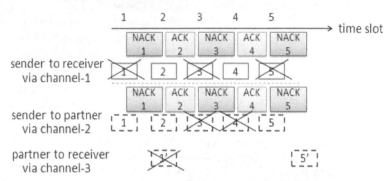

Fig. 2. A typical example of a fast HARQ with *policy_0*. When any original link packet is found failed at the receiver, only the partner will retransmit a complementary link packet if a link packet is successfully received by the partner. Link packet i' means the complementary packet of link packet i.

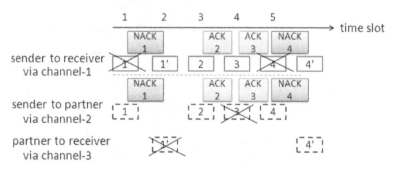

Fig. 3. A typical example of fast HARQ with *policy_1*. When any original link packet is found failed in the receiver, both the sender and the partner will retransmit a complementary link packet. Link packet i' is the complementary packet of link packet i.

3.1.3 Cooperative diversity with fast HARQ scheme

Two codes, a block code C_0 and a convolutional code C_1, are together used as the coding mechanism employed in each user (see [29], [36] for examples). For more detail description, C_1 is a rate-1/2 convolutional code with constraint length c consisting of a c-stage shift register and two generator polynomials $G_1(x)$ and $G_2(x)$. This code C_1 is used as an inner code for error detection and correction. Next, the outer code C_0 is a high rate $(n-(c-1), n-(c-1)-r)$ block code used for error detection only, where n is the length of each link packet that is transmitted in this scheme, and r is the length of parity-check bits for error detection.

When an APDU arriving at the sender, there will be s new information sequences $I_i(x)$, $1 \le i \le s$, each length $(n-(c-1)-r)$, generated in sequence. They are encoded into $J_i(x)$ with C_0 and then encoded into $V_i(x) = J_i(x)G_1(x)$ with C_1 in sequence. Each of them will be broadcasted in sequence only one time to the partner and the receiver.

Let $\tilde{V}_i(x)$ and $\hat{V}_i(x)$ be the noisy versions of $V_i(x)$ arriving at the receiver and the partner, respectively. For the receiver, the syndrome of $\tilde{V}_i(x)$ is checked in two steps. In step_1, $\tilde{V}_i(x)$ is regarded as a noisy version of a codeword in the $(n, n-(c-1))$ shortened cyclic code generated by $G_1(x)$. In step_2, an estimate $\tilde{J}_i(x)$ of $J_i(x)$ is then checked in the high rate $(n-(c-1), n-(c-1)-r)$ block code. If the syndromes are all zero in any step, an estimate $\tilde{I}_i(x)$ of $I_i(x)$ is obtained and delivered to the *receiving buffer*, which is a buffer used for waiting other link packets back for an APDU. Subsequently, an ACK packet will be sent to the sender and the partner. However, if the aforementioned syndrome check in any step is not zero, $\tilde{V}_i(x)$ is stored in the receiver buffer, and a NACK packet will be transmitted to the partner and the sender for possible retransmission. For the partner, if $\hat{V}_i(x)$ is error-free, then either *policy_0* or *policy_1* is adopted; otherwise, it is directly dropped.

Under *policy_0*, if $\hat{V}_i(x)$ is error-free, $\hat{V}_i(x)$ will be decoded to $V_i(x)$ and then re-encode via $G_2(x)$ to $V_i'(x)$. Thereafter, $V_i'(x)$ is transmitted to the receiver. Let the noisy version of $V_i'(x)$ be denoted as $\hat{V}_i'(x)$. $\hat{V}_i'(x)$ will be checked in the same way as in the first transmission. If $\hat{V}_i'(x)$ is found failed, $\hat{V}_i'(x)$ shall be combined with $\tilde{V}_i(x)$, to form a combined codeword, which is decoded by the Viterbi decoding. The result is checked in the high rate $(n-(c-1), n-(c-1)-r)$ block code. If the syndrome is zero, it is claimed as a correct result. Its information sequence is then estimated and delivered to the *receiving buffer*; otherwise, it is discarded and the retransmission for this link packet is stopped.

Under *policy_1*, a complementary link packet $V_i'(x)$ via $G_2(x)$ of $V_i(x)$ will be transmitted from the sender to the receiver, and let the noisy version be denoted as $\tilde{V}_i'(x)$. Meanwhile, if $\hat{V}_i(x)$ is error-free, $\hat{V}_i(x)$ will be decoded to $V_i(x)$ and then re-encode via $G_2(x)$ to $V_i'(x)$. Let the noisy version of $V_i'(x)$ be denoted as $\hat{V}_i'(x)$. Following that, $\tilde{V}_i'(x)$ and $\hat{V}_i'(x)$ will be checked via the two-step decoding procedure, respectively. If the syndrome of any one is zero, it is claimed as a correct result, its information sequence is then estimated and delivered to the *receiving buffer*; otherwise, $\tilde{V}_i(x)$ shall be combined with $\hat{V}_i'(x)$ and $\tilde{V}_i'(x)$, respectively, to form two combined codewords, which are decoded by the Viterbi decoding. If any result is successful, its information sequence is then estimated and delivered to the *receiving buffer*; otherwise, a new codeword, $\tilde{V}_{i,mrc}(x)$, will be further generated based on the Maximal Ratio Combining (MRC) (see [31] for more details) technique via $\hat{V}_i'(x)$ and $\tilde{V}_i(x)$. The syndrome of the new codeword is checked in the same concept of the two-step decoding procedure. If the result is successful, the estimated information sequence will be delivered to the *receiving buffer*; otherwise, it is discarded and the retransmission for this link packet is stopped.

3.2 Throughput analysis

Performances of the application layer throughput for the present scheme will be first analyzed in detail in this Section. It is assumed that the time axis is partitioned into equal size slot. In each time slot, it is separated into two parts. That is to say, the main part of a time slot is used for link packets transmissions and the rest of the time slot is reserved for ACK/NACK packets transmissions. Here, the SNR is assumed staying constant in a time slot. In addition, the M-ary, $M = 2^b$ where b is even, the Quadrature Amplitude Modulation (QAM) scheme is assumed in the OFDM subchannels of the proposed model.

3.2.1 Link packet error probability

For M-ary QAM in Nakagami-m slow-fading channels, the average BERs for channel-j, j=1,2,3, denoted as $\bar{\varepsilon}_j$, can be derived by

$$\bar{\varepsilon}_j = \int_0^\infty p_j(\gamma_j)\varepsilon_{ins,j}(\gamma_j)d\gamma_j \, , j=1,2,3, \tag{1}$$

where, in channel-j, $\varepsilon_{ins,j}(\gamma_j)$ is the instantaneous BER conditional on γ_j for M-ary QAM, γ_j is the instantaneous SNR per bit, $\gamma_j > 0$, $p_j(\gamma_j) = m^m \gamma_j^{m-1} e^{-m\gamma_j/\bar{\gamma}_j} / \bar{\gamma}_j^m \Gamma(m)$ is the probability density function (pdf) of γ_j in Nakagami-m fading given in [28], $m \geq 1/2$, $\Gamma(\cdot)$ is the gamma function, and $\bar{\gamma}_j$ is the average SNR per bit. The instantaneous BER $\varepsilon_{ins,j}(\gamma_j)$ was previously derived in [30], [32] as

$$\varepsilon_{ins,j}(\gamma_j) =$$
$$\frac{1}{\sqrt{M}\log_2\sqrt{M}} \sum_{z=1}^{\log_2\sqrt{M}} \sum_{t=0}^{f(z,M)} \left\{ erfc((2t+1)\sqrt{g\gamma_j})f(t,z,M) \right\}, \, j = 1,2,3, \tag{2}$$

where $f(z,M) = (1-2^{-z})\sqrt{M} - 1$, $g = 3\log_2 M/(2M-2)$, $f(t,z,M) = (-1)^{\lfloor t2^{z-1}/\sqrt{M} \rfloor} \left(2^{z-1} - \lfloor t2^{z-1}/\sqrt{M} + 1/2 \rfloor\right)$, and $erfc(\cdot)$ is the error function.

The average link packet error probability in a single transmission in channel-j, j=1,2,3, denoted as $\bar{P}_{j,e}$, can be given by

$$\bar{P}_{j,e} = \int_0^\infty p_j(\gamma_j)(1-(1-\varepsilon_{ins,j}(\gamma_j))^n)d\gamma_j \, , j=1,2,3. \tag{3}$$

Furthermore, the average link packet error probability after the Viterbi decoding conditional on the event that both $\hat{V}_i'(x)$ and $\tilde{V}_i(x)$ are corrupted, denoted as $\bar{P}_{f,0}$, can be approximately by (see eq. (28) in [29])

$$\bar{P}_{f,0} \cong 1-(1-p_b)^{n-(c-1)}, \tag{4}$$

where p_b is the corresponding bit error probability obtained via the Viterbi decoding. As shown in [29], p_b is bounded by

$$p_b \leq \frac{1}{2} \frac{\partial T(X,Y)}{\partial Y}\bigg|_{X=2\sqrt{\varepsilon'(1-\varepsilon')},Y=1}, \tag{5}$$

where ε', the upper bound of the conditional BERs given that the two-step (mentioned in Section 3.1.3) decoding syndromes in channel-1 and channel-3 are non-zero, is given by

$$\varepsilon' = \max\left\{\frac{\varepsilon_1}{1-(1-\varepsilon_1)^n}, \frac{(1-\varepsilon_2)\varepsilon_3}{1-(1-(1-\varepsilon_2)\varepsilon_3)^n}\right\}, \tag{6}$$

and $T(X,Y)$ is the generating function of the convolutional code. In addition, $\overline{P}_{f,1}$, the average link packet error probability after the Viterbi decoding conditional on the event that both $\tilde{V}_i(x)$ and $\hat{V}_i(x)$ are corrupted, can be given by (4)-(6) together with ε' replaced by $\varepsilon = \varepsilon_1/(1-(1-\varepsilon_1)^n)$, which is the conditional BER given that the two-step decoding syndrome in channel-1 is non-zero.

Last, but not least, the average link packet error probability after the MRC decoding, under *policy_1*, denoted as $\overline{P}_{mrc}$, can be given by

$$\overline{P}_{mrc} = \int_0^\infty \tilde{p}(\gamma_b)(1-(1-\varepsilon_{mrc}(\gamma_b))^n)d\gamma_b, \tag{7}$$

where γ_b is the instantaneous SNR per bit at the output of the MRC decoder, $\tilde{p}(\gamma_b)$ represents the pdf of γ_b, $\gamma_b > 0$, and $\varepsilon_{mrc}(\gamma_b)$ is the instantaneous BER conditional on γ_b for M-ary QAM after the MRC decoding. According to [28], [30], $\tilde{p}(\gamma_b) = m^{2m}\gamma_b^{2m-1}e^{-m\gamma_b/\overline{\gamma}_c}/\overline{\gamma}_c^{2m}\Gamma(2m)$, where $\overline{\gamma}_c$ means the equivalent average SNR for each channel. The instantaneous BER $\varepsilon_{mrc}(\gamma_b)$ can be given by (2) with γ_j replaced by γ_b.

3.2.2 Throughput

For the fast HARQ scheme with *policy_0*, the application layer throughput in APDU/slot, denoted as T_0, can be derived as

$$T_0 = \frac{1}{s}(1-\overline{P}_{1,e}\overline{P}_{2,e} - \overline{P}_{1,e}(1-\overline{P}_{2,e})\overline{P}_{3,e}\overline{P}_{f,0})^s, \tag{8}$$

where $1/s$ represents the average number of the APDUs transported per slot, and the second term indicated the success probability of an APDU transmission.

Next, for the fast HARQ scheme with *policy_1*, the application layer throughput in APDU/slot, denoted as T_1, can be derived as

$$T_1 = \frac{\alpha}{s}(1-\overline{P}_{1,e}^2\overline{P}_{2,e}\overline{P}_{f,1} - \overline{P}_{1,e}^2(1-\overline{P}_{2,e})\overline{P}_{3,e}\overline{P}_{f,0}\overline{P}_{f,1}\overline{\overline{P}}_{mrc})^s, \tag{9}$$

where α / s represents the average number of the APDUs transported per slot, and similar to the concept in (8), the second term means the success probability of an APDU transmission. In (9), $\bar{P}_{mrc}$ is the average link packet error probability after the MRC decoding conditional on $\hat{V}_i'(x)$ and $\tilde{V}_i(x)$ all found failed. Since $\bar{P}_{mrc}$ is the unconditional probability of a link packet error after the MRC decoding and the result will be correct after the MRC decoding as long as there is at least a link packet that is correct, $\bar{\bar{P}}_{mrc}$ can be derived as $\bar{\bar{P}}_{mrc} = \bar{P}_{mrc} / (\bar{P}_{1,e}^2 (1 - \bar{P}_{2,e}) \bar{P}_{3,e})$ by the definition of conditional probability [35]. Moreover, α in (9) can be obtained via the equality

$$\bar{P}_{1,e}\alpha = 1 - \alpha \tag{10}$$

since the average number of retransmission link packets generated per slot should equal the average number of retransmission completed, after normalization.

Last, but not least, for delay-sensitive flows, the maximum air-transmission delay of an APDU allowed is usually subject to a specific QoS requirement. In this case, based on the similar derivation and argument in [18], one can appropriately tune the key parameter, namely, s, in the system to achieve the highest effective throughput under a given delay constraint.

3.3 Analytical and simulation results

In this section, the considered CD environment with a sender, a partner, and a receiver remains the same as shown in Fig. 1. We assume that an APDU is composed of 5 link packets. A 16 QAM modulation scheme is adopted. The coding mechanism is referred to Section 3.1.3. Also, we set $r = 6$ bytes, $c = 9$, and $n = 257$ bytes. The ACK/NACK packet size for ARQ related schemes is set equal to 25 bytes and for HARQ related schemes is set equal to 26 bytes. The link speed is set equal to 10Mbps. Besides, excluding the error-correcting codes, the ratio of the additional header overhead associated with the lower layer protocols from the application one is set equal to 0.04.

First, we will evaluate and compare the performance results among all schemes to see main potential insights of our proposed scheme by considering the ideal case that the channel between the sender and the partner is error-free. Next, we further investigate the impact on the system performance when there is an error probability on the channel between the sender and the partner.

3.3.1 With an error-free channel-2

For the fast retransmission scheme, based on (8)-(9), analytical results of application throughputs under $\bar{\varepsilon}_1$, with $\bar{P}_{3,e} = 0.9$ and $\bar{P}_{3,e} = 0.1$, in the Nakagami-3 slow-fading environment, are depicted in Fig. 4 and Fig. 5, respectively. In Fig. 4, it can be found that if $\bar{\varepsilon}_1 \leq 0.4 \times 10^{-2}$, the optimal throughput can be achieved with only 1 channel for retransmission (via the partner); if $\bar{\varepsilon}_1 \geq 0.4 \times 10^{-2}$, it can be achieved by parallel retransmissions via 2 channels (via both sender and partner). However, in Fig. 5, it is seen that the throughput of *policy_0* is always better than that of *policy_1*. Because the average link packet error probability of channel-3 is small, the retransmission of duplicated link packet on channel-1 via *policy_1* will waste bandwidth.

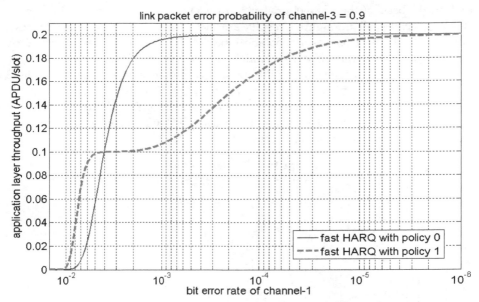

Fig. 4. Application throughputs under typical $\bar{\varepsilon}_1$ with $\bar{P}_{3,e} = 0.9$ under a fast HARQ scheme.

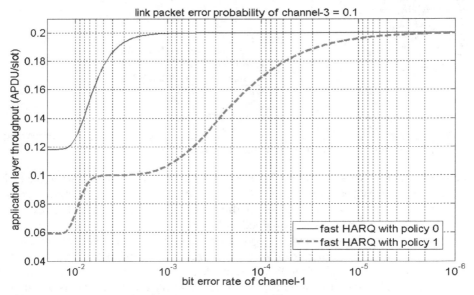

Fig. 5. Application throughputs under typical $\bar{\varepsilon}_1$ with $\bar{P}_{3,e} = 0.1$ under a fast HARQ scheme.

In what follows, we compare the throughput of our proposed scheme with that of the previous work [26] and non-CD HARQ scheme under $\bar{\varepsilon}_1$, with $\bar{\varepsilon}_3 = 10^{-3}$ in Nakagami-3 slow-fading channels, as shown in Fig. 6. Notice that in Fig. 6, the *CD with optimized fast HARQ* scheme represents the case that the retransmission policy is adaptively adjusted to be optimal on the basis of the channel quality in the CD environment. For a fair comparison among all schemes, all throughput results are in bit/second, and other 2 schemes are modified to allow only 1 retransmission and time slots for those discarded retransmissions are then used for new transmissions. Notice due to this modification, their throughput formulas are modified versions of (9) with the unused parameters removed. In details, one should set $\bar{P}_{f,1} = 1$ and replace the parameter n in (3) by n-$(c$-1$)$ for the scheme in [26], and set $\bar{P}_{2,e} = 1$ for the non-CD HARQ scheme. With the help of Fig. 6, it can be found that better performance is achieved by the optimized fast HARQ scheme except when $\bar{\varepsilon}_1$ is extremely small due to the additional overhead of the HARQ.

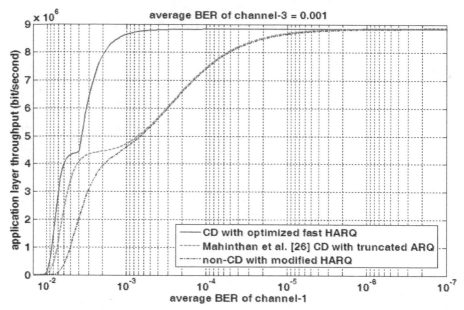

Fig. 6. Application layer throughput (in bit/sec.) comparison among various schemes under Nakagami-3 slow-fading channels when channel-2 is error-free.

Notice that, generally speaking, BER=0.001 is fairly high (in our parameter setting, which is about equal to the packet error rate =0.9 when without employing any error correcting mechanism), that is to say, the channel condition is extremely bad. Here, the reasons for setting channel-3's BER=0.001 are explained as follows. Although the sender would like to select the neighboring partner having good channel condition for helping transmission, in the worst case when the partner is far away from the receiver and both of them are at the edge of a cell such that the channel-3's condition degrades. In this case, with the validation of analytical and simulation results, the performance result of our scheme is much better than that of other schemes. It means our scheme is very powerful. Thus, it can be easily reasoned that when channel-3's BER decreases, our scheme still remains the best although the performance results for these schemes will all be improved.

Furthermore, taking $\bar{\varepsilon}_1 = 2 \times 10^{-3}$ and $\bar{\varepsilon}_3 = 10^{-3}$ as an example, effective throughput performances of these schemes under different Nakagami-m, m=1/2, 1, 3, slow-fading channels in the CD environment are compared, as listed in Table 1. From Table 1, both analytical results based on (8)-(9) and simulation results show that the optimized fast HARQ scheme always achieves better throughput performance than other schemes since the optimized fast HARQ scheme can adaptively adjust the retransmission policy according to the channel quality. Again in Table 1, it is found that the analytical results are slightly lower than the simulation ones for both the optimized fast HARQ scheme and the non-CD HARQ scheme since the Viterbi decoding mechanism via (5) is employed for them. Note that the upper bound in (5) is tight and can be regarded as an excellent approximation when the BER is lower than 10^{-2} [36].

Environment	Scheme Effective Throughput $\times 10^6$	CD with optimized fast HARQ	[26] CD with truncated ARQ	Non-CD with modified HARQ
Nakagami-1/2	analytical result	7.765	4.308	3.704
	simulation result	7.853	4.356	3.767
Nakagami-1	analytical result	7.889	4.413	3.822
	simulation result	7.968	4.462	3.887
Nakagami-3	analytical result	7.997	4.512	3.998
	simulation result	8.077	4.567	4.068

Table 1. Comparisons of the application layer throughputs (in bit/second) at $\bar{\varepsilon}_1 = 2 \times 10^{-3}$ and $\bar{\varepsilon}_3 = 10^{-3}$ among various schemes under different Nakagami-m, m=1/2, 1, 3, slow-fading channels.

3.3.2 With a non-error-free channel-2

Due to the fundamental physical characteristics of wireless channels, there often exists an error probability for each transmission channel in the real-world environment. However, in order to take the advantage of CD, the sender usually selects the neighboring partner having good channel condition between them. Thus, we herein set $\bar{\varepsilon}_2 = 10^{-4}$ for demonstrating performance results. The throughput comparisons of various schemes under $\bar{\varepsilon}_1$, with $\bar{\varepsilon}_2 = 10^{-4}$ and $\bar{\varepsilon}_3 = 10^{-3}$ in Nakagami-3 slow-fading channels, are shown in Fig. 7.

It is found in Fig. 7 that the performance of the optimized fast HARQ scheme obviously degrades when the BER of channel-1 is smaller than 3×10^{-3} when compared with that in Fig. 6. Because there exists an error probability on channel-2 and *policy_0* only uses the cooperative path (i.e., channel-2 together with channel-3) for retransmissions, the power of *policy_0* decreases. However, the throughput result of the optimized fast HARQ scheme in Fig. 7 is also shown better than that of the other 2 schemes. In addition, it can be observed that when $\bar{\varepsilon}_1 \geq 3 \times 10^{-3}$, the performance results of the first 2 good schemes are almost the same as those in Fig. 6 due to the fact that MRC is much powerful.

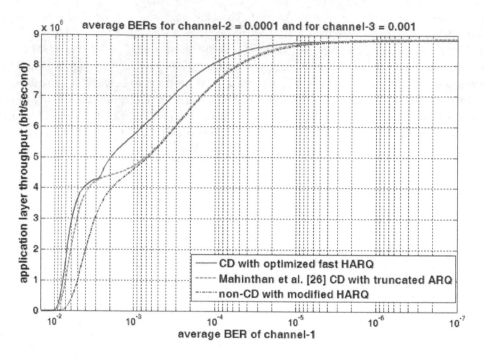

Fig. 7. Application layer throughput (in bit/sec.) comparison among various schemes under Nakagami-3 slow-fading channels when channel-2 is not error-free.

Last, for completeness, we take $\bar{\varepsilon}_1 = 2 \times 10^{-3}$, $\bar{\varepsilon}_2 = 10^{-4}$, and $\bar{\varepsilon}_3 = 10^{-3}$, as an example, to illustrate the effective throughput results for various schemes under different Nakagami-m, m=1/2, 1, 3, slow-fading channels, and summarize the results in Table 2. From Table 2, it can be found that both analytical results based on (8)-(9) and simulation results of the optimized fast HARQ scheme also always have better throughput results than those of other schemes as in Table1. The results of the non-CD HARQ scheme for both Table 1 and Table 2 are the same since its performance only depends on channel-1's BER. We also notice that throughput improvement of our scheme is significant even with $\bar{\varepsilon}_2 = 10^{-4}$ in the sender-to-partner channel.

In summary, based on Figs. 6 and 7 and Tables 1 and 2, we can thus conclude that the fast HARQ scheme is an excellent approach for transporting delay-constrained streaming-type or real-time multimedia flows in CD environments even when there is an error probability on the cooperative path. It is for the reasons that the retransmission strategy can be adaptively adjusted according to the channel condition and that the decoding procedure involving MRC and the Viterbi decoding are appropriately designed.

Scheme Environment	Effective Throughput $\times 10^6$	CD with optimized fast HARQ	[26] CD with truncated ARQ	Non-CD with modified HARQ
Nakagami-1/2	analytical result	4.805	4.258	3.704
	simulation result	4.848	4.258	3.767
Nakagami-1	analytical result	4.902	4.351	3.822
	simulation result	4.951	4.351	3.887
Nakagami-3	analytical result	5.017	4.456	3.998
	simulation result	5.067	4.456	4.068

Table 2. Comparisons of the application layer throughputs (in bit/second) at $\overline{\varepsilon}_1 = 2 \times 10^{-3}$, $\overline{\varepsilon}_2 = 10^{-4}$, $\overline{\varepsilon}_3 = 10^{-3}$ among various schemes under different Nakagami-m, m=1/2, 1, 3, slow-fading channels.

4. Conclusions

A fast HARQ packet retransmission scheme has been successfully proposed to transport delay-sensitive flows in a general CD environment. The presented scheme incorporates 2 retransmission policies, and these 2 policies can be selected adaptively by the channel SNRs and the APDU sizes. In ideal conditions, the best retransmission policy can always be selected to achieve optimized performance.

In this case study, our cooperative fast retransmission scheme has been shown to be an excellent approach for improving the effective throughput in transporting delay-sensitive flows in CD environments. Numerical results verified via simulations, show that when optimized, the proposed scheme can achieve effective throughput much better than that of other ARQ schemes (such as [26]) and non-CD HARQ schemes, especially when the sender-to-partner channel condition is good. The performance improvement is still significant even when there is an error probability (e.g. BER $\leq 10^{-4}$) in the sender-to-partner channel. Moreover, in the aspect of the battery saving, the presented scheme should save much more power than that of other schemes due to the one-time retransmission design. It is thus concluded that the proposed fast HARQ retransmission scheme is an excellent ARQ candidate for the multimedia or real-time transport in CD environments, when the time-constraint is imposed.

5. Summaries and future works

The issues of improving fast retransmission schemes under CD environments can be essential to many delay-sensitive applications. This chapter has widely covered the conceptual description of many representative retransmission schemes under various environments and presented a novel fast packet retransmission scheme intended for effectively transporting delay-sensitive flows in a general CD environment. The presented retransmission scheme and other related works should have provided a sufficient collection of schemes and analysis methodologies for designing further wireless communication systems with similar requirements.

Furthermore, due to the fact that in most practical scenarios the terminals are battery-powered, the design of the energy-efficiency transmission satisfying the respectively specific QoS requirements of these users in the network is very crucial to prolonging the battery life of these terminals. Consequently, the issue concerning the energy consumption has been increasingly paid much attention. We suggest incorporating such a concern with the present work to design an efficiently power-saving fast packet retransmission scheme in a general CD environment for delay-sensitive flows in the future. Additionally, the well design of an efficient retransmission scheme to be employed in such a CD environment simultaneously considering the issue of effective throughput, QoS, fairness, complexity, and power saving is still an open issue for research.

6. Acknowledgements

The authors wish to express their sincere appreciation for financial support from the National Science Council of the Republic of China under Contracts NSC 98-2219-E-002-002 and NSC 99-2219-E-002-002.

7. References

[1] V. Mahinthan, J. W. Mark, and X. Shen, "A cooperative diversity scheme based on quadrature signaling," *IEEE Trans. Wireless Commun.*, vol. 6, no. 1, pp. 41-45, January 2007.

[2] V. Mahinthan *et al.*, "Maximizing cooperative diversity energy gain for wireless networks," *IEEE Trans. Wireless Commun.*, vol. 6, no. 7, pp. 2530-2539, July 2007.

[3] V. Mahinthan *et al.*, "Partner selection based on optimal power allocation in cooperative diversity systems," *IEEE Trans. Veh. Technol.*, vol. 57, no. 1, pp. 511-520, January 2008.

[4] J. Chen and K. Djouani, "A multi-user cooperative diversity for wireless local area networks," *Int. J. Communications, Network and System Sciences*, vol. 3, pp. 207-283, 2008.

[5] K. J. Ray Liu, A. K. Sadek, W. Su, and A. Kwasinski, *Cooperative communication and networking*, Cambridge University Press, 2009.

[6] F. H. P. Fitzek and M. D. Katz, *Cooperation in wireless networks: principles and applications*, Springer, Netherlands, 2006.

[7] W. M. Jang, "Quantifying performance of cooperative diversity using the sampling property of a delta function," *IEEE Trans. Wireless Commun.*, vol. 10, no. 7, pp. 2034-2039, July 2011.

[8] S. Lin, et al., "Automatic-repeat-request error control schemes," *IEEE Commun. Mag.*, pp. 5-16, December 1984.

[9] J. So and N. H. Vaidya, "Multi-channel MAC for ad hoc networks: handling multichannel hidden terminals using a single transceiver," in *Proc. ACM MobiHoc*, pp. 222-233, Roppongi Hills, Tokyo, Japan, May 2004.

[10] R. Vedantham *et al.*, "Component based channel assignment in single-radio multichannel ad hoc networks," in *Proc. ACM MobiCom*, pp. 378-389, Los Angeles, CA, USA, May 2006.

[11] W. H. Tam and Y. C. Tseng, "Joint multi-channel link layer and multi-path routing design for wireless mesh networks," in *Proc. IEEE INFOCOM*, pp. 2081-2089, May 2007.

[12] P. Kyasanur and N. H. Vaidya, "Routing and link-layer protocols for multi-channel multi-interface ad hoc wireless networks," *ACM Mobile Computing and Comm. Rev,* pp. 31-43, vol. 10, no.1, January 2006.

[13] A. K. Jeng and R. H. Jan, "Optimization on hybrid channel assignment for multi-channel multi-radio wireless mesh networks," in *Proc. IEEE GLOBECOM,* November 2007.

[14] H. Su and X. Zhang, "Modeling throughput gain of network coding in multi-channel multi-radio wireless ad hoc networks," *IEEE Journal on Selected Areas in Commun.,* vol. 27, no. 5, June 2009.

[15] N. Shacham and B. C. Shin, "A selective-repeat-ARQ protocol for parallel channels and its resequencing analysis," *IEEE Trans. Commun.,* vol. COM-40, pp. 773-782, April 1992.

[16] J. F. Chang and T. H. Yang, "Multichannel ARQ protocols," *IEEE Trans. Commun.,* vol. COM-41, pp. 592-598, April 1993.

[17] Z. Ding and M. Rice, "ARQ error control for parallel multichannel communications," *IEEE Trans. Wireless Commun.,* vol. COM-5, no. 11, pp. 3039-3044, November 2006.

[18] Y.-L. Chung and Z. Tsai, "Performance analysis of two multichannel fast retransmission schemes for delay-sensitive flows," *IEEE Trans. Veh. Technol.,* vol. 59, no. 7, pp. 3468-3479, September 2010.

[19] W. Luo, K. Balachandran, S. Nanda, and K. K. Chang, "Delay analysis of selective-repeat ARQ with applications to link adaptation in wireless packet data systems," *IEEE Trans. Wireless Commun.,* vol. 4, no. 3, pp. 1017-1029, May 2005.

[20] L. Badia, M. Rossi, and M. Zorzi, "SR ARQ packet delay statistics on Markov channels in the presence of variable arrival rate," *IEEE Trans. Wireless Commun.,* vol. 5, no. 7, pp. 1639-1644, July 2006.

[21] L. B. Le, E. Hossain, and M. Zorzi, "Queueing analysis for GBN and SR ARQ protocols under dynamic radio link adaptation with non-zero feedback delay," *IEEE Trans. Wireless Commun.,* vol. 6, no. 9, pp. 3418-3428, September 2007.

[22] L. Badia, M. Levorato, and M. Zorzi, "Markov analysis of selective repeat type II hybrid ARQ using block codes," *IEEE Trans. Commun.,* vol. 56, no. 9, pp. 1434-1441, September 2008.

[23] T. V. K. Chaitanya and E. G. Larsson, "Outage-optimal power allocation for hybrid ARQ with incremental redundancy," *IEEE Trans. Wireless Commun.,* vol. 10, no. 7, pp. 2069-2074, July 2011.

[24] H. Boujemâa, "Delay analysis of cooperative truncated HARQ with opportunistic relaying," *IEEE Trans. Veh. Technol.,* vol. 58, no. 9, pp. 4795-4803, November 2009.

[25] L. Le and E. Hossain, "An analytical model for ARQ cooperative diversity in multi-hop wireless networks," *IEEE Trans. Wireless Commun.,* vol. 7, no. 5, pp. , 1786-1791, May 2008.

[26] V. Mahinthan *et al.,* "Cross-layer performance study of cooperative diversity system with ARQ," *IEEE Trans. Veh. Technol.,* vol. 58, no. 2, pp. 705-719, February 2009.

[27] T. Issariyakul, and V. Krishnamurthy, "Amplify-and-forward cooperative diversity wireless networks: models, analysis, and monotonicity properties," *IEEE/ACM Trans. Networking,* vol. 17, no. 1, pp. 225-238, February 2009.

[28] M. Nakagami, "The m-distribution - a general formula of intensity distribution of rapid fading," in *Statistical Methods of Radio Wave Propagation,* pp. 3-36, W. C. Hoffman Ed. Elmsford, Pergamon Press, New York, 1960.

[29] L. R. Lugand, *et al.*, "Parity retransmission hybrid ARQ using rate-½ convolutional codes on a nonstationary channel," *IEEE Trans. Commun.*, Vol. 37, no. 7, pp. 755-765, July 1989.

[30] M. S. Patterh, *et al.*, "BER performance of MQAM with L-branch MRC diversity reception over correlated Nakagami-m fading channels," *International Journal of Wireless Communications and Mobile Computing*, vol. 3, pp. 397-406, May 2003.

[31] D. G. Brennan, "Linear diversity combining techniques," *Proc. IRE*, vol. 47, pp. 1075-1102, June 1959.

[32] D. Yoon, K. Cho, and J. Lee, "Bit error probability of M-ary quadrature amplitude modulation," in *Proc. the IEEE VTC*, Boston, MA, September 2000.

[33] Y.-L. Chung and Z. Tsai, "On the effective throughput gain of cooperative diversity with a fast retransmission scheme for delay-sensitive flows," *IEICE Trans. Commun.*, vol. E94-B, no.12, pp.-, December 2011.

[34] Y.-L. Chung and Z. Tsai, "Cooperative diversity with fast HARQ for delay-sensitive flows," in *Proc. IEEE 71st Veh. Technol. Conf. (IEEE VTC)*, Taipei, Taiwan, May 2010.

[35] A. Leon-Garcia, *Probability, statistics, and random processes for electrical engineering*, 3rd ed., Prentice Hall, 2008.

[36] A. J. Viterbi and J. K. Omura, *Principles of digital communication and coding*, McGraw-Hall, 1979.

5

Intelligent Transport Systems: Co-Operative Systems (Vehicular Communications)

Panagiotis Lytrivis and Angelos Amditis
Institute of Communication and
Computer Systems (ICCS)
Greece

1. Introduction

The term *Intelligent Transport Systems* (ITS) is used to illustrate the application of information and communication technologies in the transport domain. The intention of ITS is to enhance road safety and traffic efficiency, minimize environmental impact and in general maximize the benefits for the road users (Zhou et al., 2010; Popescu-Zeletin et al., 2010; Hartenstein & Laberteaux, 2010).

In turn, *Co-operative Systems* are the most promising technology within the ITS framework. The word "co-operative" indicates that vehicles are collaborating with each other and with the infrastructure, exploiting wireless communications, in order to increase their awareness about the road environment. There are two types of communication in co-operative systems, namely vehicle-to-vehicle (V2V) and vehicle-to-infrastructure (V2I).

The scope of this chapter is to highlight the significant role of vehicular communications in future ITS. Standalone sensors and sensor systems can support the drivers in certain cases (e.g. maintain a safe speed and safe distance from the vehicle ahead, avoid a possible rear-end collision etc.) but are not sufficient enough. Vehicles exchanging real-time messages and sharing information about the perception of the road environment could significantly extend the benefits of the abovementioned standalone systems and also satisfy the requirements of a large number of applications (see Figure 1).

Over the past years significant efforts have been performed for the bandwidth allocation and the standardization of vehicular communications worldwide. The Federal Communication Commission (FCC) decided the allocation of a frequency spectrum for vehicular applications. In Europe under the European Commission Decision 676/2002/EC the radio spectrum dedicated to ITS is in the 5.8 GHz frequency band. ETSI and CEN have formed working groups and technical committees dedicated to the ITS domain.

Although the benefits from the use of co-operative systems in transport are numerous there are also some difficulties. Some of the concerns are the following: wide uptake of such systems, market penetration, standards finalization and consensus among different standardization organizations, all the inherent problems of wireless technologies (multi-path propagation, security issues etc.).

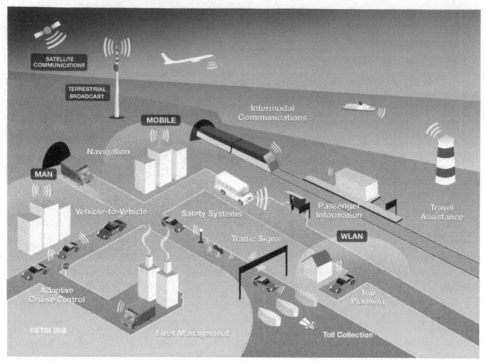

Fig. 1. Indicative ITS applications (ETSI, 2011).

The remainder of this chapter is organized as follows. In the next section, the architecture of co-operative systems is described. In the following wireless technologies used within the co-operative systems framework are outlined. The applications of vehicular networks and their corresponding categories are highlighted. Emphasis is given on hot research topics concerning co-operative systems such as data fusion, routing, security and privacy. Eventually, conclusions are drawn.

2. Architecture

In co-operative systems the specification of a unified communication architecture plays a central role for further deployment. As a result of the deployment of co-operative systems the road users will benefit from improved safety, reduced traffic congestion, environmental friendly driving and much more. The key to achieving these benefits lies in the specification of a common and standardized communication architecture among the various components of such systems. This architecture comprises four main components which can be composed arbitrarily to form a co-operative intelligent transport system. To form such a system there is no need to have all these four components available but a subset of them is sufficient. The components can communicate with each other either directly within the same communication network or indirectly across several communication networks. These four components are depicted in Figure 2 and are briefly described in the following. For a more detailed description one can refer to (Bechler et al., 2010; ETSI, 2010).

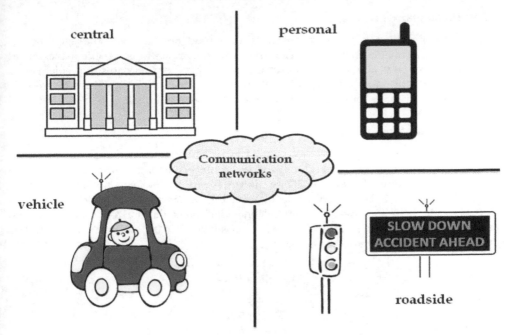

Fig. 2. Communication Architecture Components.

2.1 Vehicle component

The vehicle is equipped with communication capabilities (i.e. a router, embedded PCs) to establish communication with other vehicles and the roadside infrastructure. These modules have access to the CAN network of the vehicle as well as to other vehicle data which they collect, process and communicate to other vehicles, the roadside units or the central system. The exact HW solution is not strictly defined and it can be a unique HW unit or several units which form a LAN inside the vehicle.

2.2 Roadside component

The roadside component includes variable message signs, traffic lights and other units which are equipped with communication capabilities. This way, the roadside component can communicate with other vehicles by sending them information or acting as a relay station supporting multi-hop communication. Moreover, this component can communicate with other roadside units and the central system and therefore forward information received from vehicles. The roadside component can be also connected to the Internet.

2.3 Personal component

The personal component is actually a nomadic device, that is a personal navigator or a smartphone, which can host a variety of ITS applications. These devices can also support co-operative ITS applications based on communication with other road users or the road infrastructure.

2.4 Central component

The central component is a public authority or a road operator who manages the co-operative applications or services. An example of such component is a traffic management center which uses roadside units to inform the drivers about traffic status or accidents in a specific road network and suggests alternative routes. The central component can receive information from vehicles or roadside units and in turn send information to them.

2.5 Reference protocol stack

Each one of the above components contains an ITS station which in turn comprises a number of ITS specific functions and a set of devices implementing these functions. From a communication's point of view an ITS station is based on the reference protocol stack depicted in Figure 3. This protocol stack follows the ISO/OSI reference model and consists of four horizontal layers and two vertical ones that flank the horizontal stack. Access, networking and transport, facilities and applications layers are the horizontal ones, whereas management and security are the vertical ones.

Fig. 3. Reference protocol stack of an ITS station.

3. Wireless technologies

The wireless technologies used for the continuous communication among different vehicles and between vehicles and the road infrastructure are the cornerstone of co-operative systems. These technologies concern the *networking & transport* and *access* layers of the reference protocol stack of an ITS station (Figure 3) and can be divided into two main categories: general and vehicular specific communication technologies.

Therefore, for the connection among vehicles and the road infrastructure a mixture of general and vehicular specific technologies is needed. Some of these technologies are already in use, while some others are still under development.

3.1 General communication technologies

This category comprises well known wireless communication technologies such as cellular networks, WiMAX, WiFi, infrared, bluetooth, DVB/DAB etc. which are not specifically developed for vehicular networks but play a significant role for future deployment of co-operative systems. Currently, in ITS the main focus is on cellular networks and WiMAX and for this reason only those two will be analyzed below.

3.1.1 Cellular networks

Cellular networks are evolving rapidly to support the increasing demands of mobile networking. Although these networks are designed for voice data exchange they can be applied also to vehicular networks especially for information and entertainment applications. Nowadays, cellular networks migrate from GPRS, to UMTS, to LTE standards, increasing bandwidth and reducing delay times making these networks appropriate also for other kind of applications such as efficiency and trip planning.

Cellular networks have several characteristics suitable for co-operative systems like large scale usage and long range communication. However, some drawbacks of cellular networks which are relevant to vehicular connectivity are summarized below:

- Increased latency (i.e. voice data higher priority than text data, data sent via base stations)
- No broadcasting capabilities, only support of point-to-point communication
- Operation fees (e.g. internet access, roaming)

However, despite all the above disadvantages, cellular networks can be used for ITS applications which require moderate delay, long range communication, and low data rate. With the migration from 3G towards 4G (such as LTE) the focus remains on technologies that can serve a circular area with Internet connectivity, with no special provision for following the road infrastructure and optimising for connected car services.

3.1.2 WiMAX

WiMAX (Worldwide Interoperability for Microwave Access) is based on IEEE 802.16 standard and aims at providing wireless data over long distances in a variety of ways, from point-to-point links to full mobile cellular type access. WiMAX provides support for mobility and it will fill the gap between 3G and WLAN standards. It offers high data rates (<40Mb/sec), portable connectivity at low speeds (<60km/h) and wide area coverage (<10km) required to deliver high speed internet access to mobile clients. WiMAX provides a wireless alternative to cable and xDSL for last mile broadband access and can be used for V2I or I2I long range communication. At this point it should be mentioned that WiMAX supports several service levels including guaranteed Quality of Service (QoS) for delay sensitive applications and an intermediate QoS level for delay tolerant applications that require a minimum guaranteed data rate.

3.2 Vehicular specific communication technologies

This category includes communication technologies which are dedicated to vehicular applications and actually were the result of additional communication requirements posed by ITS applications. Dedicated communication standards are in development for co-operative systems. At the access layer, a convergence towards the IEEE 802.11p standard can be observed, while standardisation on the network and transport layer is still in progress. Several prototype implementations exist and are used in demonstrations and pilots. IP communication (focus is on IPv6) can be used on top of 802.11p, but due to the highly dynamic character of the network (i.e. movement of vehicles, relatively short communication distances) dedicated standards have been developed and are being standardised by ISO and ETSI. The most important of them, namely DSRC, WAVE and CALM, are illustrated below.

3.2.1 Dedicated Short Range Communications

Dedicated Short Range Communications (DSRC) is a short to medium range communications service that supports both public safety and private operations in V2V and V2I communication environments (DSRC, 2003). DSRC is meant to be a complement to cellular communications by providing very high data transfer rates in circumstances where minimizing latency in the communication link and isolating relatively small communication zones are important.

DSRC is designed for vehicular wireless communications and operates on radio frequencies in the 5.725 to 5.875 GHz (Industrial, Scientific and Medical - ISM) band in Europe and in the 5.850 to 5.925 GHz band in the United States. DSRC systems consist of Road Side Units (RSUs) and On Board Units (OBUs) with transceivers and transponders. The DSRC standards specify the operational frequencies and system bandwidths, but also allow for optional frequencies which are covered (within Europe) by national regulations.

The range of communication using DSRC is up to 1000m with data rates of 6–27 Mb/s, where vehicles may be moving at speeds up to 140 km/h. As mentioned previously, DSRC is divided into two types of communication, namely V2V and V2I. V2V communication is used when vehicles need to exchange data among themselves in order for co-operative applications to work properly, whereas V2I communication is used when roadside units are part of the co-operative application. In co-operative systems, some applications are required to send messages periodically (e.g. every 100ms), whereas other applications send messages when an event occurs.

At this point it should be highlighted that DSRC systems are used in the majority of European Union countries, but these systems are currently not totally compatible. Therefore, standardization is essential in order to ensure pan-European interoperability, particularly for applications such as electronic fee collection, for which the European Union imposes a need for interoperability of systems.

Standardization will also assist with the provision and promotion of additional services using DSRC, and help ensure compatibility and interoperability within a multi-vendor environment.

3.2.2 Wireless Access in Vehicular Environments

The design of an efficient communication protocol in the automotive sector that deals with privacy, security, multi-channel operation and management of resources is a difficult task, which is under intensive scientific investigation. This task is assigned to a special IEEE working group and the ongoing suite of protocols is the IEEE 1609, mostly known as Wireless Access in Vehicular Environments or simply WAVE (WAVE, 2007).

The WAVE standards define an architecture and a complementary, standardized set of services and interfaces that collectively enable secure V2V and V2I wireless communications. Together these standards provide the foundation for a broad range of applications in the transportation environment, including vehicle safety, automated tolling, enhanced navigation, traffic management and many others.

The architecture, interfaces and messages defined in WAVE support the operation of secure wireless communications among vehicles and between vehicles and the road infrastructure. Applications can use these standards in conjunction with equipment operating at 5.9 GHz to provide, for example, services for drivers, road operators, facilities operators and maintenance staff.

The IEEE 1609 Family of Standards for WAVE consists of four trial use standards which have full use drafts under development and two unpublished standards under development:

- *IEEE P1609.0 - Draft Standard for Wireless Access in Vehicular Environments (WAVE) – Architecture*
  This standard describes the WAVE architecture and services necessary for multi-channel DSRC/WAVE devices to communicate in a mobile vehicular environment.
- *IEEE 1609.1-2006 - Trial Use Standard for Wireless Access in Vehicular Environments (WAVE) - Resource Manager*
  This standard specifies the services and interfaces of the WAVE Resource Manager application. It describes the data and management services offered within the WAVE architecture. It defines command message formats and the appropriate responses to those messages, data storage formats that must be used by applications to communicate between architecture components, and status and request message formats.
- *IEEE 1609.2-2006 - Trial Use Standard for Wireless Access in Vehicular Environments (WAVE) - Security Services for Applications and Management Messages*
  This standard defines secure message formats and processing. This standard also defines the circumstances for using secure message exchanges and how those messages should be processed based upon the purpose of the exchange.
- *IEEE 1609.3-2007 - Trial Use Standard for Wireless Access in Vehicular Environments (WAVE) - Networking Services*
  This standard defines network and transport layer services, including addressing and routing, in support of secure WAVE data exchange. It also defines WAVE Short Messages, providing an efficient WAVE-specific alternative to IPv6 (Internet Protocol version 6) that can be directly supported by applications. Further, this standard defines the Management Information Base (MIB) for the WAVE protocol stack.
- *IEEE 1609.4-2006 - Trial Use Standard for Wireless Access in Vehicular Environments (WAVE) - Multi-Channel Operations*
  This standard provides enhancements to the IEEE 802.11 Media Access Control (MAC) to support WAVE operations.
- *IEEE P1609.11 Over-the-Air Data Exchange Protocol for Intelligent Transportation Systems (ITS)*
  This standard will define the services and secure message formats necessary to support secure electronic payments.

Additionally, the IEEE 1609 standards rely on IEEE P802.11p. This proposed standard specifies the extensions to IEEE 802.11 that are necessary to provide wireless communications in a vehicular environment.

3.2.3 Communications Access for Land Mobiles

The Communications Access for Land Mobiles (CALM) framework is an ISO TC204 initiative that specifies a common architecture, network protocols and communication interface definitions for wired and wireless communications using various access technologies including cellular 2nd generation, cellular 3rd generation, satellite, infra-red, 5 GHz micro-wave, 60 GHz millimetre-wave, and mobile wireless broadband (CALM, 2007). These and other access technologies that can be incorporated are designed to provide broadcast, unicast and multicast communications between mobile stations, between mobile and fixed stations and between fixed stations in the ITS sector.

The CALM concept is therefore developed to provide a layered solution that enables continuous or quasi continuous communications between vehicles and the infrastructure, or between vehicles, using such (multiple) wireless telecommunications media that are available in any particular location, and have the ability to migrate to a different available media where required. Media selection is at the discretion of user determined parameters.

The motivations behind this standardization effort are the following:

- different countries use different ITS media,
- different ITS applications have different requirements, therefore it is impossible to use a single carrier to support all types of applications.

The following communication types are supported by CALM:

- *Vehicle-to-Infrastructure*: Multipoint communication parameters are automatically negotiated and subsequent communication may be initiated by either roadside or vehicle.
- *Infrastructure-to-Infrastructure*: The communication system may also be used to link fixed points where traditional cabling is undesirable.
- *Vehicle-to-Vehicle*: A low latency peer-to-peer network with the capability to carry safety related data such as collision avoidance and other vehicle-vehicle services such as ad-hoc networks linking multiple vehicles.

At a high level, on the one side there are multiple services possibly operating simultaneously all requesting communications services, whereas on the other side there is a possibility of multiple communications media opportunities in the vehicle to handle the transaction. In the middle CALM is located managing quasi continuous communications using the available media, to satisfy the needs of one or multiple applications. It is important to understand that the vehicle may be maintaining multiple simultaneous sessions.

Finally, it is important to highlight that the specifications and standards of CALM are not a physical piece of equipment. While CALM may indeed operate through a "box" designed to achieve its tasks, it is actually a set of protocols, procedures and management actions. The implementation is actually a commercial decision.

4. Applications

Together with the evolution of vehicular networks numerous novel ITS applications have emerged. Typical examples of co-operative applications include remote diagnostics, collision avoidance, online navigation, map update, congestion avoidance for the driver, internet in the vehicle for passengers (e.g. gaming, downloading videos, reading the news etc.). These applications can be divided into three major categories, namely safety, efficiency and infotainment. In the following some indicative applications in each category are selected and will be described briefly.

4.1 Safety

One of the main goals of transport authorities is the minimization of traffic accidents and the increase of road safety. The exploitation of wireless technologies will be a significant asset towards this direction as it has been obvious from the results of the SAFESPOT project (SAFESPOT, 2006-2010). Examples of co-operative safety related applications will be given below.

4.1.1 Frontal collision warning

Frontal collisions represent a major proportion of accidents worldwide. Typical causes of such accidents are the distraction of the driver, sudden braking of a vehicle ahead, the presence of a stationary obstacle in front of the vehicle (e.g. right after a turn) etc. Conventional collision warning systems are based on sensors installed in the vehicle. These sensors could be long range radars for adaptive cruise control, camera sensors for objects detection, cameras covering the blind spot area, laser scanners for both detecting and classifying objects. This way a vehicle can be informed about events and targets which are within range of the detection sensors. Figure 4 shows a frontal collision warning application where the vehicle in front brakes while other vehicles are following.

The reliability and accuracy of a conventional collision warning system is based on the number and the type of sensors used, as well as the type of the environment (i.e. urban, inter-urban, highway) around the vehicle. The occlusion of sensors from obstacles, the limited range of sensors and other physical constraints, reduce system's range and degrade its performance. Apart from these factors, a collision warning system to function properly needs a multitude of sensors to cover the entire area around the vehicle, which makes such a system extremely expensive.

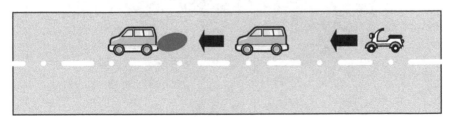

Fig. 4. Frontal collision warning application.

The collision warning system can be much more effective if other neighboring vehicles communicate with the subject vehicle, extending thus the perception of the driver in relation to the limited perception based only on sensors installed in the vehicle. Actually, this is the principle of co-operative collision warning systems. While driving, equipped vehicles anonymously share relevant information, including their position, speed and direction. This way each vehicle monitors the intentions of other drivers and the location and behavior of all vehicles in the neighborhood. When a vehicle detects a critical situation, the system warns the driver with a visual, audible and/or haptic manner. Thus, the driver has enough time to intervene and avoid a collision.

In time critical situations, immediate intervention to avoid a collision is feasible with the use of communication. This would not be possible in case only onboard sensors were used because of the delay in detecting and classifying objects and analyzing the ongoing situation. The co-operative approach also has great influence on the classification of objects. If vehicles are equipped with wireless communication they can directly exchange information about their type (e.g. truck, car, motorbike).

4.1.2 Intersection safety

A significant number of accidents in urban areas occur at intersections. The reasons for this are the significant burden of the driver from the complex situations that can occur at intersections due to many vehicles that are flooding them from different directions, the variety of road users (cars, trucks, pedestrians, cyclists etc.) as well as buildings and walls that limit the visibility of the driver.

In order for an accident or a dangerous situation to occur at an intersection it is supposed that there should be a violation of traffic rules such as traffic light or "STOP" sign violation. But intersections are complex road environments and accidents are likely to occur even if the rules are obeyed (e.g. abrupt braking while the traffic light turns from green to red). A simplified example of an intersection safety application is shown in Figure 5.

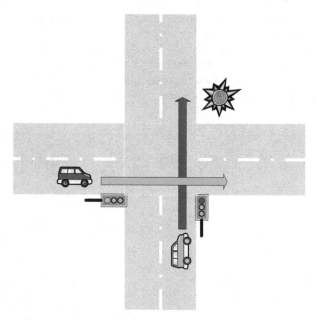

Fig. 5. Intersection safety application.

The benefits from the exploitation of wireless communication for safety reasons at intersections are significant. First of all, the traffic lights and other traffic signs can emit their status, together with information concerning the instant that this status is going to change (e.g. traffic light that turns from red to green), to the interested drivers informing them about the real ongoing situation at the intersection. In addition, vehicles can emit their position and dynamic state (speed, acceleration, steering angle) and thus to inform other drivers at the intersection about their presence. Otherwise it would be impossible for other drivers to be aware of their presence because of the occlusion of the surrounding buildings, walls and other obstacles. Finally, some sensors, such as laser scanners, could be installed at critical points at an intersection to detect pedestrians and cyclists (vulnerable road users) and inform the drivers for their presence through wireless communication.

4.1.3 Slippery road detection

This application informs the driver about the status of a road segment where there is a possible risk. The risk is primarily related to a slippery road surface that may be due to adverse weather conditions (e.g. ice, rain) or to any extraordinary event (e.g. an oil leak of the vehicle ahead). The detection of the slippery road segment can be carried out by a vehicle either directly by specific sensors or indirectly by activation of the ABS or ESP system. Even though this information, about the dangerousness of the road, is transmitted to the driver directly, it may be too late to take action because it is highly likely that the vehicle has slipped already since the ESP or ABS has been activated. A slippery road detection application is depicted in Figure 6.

As it is obvious, this application has almost no interest without the use of wireless communication. By taking advantage of wireless communication the vehicular network can share information related to hazardous road segments such as slippery roads. For example, a vehicle that its ESP system is activated associates this data with its current location and notifies other nearby vehicles and possibly a RSU if it is in communication range. The neighboring vehicles receiving such information shall promptly inform the drivers about the potential risk, while retransmitting this information to other nearby vehicles (multi-hop communication). As an alternative, a RSU can be equipped with some special sensors which can detect the hazardous road conditions and inform the drivers approaching this area.

Fig. 6. Slippery road detection application.

As a conclusion, there is an apparent advantage in using wireless communication to broadcast information about the road conditions to the approaching drivers who will then have sufficient time to react. With this co-operative approach the RSU has the ability to transmit road condition information to the traffic management center which then can be analyzed and checked for accuracy and quality and transmitted to other vehicles that drive on this road segment.

4.2 Efficiency

The climate change that has been observed in recent years has also affected the priorities in the transport agenda. Nowadays minimization of CO_2 emissions related to transport and environmental friendly driving comprises a top priority. The results of the CVIS project (CVIS, 2006-2010) have shown the added benefit from the use of wireless communication to enhance efficiency in transport. Examples of co-operative efficiency related applications will be highlighted in the following.

4.2.1 Enhanced route planning

In this application the infrastructure continuously collects information related to traffic density and makes forecasts for road segments with potential traffic jams. Then when an equipped vehicle drives next to a RSU the traffic density information on the neighboring area as well as driving instructions are transmitted to it. This information is processed by the vehicle and then the driver is informed about possible delays and alternative routes to avoid traffic jams. This way a significant number of drivers could be guided around congested areas so the entire transport system to become more efficient. A significant side effect of this application will be the reduction of pollution deriving from vehicles got stack in congested highways.

Figure 7 shows an enhanced route planning scenario in which the RSU informs the driver of the red vehicle about heavy traffic ahead and suggests a faster alternative route to the driver's final destination. It is obvious from the description of this application that for its implementation a RSU with wireless communication is essential to provide the necessary information. A vehicle equipped only with some perception sensors, without communication capabilities, could make a rough estimation about the traffic density in the road segment it is currently driving, but without any information on alternative routes.

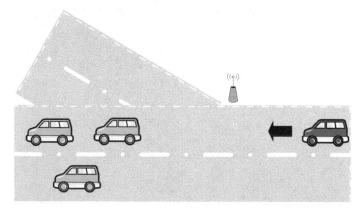

Fig. 7. Enhanced route planning application.

4.2.2 Optimal speed advice

The aim of this application is to provide the needed information to the driver in an effort to make driving smoother and reduce significantly start and stop situations. As the vehicle approaches a signalized intersection it receives information about the exact position of the traffic light and the duration of its current status (e.g. the remaining seconds for the traffic light to become red).

Based on this information, that is using the distance from the intersection and the time until the traffic light turns to green, the approaching vehicle calculates the speed that it should follow to avoid stopping at the intersection. Then this "optimal" speed is provided to the driver and if this suggestion is followed it is very likely that the traffic light will turn into green as soon as the vehicle reaches the intersection and there is no need for it to stop. Fewer stops result in increased traffic flow and reduced fuel consumption for the equipped vehicles.

Figure 8 shows an example of this application in which the driver follows the optimal speed suggestion and thus will not have to stop at the intersection. For such kind of application to become a reality it is essential that the traffic light should be equipped with wireless communication capabilities.

Fig. 8. Optimal speed advice application.

4.2.3 Traffic merge assistant

The traffic merge assistant application allows the merging of cars in a joint traffic flow without the need to interrupt their smooth flow. When a vehicle enters a highway from an entrance communicates this intention to the neighboring vehicles. The vehicle entering the highway requires specific maneuvers by surrounding vehicles to adapt in a safe and continuous way to the traffic. If there are no objections from other drivers, then either the traffic will be adjusted automatically or advice will be given to the drivers on how to act. In this way the vehicle entering the highway can adjust smoothly into the flow of traffic without causing major disruptions to it. This application can be enhanced by using a RSU which can determine the movements of each participant.

This application, as shown in Figure 9, requires a significant number of vehicles to be equipped with wireless capabilities. Moreover, it is one of the most difficult and complex applications which is based on the harmonious and trustworthy co-operation among highly dynamic vehicles.

Fig. 9. Traffic merge assistant application.

4.3 Infotainment

The entertainment and information provision to the driver and the passengers might not be life critical but is still very important in today's society, where internet and exchange of information dominate. Examples of co-operative applications related to information and entertainment (infotainment) will be outlined in this section.

4.3.1 Points of interest

The Points of Interest (PoI) application allows local businesses, touristic attractions, restaurants, gas stations and so on to advertise their availability to nearby vehicles. In this case, a RSU transmits information about a PoI, such as its location, hours of operation and pricing. Then this information is filtered by the vehicles dynamically, depending on the case, and the relevant information is presented to the driver. For example, if the fuel level is low, the vehicle could show to the driver the locations and prices of gas stations in the surrounding area. The benefit of this application is that advertising gets more effective as the driver moves within the geographical area where the PoI is located and it is more likely to visit it rather than if he listened about it to a radio station or found it on the web. Moreover, another benefit is that consumers receive up to date information directly from a business in the neighborhood.

4.3.2 Internet access

This application allows drivers and passengers to access the Internet. This in turn means the use of all types of services based on the IP protocol inside the vehicle. Therefore, a multi-hop route from a RSU to the relevant vehicle is installed and maintained to act as a gateway to the Internet. This path of multiple hops takes place transparently to the upper layers of the protocol stack and enables almost any service based on IP protocol to be used inside vehicles. Finally, this application allows access to the driver and passengers in any type of information available on the Internet (e.g. downloading of updated digital maps).

4.3.3 Remote diagnostics

This application allows an authorized service station to assess the condition of a vehicle without needing a physical connection with it. When a vehicle enters the parking building of the station, remote diagnostics system may ask the vehicle about relevant information to support the diagnosis of the problem reported by the client. Moreover, as the vehicle is approaching its service history and the necessary customer information can be retrieved from a database and be ready for use by the technician. If software updates are necessary, the system can install the updates without a physical connection. This application can reduce the amount of time required for a customer during a visit to an authorized service station. This fact will also reduce both the repair cost and the waiting time for the customers.

5. Research topics

Co-operative systems have received particular attention, with respect to the research activities in the field of ITS, the last decade. The recent advances in information and communication technologies have enabled the deployment of co-operative systems as an exciting platform for developing new and useful vehicular applications. The research activities focus mainly on data fusion, routing as well as on privacy and security issues which will be analyzed in the following.

5.1 Data fusion

Data fusion plays an important role in co-operative systems. A stand alone sensor or several sensors installed in a vehicle cannot overcome certain physical limitations as, for example, the limited range and field of view. Therefore combing information coming from both onboard sensors and wireless messages, encompassing information from other vehicles, broadens the awareness of the driver and increases the reliability of the whole system in case of sensor failure. However, fusing information from highly mobile vehicles, forming a wireless network, is a challenging task (Ahlers & Stimming, 2008; Lytrivis et al., 2008).

The Joint Directors of Laboratories (JDL) functional model, which is the most prevalent in data fusion community, is depicted in Figure 10. According to this model the data processing is divided to the following levels: signal, object, situation and application. All these levels communicate and exchange data through a storage and system manager (Liggins et al., 2008).

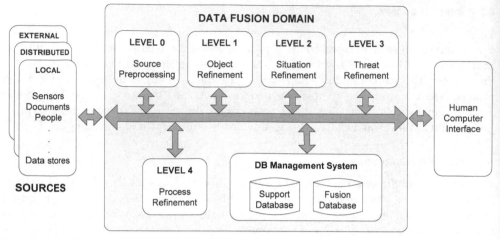

Fig. 10. Joint Directors of Laboratories (JDL) model.

In the data fusion process the main focus is on object and situation refinement levels, which refer to the state estimation of objects and the relations among them, correspondingly. The discrimination between these levels is also made by using the terms low and high level fusion instead of object and situation refinement. The different levels of the JDL model are summarized below:

- *Level 0*: Preprocessing of sensor measurements (pixel/signal-level processing).
- *Level 1*: Estimation and prediction of entity states on the basis of inferences from observations.
- *Level 2*: Estimation and prediction of entity states on the basis of inferred relations among entities.
- *Level 3*: Estimation and prediction of effects on situations of planned or estimated/predicted actions by the participants.
- *Level 4*: Adaptive data acquisition and processing related to resource management and process refinement.

In the past decade the advances in autonomous sensor technologies and the major objective of the European Union to reduce to a half road accidents and fatalities by 2010, led to the development of advanced driver assistance systems (ADAS). The fusion of data coming from different advanced in-vehicle sensors was initially in the centre of this attempt. However, this approach suffers from serious limitations. Specifically:

- the perception environment of the vehicle cannot go beyond the sensing range,
- the sensor systems cannot perform well in all environments (the urban roads comprise a major challenge),
- in several cases the system is not able to perceive the situation in time in order to warn the driver and suggest a corrective action,
- the cost of the sensor systems is too high and so their installation is feasible only at luxurious vehicles.

Recently the focus of research activities on co-operative systems is driven by the attempts to overcome all the above limitations. The limited bandwidth, security issues, privacy, reliability and propagation are some of the emerging disadvantages of the wireless connectivity in vehicles. All these issues poses additional challenges to the data fusion process. The association and synchronization of data from on-board sensors together with the wireless network data is the main challenge. Moreover, the manipulation of delayed information and the reliability of the information transferred via the network are other important issues.

5.2 Routing

Routing is the process of finding a path from a source node to a destination node. In this section the word "node" will be used interchangeably with the word "vehicle" because a vehicle is actually a node of a vehicular network. Since each node has limited transmission range, messages often need to be forwarded by other nodes in the network to reach their final destination (i.e. multi-hop communication).

Despite the fact that there are already some routing protocols available, which are mainly derived from the Mobile Ad-hoc Network (MANET) domain, it is an intensive scientific research area due to the highly dynamic nature of vehicular networks.

The routing protocols designed specifically for co-operative systems (Lee et al., 2010; Li & Wang, 2007) can be divided into two broad categories: *topology-based* routing and *location-based* routing. The former use information about the existing links of the network to forward the relevant messages. In the latter forwarding decisions are based on the location of the nodes. Moreover, position based routing protocols can be further divided into *proactive* and *reactive*.

Proactive algorithms are using classical routing strategies such as distance-vector routing or link-state routing. Proactive algorithms maintain routing information about the available paths in the network even if these paths are not currently used. The main disadvantage of this approach is that the maintenance of unused paths occupies a significant part of the available bandwidth if the network topology changes frequently.

In response to the problem of maintaining the paths of proactive protocols, reactive routing protocols were created. Reactive protocols maintain only routes that are in use, thereby reducing the load on the network when only a small subset of available paths are used.

In location-based routing, forwarding decisions are based on the location of the node that forwards the message according to the location of the source and destination nodes. In contrast to pure ad hoc approaches which are based on topology-based routing, here it is not necessary to setup or maintain a path since packets are forwarded directly. Location-based routing protocols consist of location services and geographical forwarding.

Geographical forwarding takes advantage of a topological assumption which works well for wireless ad hoc networks: nodes that are physically close are likely to be close in the network topology too. Each node is aware of its location using technologies such as GPS and periodically broadcasts its presence, location and speed to its neighbors. Thus, each node maintains a table with the identities and locations of its current neighbors. When one node needs to forward a packet it includes the identifier of the destination-node and its geographical location into the header of the packet. Each node along the forwarding path consults its list of neighbors and forwards the packet to the neighbor closest to the destination in terms of physical location, until it reaches its final destination.

Although the geographical forwarding works well for networks where nodes are uniformly distributed, perhaps cannot find a route to a packet's destination when the packet has to travel around a topology "hole" - that is, when an intermediate forwarding node has no neighbors who are closer than itself to the destination of the packet.

An overview of some *topology-based* routing algorithms is given below:

- **Ad Hoc On Demand Distance Vector (AODV)** is a routing algorithm where the nodes of the network upon receiving a broadcast query they record the address of the querying node to their routing table. The process of recording the previous hop is called backward learning. When a packet reaches its destination a reply packet is sent back to the source through the full path retrieved from the process of backward learning. At every node of the path, the previous hop should be recorded, creating this way the forward path from the source. The query together with the response create a complete bidirectional path. After setting the path, it is maintained as long as the source uses it. A failure on a link will be reported recursively to the source and in turn this will trigger another query-response process for finding the new route. More details about AODV one can find in (Perkins & Royer, 1999).
- **Dynamic Source Routing (DSR)** is an algorithm that uses source routing, that is the source indicates to a data packet the sequence of intermediate nodes on the routing path. In DSR, the query packet copies in its header the identities of the intermediate nodes it has already visited. Afterwards, the destination uses the query packet to retrieve the entire path to respond to the source. As a result, the source can establish a path to the destination. If the destination node is allowed to send multiple routes responses, the source node may receive and store these multiple routes. An alternative route can be used in case a link of the current path is broken. In a low mobility network DSR has the advantage over AODV in case the alternative route can be tested before the DSR initiates another query to discover the route. There are two major differences between AODV and DSR. The first is that in AODV data packets carry the destination address, while in DSR data packets carry all the routing information. This means that DSR has probably more routing burden than AODV. Moreover, as the diameter of the network increases, the burden on the data packet will continue growing. The second difference is that in AODV route response packets carry the destination address and the sequence number, while in DSR they carry the address of each node along the route. The interested reader in DSR can refer to (Johnson & Maltz, 1996).

A brief description of some *location-based* routing algorithms is given below:

- **Connectivity-Aware Routing (CAR)** is a routing algorithm which derives from the work performed by the Preferred Group Broadcast (PGB) to reduce the broadcasted packets during the discovery of the AODV route taking also into account the mobility of the nodes. CAR uses the route discovery of AODV to find routes with reduced broadcasting from PGB. However, the nodes forming the route record neither their previous node from the backward learning nor their previous node which forwards the response route packet from the destination. Only anchor points, which are nodes near an intersection or a curve of the road, are recorded in the route discovery packet. A node defines itself as an anchor point if its velocity vector is not parallel to the velocity vector of the previous node in the packet. The destination may receive multiple route discovery packets. If this happens it chooses the path that provides the best connectivity and the shortest delays. More details about CAR can be found in (Naumov & Gross, 2007).

- **Geographic Source Routing (GSR)** is based on the availability of a map. It calculates the shortest Dijkstra path of the cascading graph where vertices are intersection nodes and edges are the roads connecting these vertices. The sequence of intersections is setting up the route to the destination. Then the packets are greedily forwarded between intersections. GSR does not take into account the connectivity between two intersections, so the route might not be fully connected. In case such a situation occurs a recovery with greedy forwarding takes place. The most significant difference between the GSR and CAR is that CAR does not use a map and uses proactive discovery of anchor points that indicate a turn at an intersection. More details about GSR can be found in (Lochert et al., 2003).

5.3 Security and privacy

Security in V2V and V2I communications is a prerequisite for future development of co-operative systems and actual deployment in the real world. Co-operative systems have to ensure that data transmission derives from a trusted source and has not been counterfeited. For example, in a red light violation warning application, the in-vehicle system receives data from the equipment which is installed in the traffic light and then decides to issue or not a warning to the driver. An incorrect transmission from a malfunctioning or compromised unit might jeopardize vehicle's safety as well as others' safety in the vicinity. Similarly, the future development of safety applications is jeopardized without securing that transmissions are coming from a trusted source.

Privacy and anonymity are primary issues that also have to be addressed. In co-operative applications vehicles are broadcasting messages about their current location, speed and heading. It is desirable for the users to maintain their privacy since they fear that such a system could be used to build tracking mechanisms which would allow harassment, automatic issue of tickets for speeding or otherwise act in an undesirable way for them.

Unfortunately, on the other hand anonymity may be abused. Some examples are sending fake information or spamming. If the system ensures accountability[1] then the users know that there will be consequences for others if their data is abused. The challenge here is ensuring anonymity and at the same time accountability, as they seem to be conflicting.

There are many ongoing research activities on security and privacy in co-operative systems. Some ideas that have been proposed for solving such issues include public key certificates or digital signatures. For more information the interested reader can refer to (Fischer et al., 2007; Raya & Hubaux, 2007).

---

[1]Accountability is the ability to attribute actions to the entity that caused those actions.

6. Conclusion

An overview of co-operative systems and their importance in future transportation systems has been presented in this chapter. The chapter started with a short introduction about the historical background and the purpose of co-operative systems. Emphasis was given to the communication architecture, including its components, which is mainly the outcome of the efforts carried out so far in Europe. Also the importance of the definition of a common architecture for further deployment of co-operative systems was stressed.

In the following, the focus was on the wireless technologies used within the co-operative systems framework which are divided into two categories: general and vehicular specific communication technologies. These technologies are the cornerstone of co-operative systems and their objective is the continuous communication among different road users (vehicles, motorbikes, trucks, roadside units, infrastructure etc.). To achieve this continuous and seamless communication a mixture of general and vehicular specific technologies is needed. Some of these technologies are already in use, while some others are still under development.

Additionally, some co-operative applications were described which are categorized into three main groups: safety, efficiency and infotainment. The applications addressing safety and efficiency are of great importance today because the minimization of accidents and their consequences as well as the reduction of CO_2 emissions are the primary targets worldwide. Finally, emphasis was given on hot research topics concerning co-operative systems such as data fusion, routing, security and privacy. A general description highlighting each of these topics, the research challenges as well as some solutions were indicated.

Although many problems are not yet solved, the general feeling is that vehicles could benefit from evolving wireless communications in the near future, making "talking vehicles" a reality. Co-operative systems will not only provide lifesaving and environmental friendly applications, but they will become a powerful communication tool for their users.

7. References

Ahlers, F. & Stimming, C. (2008). "Cooperative Laserscanner Pre-Data-Fusion", in Proc. *IEEE Intelligent Vehicles Symposium* (IV 2008), Eindhoven, 2008, pp. 1187–1190

Bechler, M. et al. (2010). *European ITS Communication Architecture, Overall Framework, Proof of Concept Implementation*, version 3.0, COMeSafety, February 2010

CALM (2007). International Organization for Standardization, Intelligent Transport System-Continuous Air Interface Long and Medium – Medium Service Access Point, Draft International Standard ISO/DIS 21218, 2007

CVIS (2006-2010). "Cooperative Vehicle-Infrastructure Systems", Integrated Project co-funded by the European Commission, Available from http://www.cvisproject.org

DSRC (2003). Standard Specification for Telecommunications and Information Exchange Between Roadside and Vehicle Systems-5GHz Band Dedicated Short Range Communications (DSRC) Medium Access Control (MAC) and Physical Layer (PHY) Specifications, September 2003

ETSI (2010). *ETSI EN 302 665 - Intelligent Transport Systems (ITS): Communications Architecture*, v1.1.1, European Standard (Telecommunications series), September 2010

ETSI (2011). European Telecommunications Standards Institute (ETSI), Intelligent Transport Systems - ITS, Available from http://www.etsi.org/WebSite/Technologies/IntelligentTransportSystems.aspx

Fischer, L.; Stumpf, F. & Eckert, C. (2007). "Trust, Security and Privacy in VANETs - A Multilayered Security Architecture for C2C-Communication", *In VDI/VW-Gemeinschaftstagung: Automotive Security*, Wolfsburg, Germany, November 2007

Hartenstein, H. & Laberteaux, K. (2010). *VANET Vehicular Applications and Inter-Networking Technologies* (Intelligent Transport Systems), Wiley, 1st edition, March 2010

Johnson, D. & Maltz, D. (1996). "Dynamic Source Routing in Ad Hoc Wireless Networks," Mobile Computing, T. Imielinski and H. Korth, Eds., Ch. 5, Kluwer, 1996, pp. 153– 81

Lee, K.; Lee, U. & Gerla, M. (2010). *Survey of Routing Protocols in Vehicular Ad Hoc Networks*, Advances in Vehicular Ad-Hoc Networks, M.Watfa Book Editor, IGI Global

Li, F. & Wang, Y. (2007). "Routing in vehicular ad hoc networks: A survey", *IEEE Vehicular Technology Magazine*, June 2007, Vol. 2, Issue 2, pp. 12-22

Liggins, M.; Hall, D. & Llinas J. (2008). *Handbook of Multisensor Data Fusion: Theory and Practice*, Second Edition, CRC Press

Lochert, C.; Hartenstein, H.; Tian, J.; Fussler, H.; Hermann, D. & Mauve, M. (2003). "A routing strategy for vehicular ad hoc networks in city environments," *IEEE Intelligent Vehicles Symposium 2003*, 9-11 June 2003, pp. 156-161

Lytrivis, P.; Thomaidis, G. & Amditis, A. (2008). "Cooperative Path Prediction in Vehicular Environments," in Proc. *11th Int. IEEE Conf. on Intelligent Transportation Systems* (ITSC 2008), Beijing, 2008, pp. 803–808

Naumov, V. & Gross, T. (2007). "Connectivity-Aware Routing (CAR) in Vehicular Ad-hoc Networks," *26th IEEE International Conference on Computer Communications* (INFOCOM 2007), May 2007, pp.1919-1927

Perkins, C. & Royer, E. (1999). "Ad-Hoc On-Demand Distance Vector Routing," *Proc. IEEE WMCSA '99*, New Orleans, LA, Feb. 1999, pp. 90–100

Popescu-Zeletin, R.; Radusch, I. & Rigani, M. A. (2010). *Vehicular-2-X Communication: State-of-the-Art and Research in Mobile Vehicular Ad hoc Networks*. Springer, 1st Edition, May 2010

Raya, M. & Hubaux, J.P. (2007). "Securing vehicular ad hoc networks", *Journal of Computer Security 15* (2007), IOS Press, pp. 39–68

SAFESPOT (2006-2010). "Cooperative vehicles and road infrastructure for road safety", Integrated Project co-funded by the European Commission, Available from http://www.safespot-eu.org

WAVE (2007). IEEE Standards Association, IEEE P1609.1 – Standard for Wireless Access in Vehicular Environments (WAVE) – Resource Manager, IEEE P1609.2 – Standard for Wireless Access in Vehicular Environments (WAVE) – Security Services for Applications and Management Messages, IEEE P1609.3 – Standard for Wireless Access in Vehicular Environments (WAVE) – Networking Services, IEEE P1609.4 – Standard for Wireless Access in Vehicular Environments (WAVE) – Multi-Channel

Operations, adopted for trial-use in 2007, IEEE Operations Center, 445 Hoes Lane, Piscataway, NJ, 2007

Zhou, M.-T.; Zhang, Y. & Yang, L. (2010). *Wireless Technologies in Intelligent Transportation Systems* (Transportation Issues, Policies and R & D), Nova Science Pub Inc, June 2010

6

User Oriented Quality of Service Framework for WiMAX

Niharika Kumar, Siddu P. Algur and Amitkeerti M. Lagare
RNSIT
BVB College of Engineering
Motorola Mobility
India

1. Introduction

IEEE 802.16 provides last mile broadband wireless access. Also called as WiMAX, IEEE 802.16 is rapidly being adopted as the technology for Wireless Metropolitan area networking (MAN). WiMAX operates at the microwave frequency and each WiMAX cell can have coverage area anywhere between 5 to 15 kilometers and provide data rates upto 70Mbps.

IEEE 802.16m has been submitted to ITU as a candidate for 4G. With data rates of 100Mbps for mobile users and 1Gbps for fixed users, IEEE 802.16m holds a lot of promise as a true 4G broadband wireless technology.

This chapter introduces a user based framework in WiMAX. In section 2, user based bandwidth allocation algorithms are introduced. In section 3, user based packet classification mechanism is explored. In section 4 user based call admission control algorithm is explored.

2. User based bandwidth allocation

IEEE 802.16 (WiMAX) provides differentiated Quality of Service (QoS) (IEEE 802.16 2004) (IEEE 802.16e 2005) (Vaughan-Nichols 2004). This is achieved by having five different types of service classes. Each of these service classes caters to specific type of data. Unsolicited Grant Services (UGS) supports real time data streams that generate fixed size packets at periodic intervals. For example Voice over IP without silence suppression, T1/E1. Extended Real Time Polling Services (eRTPS) is designed to support real-time service flows that generate variable sized data packets on periodic basis, like VoIP with silence suppression. Real Time Polling Services (RTPS) supports real time data streams that generate variable size packets on periodic basis. For example Multimedia formats like an MPEG video. Non Real Time Polling Services (nRTPS) supports delay tolerant data streams generating variable size data packets, like FTP. Best Effort(BE) supports data streams which do not require any service level. Ex Web browsing, Email etc.

User keeps generating the data. This data gets queued into one of the five service classes based on the type of data and the quality of service requirements for the data. Once the data gets queued, the device needs to request for bandwidth so that the data packets can be transmitted. Classically, the widely used bandwidth allocation algorithms have followed contention based logic. The device contends for the wireless medium. If no other device is contending for the bandwidth then the device transmits the data. Algorithms like ALOHA, Slotted ALOHA, CSMA, CSMA-CD use contention based bandwidth allocation. Even IEEE 802.11 (Wi-Fi) uses contention based bandwidth allocation mechanism called CSMA-CA.

WiMAX supports demand based bandwidth allocation mechanism. Each Mobile Station (MS) is allocated small amount of bandwidth that is used by the MS to request for additional bandwidth. Based on the availability of bandwidth and the type of service requesting for bandwidth, the Base Station (BS) allocates bandwidth. MS requests bandwidth on a per service class basis and the BS allocates bandwidth on a per-SS basis. Various types of contention based bandwidth request/allocation mechanisms have been proposed in WiMAX. Aggregate bandwidth request mechanism is proposed in (Tao & Gani, 2009). Instead of sending separate bandwidth request for each service class, a single request is sent. Service class bandwidth allocation is proposed in (Wee & Lee, 2009). Delay intolerant service classes are provided bandwidth on priority. Subsequently delay tolerant service classes are allocated bandwidth. Adaptive bandwidth request scheme is proposed in (Liu & chen, 2008). Contention free bandwidth request opportunities are provided within the contention based request opportunities for some SS. Predictive bandwidth allocation algorithm is proposed in (Peng et. al, 2007). Based on the current arrival pattern, bandwidth is requested beforehand for future packets. Channel aware bandwidth allocation algorithm is proposed in (Lin et. al, 2008). Another form of adaptive bandwidth allocation algorithm is proposed in (Chiang et. al, 2007). The TDD frame is dynamically adjusted based on the amount of uplink and downlink data. In (Park, 2009) bandwidth request algorithm is proposed that takes both the current size of the queue and the deadline assigned to each packet. CDMA bandwidth request code based bandwidth allocation mechanism is proposed in (Lee et. al, 2010). The CDMA bandwidth request code is chosen randomly, but in (Lee et. al, 2010) the bandwidth request code is intelligently chosen so that the code itself indicates the amount of bandwidth needed by the MS. This reduces the number of control message transactions between the MS and SS. In (Rong et al, 2007) two algorithms are proposed namely adaptive power allocation (APA) and call admission control (CAC). The two algorithms work in tandem to allocate bandwidth to the MS.

All the algorithms proposed above are service class based bandwidth request/allocation algorithms. MS shall send bandwidth request for all its service classes. Bandwidth is then allocated based on the service class. All UGS service classes from different users are allocated bandwidth first then the RTPS service flows are allocated bandwidth followed by eRTPS. Next, the delay tolerant service class nRTPS is allocated bandwidth. Finally BE service class is allocated bandwidth. This method of bandwidth allocation treats all MS alike. If there are 10 MS in the network and if all of them are generating BE traffic then all the BE service classes are allocated bandwidth on a first come first serve basis. Of these 10 users, there may be some users who may wish to pay more if their BE traffic is treated on priority. So, users can be segregated into different groups and bandwidth can be allotted to the users based on the group to which they belong to. In this section we shall explore three user based bandwidth allocation algorithms. Fig. 1 shows service class based bandwidth allocation mechanism.

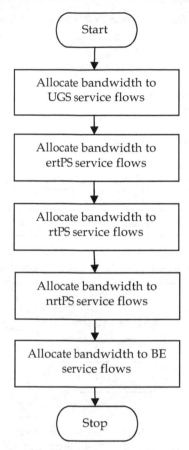

Fig. 1. Service class based bandwidth allocation mechanism.

2.1 Differentiated Bandwidth Allocation Mechanism (DBAM)

There shall be three different categories of users/MS as listed in Table 1.

User Category	Priority Value	Description
High-Priority User/MS/SS	1	Users who will receive higher priority for their traffic within each of the WiMAX service class. High-Priority users could be those users who are ready to pay more to enjoy higher QoS.
Low-Priority User/MS/SS	2	Users who will receive lower priority for their traffic for each of the WiMAX service class. Low-Priority users could be those users who wish to pay less and settle for lower quality of service.
Regular User/MS/SS	0	Users who fall in-between High-Priority and Low-Priority users.

Table 1. Classification of Users into three different categories.

Bandwidth allocation is done for all the service class for the three types of users as per fig. 2

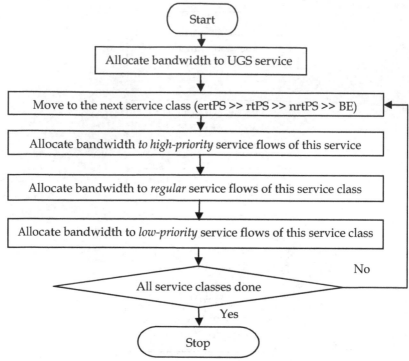

Fig. 2. DBAM Algorithm.

From the algorithm in fig.2, we see that when the BS receives bandwidth requests for BE traffic from High-Prioirty, Regular and Low-Priority users, BS shall allocate bandwidth first to the high-priority user then the regular user and finally to the low-priority user (Kumar et al. 2011a).

2.1.1 Implementation of BDAM

The WiMAX Network Reference architecture is given in the fig. 3.

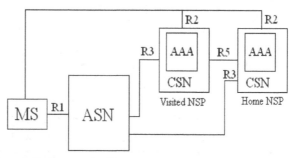

Fig. 3. WiMAX Network Reference Architecture.

Each network service provider (NSP) has a Authentication, Authorization and Accounting (AAA) server.. This server maintains the information about the users. The Access Service Network (ASN) interacts with the AAA server to obtain the information about the user.

The AAA server shall maintain a table of MAC address for the users and the priority value associated with the user. A sample state of the table could be as shown in Table-2

MAC Address	Priority Value
12:34:56:78:9a:bc	2
bc:9a:78:56:34:12	1
11:11:11:11:11:11	0
22:22:22:22:22:22	0
33:33:33:33:33:33	2
88:77:66:11:22:44	1
....	....

Table 2. Sample state table of priority values for users.

When the MS initiates the ranging process, it sends the Ranging Request (RNG-REQ). Upon receiving the ranging request, BS shall query the AAA server to obtain the priority value associated with the user. BS shall store the priority value for the user in its local cache.

Subsequently, when the MS makes bandwidth request for any of its service flows, BS shall check the priority of the MS. Based on the priority value, bandwidth shall be allotted to the service flow.

2.1.2 Analytical modeling

Table 3 lists the notations used for analytical modeling.

Symbol	Description
$ertps\_pri\_bw\_req(p)$	Bandwidth needs of pth ertPS service flow of priority SS.
$ertps\_reg\_bw\_req(p)$	Bandwidth needs of pth ertPS service flow of regular SS.
$ertps\_npr\_bw\_req(p)$	Bandwidth needs of pth ertPS service flow of low-priority SS
$ertps\_pri\_bw\_allot(p)$	Bandwidth allotted to the pth ertPS service flow of priority SS.
$ertps\_reg\_bw\_allot(p)$	Bandwidth allotted to the pth ertPS service flow of regular SS.
$ertps\_npr\_bw\_allot(p)$	Bandwidth allotted to the pth ertPS service flow of low-priority SS.
tot_bw	Total bandwidth available on the uplink for the current frame
tr	Minimum Reserved traffic rate
avl_bw	Amount of unallocated bandwidth available in the frame.
m	Number of high-priority ertPS service flows
n	Number of regular ertPS service flows
o	Number of low-priority ertPS service flows

Table 3. Notations used in Analytical Modeling.

Throughput modeling is described below. For the purpose of brevity bandwidth allocation is explained for the three types of users for eRTPS service flow. Similar equations can be derived for the other service flows.

BS allots bandwidth to the high-priority eRTPS service flows as per eqn 1.

$$ertps\_pri\_bw\_allot(p) = \begin{cases} ertps\_pri\_bw\_req(p) \\ \quad if\ ertps\_pri\_bw\_req(p) < tr \\ \quad and\ ertps\_pri\_bw\_req(p) < avl\_bw \\ tr \quad if\ tr \le ertps\_pri\_bw\_req(p) \\ \quad and\ ertps\_pri\_bw\_req(p) < avl\_bw \\ avl\_bw \quad if\ avl\_bw \le ertps\_pri\_bw\_req(p) \\ \quad and\ avl\_bw \le tr \\ tr \quad otherwise \end{cases} \tag{1}$$

Once bandwidth is allotted to a high-priority eRTPS service flow, the leftover bandwidth is calculated as per eqn 2.

$$avl\_bw = tot\_bw - \left(\sum_{j=1}^{x} ertps\_pri\_bw\_allot\,(j) \right)$$

$$x \le m, \tag{2}$$

After all the high-priority eRTPS service flows are allotted bandwidth, bandwidth is allotted to the regular eRTPS service flows as per eqn 3.

$$ertps\_reg\_bw\_allot(p) = \begin{cases} ertps\_reg\_bw\_req(p) \\ \quad if\ ertps\_reg\_bw\_req(p) < tr \\ \quad and\ ertps\_reg\_bw\_req(p) < avl\_bw \\ tr \quad if\ tr \le ertps\_reg\_bw\_req(p) \\ \quad and\ ertps\_reg\_bw\_req(p) < avl\_bw \\ avl\_bw \quad if\ avl\_bw \le ertps\_reg\_bw\_req(p) \\ \quad and\ avl\_bw \le tr \\ tr \quad otherwise \end{cases} \tag{3}$$

After allocating bandwidth to a regular eRTPS service flow, leftover bandwidth is calculated as per eqn. 4

$$avl\_bw = avl\_bw - \left(\sum_{j=1}^{x} ertps\_reg\_bw\_allot\,(j) \right)$$

$$x \le n, \tag{4}$$

Once we are through with the regular eRTPS service flows, bandwidth is allotted to the low-priority eRTPS service flows as per eqn. 5.

$$
ertps\_npr\_bw\_allot(p) = \begin{cases} ertps\_npr\_bw\_req(p) \\ \quad if \ ertps\_npr\_bw\_req(p) < tr \\ \quad and \ ertps\_npr\_bw\_req(p) < avl\_bw \\ tr \quad if \ tr \le ertps\_npr\_bw\_req(p) \\ \quad and \ ertps\_npr\_bw\_req(p) < avl\_bw \\ avl\_bw \quad if \ avl\_bw \le ertps\_npr\_bw\_req(p) \\ \quad and \ avl\_bw \le tr \\ tr \quad otherwise \end{cases} \tag{5}
$$

After allotting bandwidth to the jth low-priority eRTPS service flow, leftover bandwidth is calculated as per eqn. 6

$$
avl\_bw = avl\_bw - \left(\sum_{j=1}^{x} ertps\_npr\_bw\_allot\,(j) \right) \tag{6}
$$

$x \le 0,$

At this point bandwidth has been allotted to all the eRTPS connections. The above method of bandwidth allocation is repeated for RTPS, nRTPS and BE. This ensures that for each service flow, bandwidth is allotted to high-priority users first followed by regular users and finally the low-priority users.

2.1.3 Simulation results

In order to evaluate DBAM, simulations were carried out on NS-2. Light WiMAX module (LWX) (Chen 2008) was used to simulate the WiMAX environment in NS-2. Strict priority bandwidth allocation algorithm of LWX was modified to accommodate DBAM algorithm. Simulations were carried out with the parameters from table 4.

Parameter	Value
Uplink data rate	10 Mbps
OFDMA Frame Duration	5 ms
OFDMA symbol time	100.94 µs
eRTPS data arrival rate	1 Mbps

Table 4. Simulation parameters for DBAM.

Simulation network was setup such that at any point in time, 33% of the SS are priority SS, next 33% are regular SS and the final 1/3rd are low-priority SS. Each SS generates only eRTPS traffic. Uplink data is generated at the rate of 1Mbps. Downlink ftp traffic was also added. Downlink data is generated at the rate of 1Mbps.

Simulation results for throughput are shown in Fig. 4.

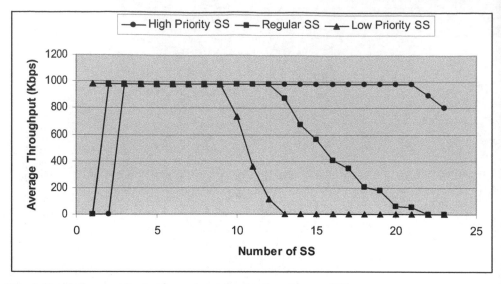

Fig. 4. Simulation results for throughput for the three types of SS.

When the number of MS is 9 each MS has sufficient bandwidth to transmit its data. But, when the number of SS is more than 9, there isn't sufficient bandwidth to support all SS. DBAM provides bandwidth to high-priroity SS first then regular SS and the leftover bandwidth is shared by low-priority SS. When the number of SS crosses 13, bandwidth for regular SS keeps reducing. Theoretical Results are shown in Figure 5.

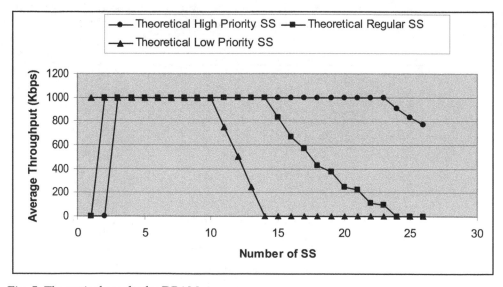

Fig. 5. Theoretical results for DBAM.

Comparing figure 4 and figure 5 we see that the simulation results closely follow the theoretical results.

By introducing DBAM we can provide graded quality of service to the users. This is a win-win situation for both users and operators. The users win because their data gets prioritized and hence they get a better quality of service. The service providers stand to gain because they get higher revenue for the same amount of data being transmitted. Its just that the order of bandwidth allocation is modified.

2.2 Enhanced Differentiated Bandwidth Allocation Mechanism (eDBAM)

In case of DBAM, the order of bandwidth allocation follows the below sequence:

High-priority RTPS > Regular priority RTPS > Low-priority RTPS > High-priority nRTPS > Regular priority nRTPS > Low-priority nRTPS > High-priority BE > Regular priority BE > Low-priority BE

Basically DBAM ensured that the order of service class is maintained and within the service class we can have graded users. However there is scope for further optimization. We can have seven different ways in which the bandwidth can be allotted. Table 5 and Table 6 list the seven different ways in which bandwidth can be allotted. Each column in the table represents a unique way of bandwidth allotment. The order of allotment is from top to bottom (Kumar et. al, 2011b).

DBAM	eDBAM Method 1	eDBAM Method 2	eDBAM Method 3
High-priority RTPS	High-Priority RTPS	High-priority RTPS	High-priority RTPS
Regular priority RTPS	High-priority nRTPS	Regular priority RTPS	High-priority nRTPS
Low-priority RTPS	High-priority BE	Low-priority RTPS	Regular priority RTPS
High-priority nRTPS	Regular priority RTPS	High-priority nRTPS	Low-priority RTPS
Regular priority nRTPS	Low-priority RTPS	High-priority BE	Regular priority nRTPS
Low-priority nRTPS	Regular priority nRTPS	Regular priority nRTPS	Low-priority nRTPS
High-priority BE	Low-priority nRTPS	Low-priority nRTPS	High-priority BE
Regular priority BE	Regular priority BE	Regular priority BE	Regular priority BE
Low-priority BE	Low-priority BE	Low-priority BE	Low-priority BE

Table 5. Method 1 to Method 3 of eDBAM.

eDBAM Method 4	eDBAM Method 5	eDBAM Method 6	eDBAM Method 7
High-priority RTPS	High-priority RTPS	High-priority RTPS	High-priority RTPS
High-priority nRTPS	Regular priority RTPS	Regular priority RTPS	Regular priority RTPS
Regular priority RTPS	High-priority nRTPS	High-priority nRTPS	Low-priority RTPS
Low-priority RTPS	Regular priority nRTPS	Regular priority nRTPS	High-priority nRTPS
High-priority BE	Low-priority RTPS	High-priority BE	Regular priority nRTPS
Regular priority nRTPS	Low-priority nRTPS	Regular priority BE	High-priority BE
Low-priority nRTPS	High-priority BE	Low-priority RTPS	Regular priority BE
Regular priority BE	Regular priority BE	Low-priority nRTPS	Low-priority nRTPS
Low-priority BE	Low-priority BE	Low-priority BE	Low-priority BE

Table 6. Method 4 to Method 7 of eDBAM.

In eDBAM (for example Method 2), low priority service class of high priority user (ex: Low-Priority BE) can be allocated bandwidth ahead of high-priority service class of regular/low-priority user (Regular/Low priority nRTPS). This out of turn allocation of bandwidth improves the throughput for even low priority service class (BE) for high-priority users.

2.2.1 Implementation

Implementation of eDBAM is similar to DBAM. The AAA server shall maintain a mapping of MAC address to the priority value associated with the MAC address. When a MS sends RNG-REQ to BS, BS shall obtain the priority value associated with the MS and allocated bandwidth based on one of the seven methods proposed for eDBAM. BS does not switch between the seven different methods of eDBAM. Each BS shall implement one of the seven methods and stick to that method throughout its operation.

2.2.2 Analytical modeling

Throughput modeling follows similar patterns as that of DBAM. Only the order of bandwidth allocation shall change. Delay modeling is explained in this section. The notations used for delay modeling are given in Table 7.

Symbol	Description
λ	Mean arrival rate
μ	Mean service rate
ρ	Service utilization
L	Mean number of packets of a service flow for a particular SS in the system.
W	Mean end-to-end delay for secure packets of a particular service flow for a particular SS.

Table 7. Delay modeling parameters.

For BE packets, Packet arrivals are assumed to have a Poisson arrival.

$$P_n(t)=\frac{(\lambda t)^n}{n!}e^{-\lambda t} \tag{7}$$

We know that, Service Utilization = Mean arrival rate / Mean service rate. i.e.

$$\rho=\frac{\lambda}{\mu} \tag{8}$$

For BE traffic (exponential distribution), mean number of packets for a service flow for a particular-SS is given in (9)

$$L=\frac{\rho}{1-\rho} \tag{9}$$

Queuing delay for a service flow for a particular SS is given in (10)

$$W=\frac{L}{\lambda}=\frac{\frac{\rho}{1-\rho}}{\lambda} \tag{10}$$

For RTPS and nRTPS we assume constant arrival pattern. So mean number of packets for a service flow for a particular SS is given in (11)

$$L=\frac{\rho(2-\rho)}{2(1-\rho)} \tag{11}$$

Hence the queuing delay for packets that have constant arrival pattern is:

$$W=\frac{L}{\lambda}=\frac{\frac{\rho(2-\rho)}{2(1-\rho)}}{\lambda} \tag{12}$$

2.2.3 Simulation of eDBAM

Simulation was carried out using NS 2.29. LWX was used to simulate wimax on top of ns2. Simulations were carried out for method-2 for eDBAM. Simulation parameters used, are given in Table-8

Parameter	Value
Data rate	10 Mbps
OFDMA Frame Duration	5 ms
OFDMA symbol time	100.94 µs
RTPS data arrival rate	333 Kbps
nRTPS data arrival rate	333 Kbps
BE data arrival rate	333 Kbps

Table 8. Simulation parameters for eDBAM.

Simulation setup was done such that at any given time the network consists of 1/3rd High-priority SS, 1/3rd Regular-SS and 1/3rd low-priority SS. Each SS is assumed to have RTPS, nRTPS and BE traffic. Downlink ftp traffic at 1 Mbps was introduced.

2.2.3.1 Throughput results

Simulation was done to compare the throughput for High-Priority, Regular and Low-Priority BE traffic. Fig. 6 shows the simulation results. A comparison with theoretical results is also provided.

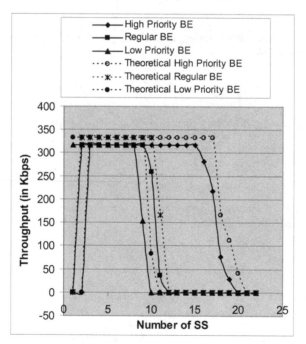

Fig. 6. Throughput for BE traffic for the three different types of user.

From fig. 6 we see that as the number of SS in the network increases, the throughput form Low-priority BE drops. Subsequently the throughput reduces for regular BE and finally the throughput for High-priority BE. Since method-2 prioritized high-priority BE ahead of Regular nRTPS and Low-priority nRTPS, simulations were carried out for the service flow. Results of simulation are shown in Fig.7.

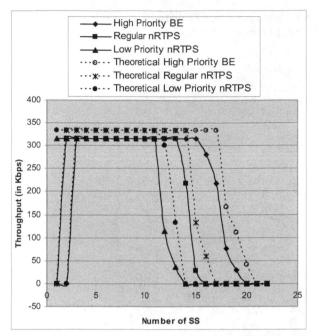

Fig. 7. Throughput for High Priority BE v/s Regular nRTPS v/s Low priority nRTPS.

2.2.3.2 Delay results

Simulations were carried out to find the delay incurred by the service flows. Fig. 8 shows the delay for High-Priority BE, Regular BE and Low-Priority BE.

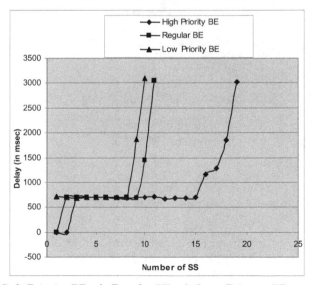

Fig. 8. Delay for High-Priority BE v/s Regular BE v/s Low-Priority BE.

Packet delay was measured for High-Priority BE, Regular nRTPS and Low-Priority nRTPS. Results of simulation are shown in Fig. 9.

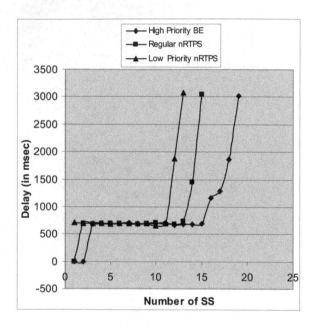

Fig. 9. Delay Results for High Priority BE v/s Regular nRTPS v/s Low Priority nRTPS.

From Fig. 8 and Fig. 9 we see that using eBBAM, packets from high-priority SS are subjected to lesser delay compared to regular and low-priority SS.

2.2.3.3 DBAM v/s eDBAM

Simulations were done to compare the throughput and delay for DBAM and eDBAM. Fig. 10 shows the throughput comparision for DBAM and eDBAM. We consider method-2 for eDBAM.

From Fig. 10 we observe that the throughput for DBAM drops down much before eDBAM. This is because in case of eDBAM, high-priority BE is allotted bandwidth ahead of regular nRTPS and low-priority nRTPS. Figure 11 shows the simuation results for delay. Again eDBAM fairs better than DBAM.

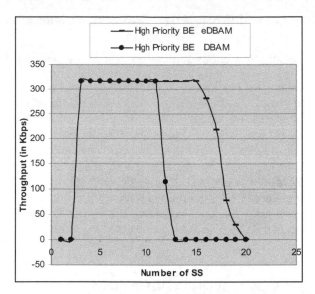

Fig. 10. Throughput comparison for eDBAM and DBAM for high priority BE.

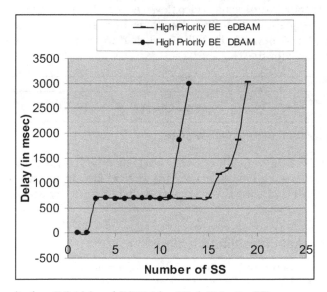

Fig. 11. Delay results for eDBAM and DBAM for High Priority BE.

2.3 Network Aware Differentiated Bandwidth Allocation Mechanism (nDBAM)

Though eDBAM improves the throughput, the algorithm is indifferent to the current network conditions. Especially, if eDBAM method-1 is implemented, it could result in delays for regular and low-priority RTPS. Users might face jitter when they are viewing videos. This might not be desirable. nDBAM takes care of current network conditions before allocating bandwidth to the different service flows. The steps for nDBAM algorithm as given below.

Step 1. Users shall be allotted bandwidth as per one of the selected Seven methods of eDBAM.

BS keeps monitoring the network condition. BS could poll the SS to know their current queue length and the average queuing delays faced for each service flow. BS and SS can use the ranging mechanism to pass the information between them.

Step 2. If the average queuing delay exceeds the QoS limits for the service class then the BS shall fallback from eDBAM to DBAM bandwidth allocation mechanism

Step 3. BS checks with the SS if the average queuing delay has reduced. If yes then BS sticks to DBAM. If the average queuing delay is still high them BS falls back to First-come-first-serve (FCFS) method of bandwidth allocation.

Step 4. BS keeps monitoring the queuing delay. If the delay reduces and stays within acceptable limits then BS moves back to eDBAM algorithm

2.3.1 Implementation

BS does ranging at periodically with the SS. Ranging process is generally done to adjust the power levels and the clock skews. During the ranging process, BS can also request for the current queue state for the different service flows. As a part of ranging response (RNG-RSP) The SS can send the queue state to BS. The information is generally sent as a TLV (Type-Length-Value) header. A new header will be required to send the queue state information. Table 9 lists an example for the TLV.

Type	Length	Value	Scope
Unused TLV type (ex: 105)	1	Average Queue delay for Service flow	RNG-RSP

Table 9. TLV header used to send Queue state.

BS receives the RNG-RSP from all the SS for each of their service class. BS then checks if the queuing delay is within the QoS limits for the service class. If not then it means that the eDBAM algorithm is introducing delay for regular and low-priority users. So, BS shifts from eDBAM to DBAM.

3. User based packet classification algorithm

We know that WiMAX supports 5 different types of service classes i.e. UGS, RTPS, eRTPS, nRTPS, BE. When a user generates data (ex: video packets) they are classified and placed into one of the 5 queues at SS (ex: Video packets are classified as RTPS packets and placed in the RTPS queue). As the user keeps generating data packets, these are classified and placed in one of the queues.

This method of classification is application specific. i.e. if the user keeps generating video packets they are always classified as RTPS packets and placed in RTPS queue and if the user generates web browsing/email packets they are generally classified as BE packets and places in BE queue. Packet classification is not user specific. i.e. there may be some users who are ready to pay more if their browsing packets are treated as high priority packets i.e. the browsing packets generated by such users are treated as RTPS packets instead of BE packets and placed in RTPS queue.

There may be some users who may wish to pay less and still enjoy broadband facility. For such users we may want to downgrade even their high priority packets like RTPS packets and treat them as low priority BE packets. A third set of users may fall in-between the high-priority and low-priority users.

There shall be 8 different ways of classifying the packets as given in Table 10 (Lagare & Das 2009).

Priority	Bit Value	Description
0	000	802.16e's existing packet classification mechanism is retained. i.e. real time packets will be placed in RTPS queue. Non real time packets are placed in nRTPS queue and delay tolerant packets are placed in BE queue.
1	001	RTPS, nRTPS and BE packets are classified as real-time packets and placed in RTPS queue.
2	010	nRTPS packets are promoted as RTPS packets and all BE packets are promoted as nRTPS packets
3	011	Only the BE packets are promoted as RTPS packets. Other Packets are placed in their respective priority queues.
4	100	Only the BE packets are promoted as nRTPS packets. Other Packets are placed in their respective priority queues.
5	101	RTPS, nRTPS and BE packets are classified as delay tolerant packets and moved to BE queue.
6	110	RTPS packets will be blocked. This priority level can be set to a certain set of users so that these users can be blocked from transmitting RTPS packets like MPEG videos.
7	111	RTPS and nRTPS packets will be blocked. This priority can be set to very low priority users.

Table 10. Eight different ways of packet classification.

3.1 Implementation

When the MS enters the network, it sends the RNG-REQ to BS. On receiving the range request, BS shall check the priority value associated with the SS. This priority value is passed to the SS in the RNG-RSP. On receiving the priority value the SS shall classify the packets as per table 10.

3.2 Simulation

Simulations were carried out to observe the improvement in throughput by implementing user based packet classification. Priority 3 scenario of table 11 was simulated. The simulation network consists of one priority MS whose packets are prioritized as per Priority

3. Other MS are regular users whose packets are prioritized as per priority 1. Table 11 lists the simulation parameters used.

Parameter	Value
Uplink Bandwidth	2Mbps
Uplink Frame Duration	1msec (2000 bits)
Number of Uplink frames per second	1000/Sec
Maximum Uplink bandwidth per SS per Frame	400 bits/frame
Minimum Reserved Traffic Rate for RTPS	240Kbps
Arrival Pattern for RTPS Traffic	Variable bit rate packets at regular interval of time
Arrival Pattern for BE Traffic	Poisson Arrival
Average arrival Rate for RTPS traffic	160Kbps
Average arrival Rate for BE traffic	72Kbps

Table 11. Simulation Parameters.

Figure 12 shows the simulation results for BE traffic when the priority MS and regular MS generate both RTPS and BE packets. For priority MS, the BE packets are classified as RTPS packets.

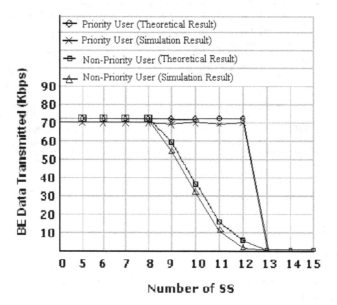

Fig. 12. BE data (in Kbps) transmitted by priority-SS compared to regular-SS.

From Fig. 12 we see that, when the number of MS in the network are less than 8, both Priority MS and non-priority MS are able to transmit all their data. When the number of MS in the network goes beyond 8, there isn't enough bandwidth to support the BE traffic for non-priority users. So the average throughput for non-priority user drops. Since priority MS request bandwidth for their BE traffic as RTPS traffic, priority MS continue to receive bandwidth. Beyond 12 SS there isn't enough bandwidth to support elevation of BE traffic as RTPS traffic. So throughput for even priority-MS drops down. Fig. 13 shows the simulation results when the network consists of only BE traffic.

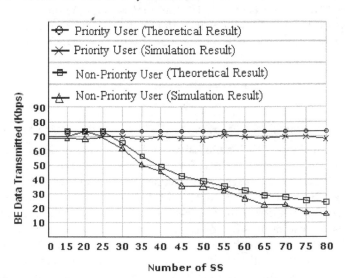

Fig. 13. BE data transmitted by priority-MS and regular-MS when only BE packets are present in the network.

In Fig. 13 we see that priority MS enjoy constant throughput of 70Kbps where as the throughput for non-priority MS keeps decreasing as the number of MS increases. This happens because BE traffic for priority MS is treated as RTPS traffic. So bandwidth is allotted to priority MS. The leftover bandwidth is shared by non-priority MS. So, by implementing priority based packet classification we can provide graded QoS to the users.

4. User based Call Admission Control (CAC) algorithm

Call admission control (CAC) plays a very important role in the IEEE 802.16 based wireless network. WiMAX networks aim to ensure that the QoS requirements for each service class are met. In order to provide QoS, the network should have a robust CAC algorithm.

When an SS/MS wants to establish a connection for a particular service class, it sends a DSA (Dynamic Service Addition) request to BS. This DSA request also contains the QoS parameters for the service class. Upon receiving the DSA request the BS decides to accept or reject the connection. If BS accepts the connection then it has to support the QoS needs of that connection.

When a BS decides to accept a connection, various factors need to be considered. For example the minimum and maximum data rates on the connection, the delay and jitter parameters for the connection etc. There can be other criteria like fairness, revenue per connection that can also play a role while admitting a connection.

Many CAC algorithms have been proposed both for wired and wireless medium. Because of the unique characteristics of wireless medium, many of the CAC algorithms of wired world cannot be applied to the wireless networks. Researches have proposed some CAC algorithms for WiMAX. In (Chen et. al, 2005) a simple bandwidth based CAC algorithm is proposed. A new connection is accepted if the bandwidth requirements for the connection can be satisfied by the BS. This algorithm does not take into consideration the deadline consideration of the connections. Once the bandwidth is allocated to the connection, the available bandwidth is calculated using the below equation:

$$BW_{avail} = BW - \sum_{s \in \{UGS, RTPS, nRTPS\}} \sum_{i=1}^{N^s} C_i^s[rate] \qquad 13$$

Where $C_i^s[rate]$ represents the data rate for the ith connection which belongs to s service class. In (Chandra & Sahoo, 2007) a QoS aware CAC is proposed. BS contains CAC queues for each service class. So there shall be 5 CAC queues (one each for UGS, RTPS, eRTPS, nRTPS and BE). When an SS makes a CAC request for a particular connection, the BS shall queue the request in one of queues based on the QoS requirements for the Class. BS then goes through each of the queues and accepts the connections. (Chandra & Sahoo, 2007) also provides criteria for call admission for each of the service class. In (Shu'aibu et. al, 2010) (Shu'aibu et. al, 2011) a partition based CAC algorithm is proposed. The total bandwidth is divided into many partitions like constant bit rate partition (CBR), variable bit rate partition (VBR) and Handover partition (HO) etc. CAC is applied to each of these partitions. CAC algorithms proposed above, are all service class based algorithms. In this section we shall look at user based CAC algorithm. The algorithm is based on (Chandra & Sahoo, 2007).

4.1 User based CAC algorithm

Fig. 14 shows the control flow at the SS when a new connection request is sent.

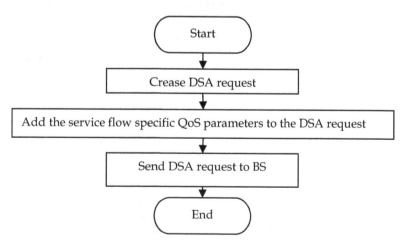

Fig. 14. User based CAC at SS.

Fig. 15 shows the classification of DSA request into different queues based on the priority of user.

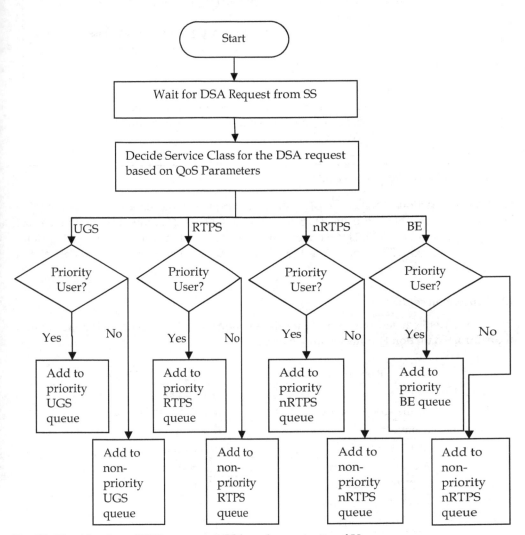

Fig. 15. Classification of DSA request at BS based on priority of User.

Because of lack of space, the control flow of eRTPS service class cannot be shown. However the logic for classifying the DSA request for eRTPS would be similar to UGS.

Once DSA requests are classified into the respective queues, BS goes through the DSA requests in each queue to admit the connection. High Level view of Admission control algorithm is given in Fig 16.

```
AdmissionControlAtBS( )
Begin

        for each service class (i.e UGS, RTPS, eRTPS, nRTPS and BE)
            if bandwidth available
                for each connection request in priority queue of service class
                    if BS can support QoS needs of connection request
                        Admit Connection.

            if bandwidth available
                for each connection request in non-priority queue of service class
                    if BS can support QoS needs of connection request
                        Admit Connection.
End
```

Fig. 16. User Based Admission control.

So, first, connections of priority UGS shall be accepted, followed by Non priority UGS. Priority RTPS and Non Priority RTPS follow next. Once RTPS connections are taken care, priority and Non priority eRTPS connection as admitted in that order. Subsequently priority and Non priority nRTPS connections are admitted. And finally priority and non priority BE connections are admitted.

4.2 Simulation results

Silulations were carried out to evaluate the performance of user based admission control algorithm. Simulation Parameters are given in Table-12.

Parameter	Value
Uplink Capacity	16 Mbps
Arrival of Connection Requests	Poisson arrival pattern
Lifetime of Connections	2 - 6 seconds
Data rate of UGS connections	256 kbps
Data rate of RTPS connections	256 kbps
Data rate of eRTPS connections	256 kbps
Data rate of nRTPS connections	256 kbps
Data rate of BE connections	256 kbps
Simulation Lifetime	200 seconds

Table 12. Simulation parameters.

Simulations were carried out to find the acceptance ratio for the connection requests for priority users and non-priority users. Acceptance ratio is defined as the ratio between the number of connections accepted to the total number of connections requested. Fig. 17 shows the simulation results for acceptance ratio when the network contains only RTPS connections.

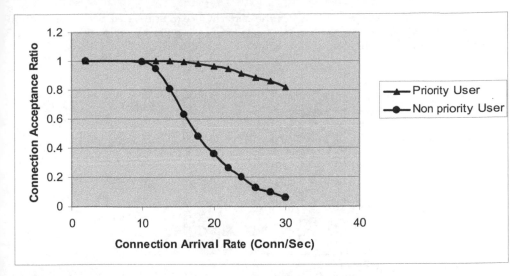

Fig. 17. Connection acceptance ratio for RTPS connections.

From Fig. 17 we can see that till the connection arrival rate is 10 connections/sec, there is enough capacity to accept both priority and non-priority connections. But beyond that the network cannot support the connection. So it starts to reject the connection. Since connection requests from priority users are processed first, the acceptance ratio for priority users would be higher compared to non-priority users.

Fig. 18 shows the simulation results for RTPS connections for different uplink capacities.

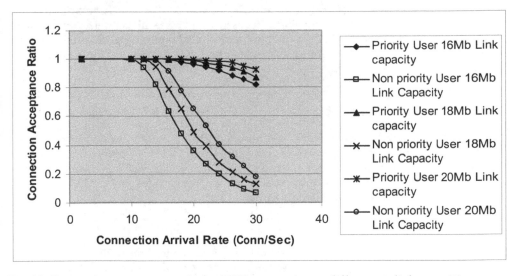

Fig. 18. Connection acceptance ratio for RTPS connection at different uplink capacities.

Simulations were also carried out to check the performance of user based admission control algorithm when BS receives connection requests for all the types of service classes i.e UGS, RTPS, nRTPS, eRTPS and BE. Fig 19. illustrates the simulation results for this scenario.

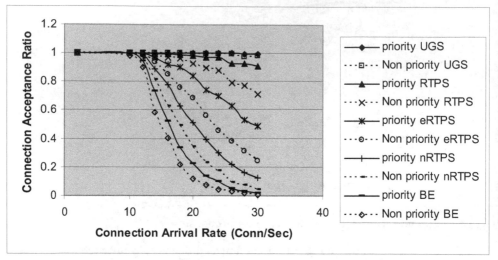

Fig. 19. Connection acceptance ratio for connections from different service classes.

From the simulation results it is clear that implementing user based admission control improves the connection acceptance ratio for priority users, there by improving the broadband experience for this section of users.

4.3 Drawback of user admission control

If, at any point in time, the network receives many connection requests from priority users then there is a chance that the non-priority users might see higher rejections of their connections. This can be tackled at the operator level. Based on the capacity of the network, a network service provider can limit the number of priority users that he can support. So when signing a new user, the operator can decide whether he wishes to provide the user the privilege of being a priority user.

5. Conclusion

In this chapter a comprehensive user based framework is proposed across various modules in WiMAX. Though operator can provide graded services by having different data rates at different price points, it does not give the flexibility that user based framework provides. Using the user based framework, graded services can be managed at the MAC layer and users can be up-graded/down-graded dynamically.

6. References

Chandra S. & Sahoo A. (2007), An efficient call admission control for IEEE802.16 networks, *in Proceedings of the 15th IEEE Workshop on Local and Metropolitan Area Networks*, pp. 188–193, ISBN 1-4244-1100-9, Princeton, NJ, USA, June 2007

Chen J.; Jiao W. & Wang H. (2005), A Service Flow Management Strategy for IEEE 802.16 Broadband Wireless Access Systems in TDD Mode, *IEEE International conference on Communication*, pp. 3422 - 3426, ISBN 0-7803-8938-7, May 16-20 2005

Chen Y. H. (2008), Light WiMAX Module, Available from http://code.google.com/p/lwx/

Chiang C.H.;Liao W.; & T. Liu (2007), Adaptive Downlink/Uplink Bandwidth Allocation in IEEE 802.16 (WiMAX) Wireless Networks: A Cross-Layer Approach, *IEEE Global Telecommunications Conference*, pp. 4775-4779, ISBN 978-1-4244-1043-9, Washington DC, US, Nov. 26-30 2007

IEEE (2004). IEEE 802.16, *Air Interface for Fixed Broadband Wireless Access Systems*, ISBN 0-7381-4070-8

IEEE (2005). IEEE 802.16e, *Amendment to IEEE Standard for Local and Metropolitan Area Networks - Part 16: Air Interface for Fixed Broadband Wireless Access Systems - Physical and Medium Access Control Layers for Combined Fixed and Mobile Operation in Licensed Bands*

Kumar N.; Murthy K. N. B. & Lagare A. M. (2011), DBAM: Novel User Based Bandwidth Allocation Mechanism in WiMAX, *2nd International Conference on Recent Trends in Information, Telecommunication and Computing*, pp. 229-236, ISBN 978-3-642-19541-9, March 2011

Kumar N.; Murthy K. N. B. & Lagare A. M. (2011), User oriented Network aware bandwidth allocation in wimax, *International Journal on Recent Trends in Engineering and Technology*, vol. 5, no. 1, pp 8-14, ISSN 2158-5555, March 2011

Lagare A.M.; Das D. (2009). Novel user-based packet classification in 802.16e to provide better performance, *IEEE 3rd International Symposium on Advanced Networks and Telecommunication Systems (ANTS)*, pp. 1 - 3, ISSN: 2153-1676, Dec 2009

Lee N.; Choi Y.; Lee S. & Kim N. (2010), A new CDMA-based bandwidth request method for IEEE 802.16 OFDMA/TDD systems, *IEEE Communications Letters*, Vol 14, Iss 2, pp 124-126, ISSN 1089-7798, Feb 2010

Lin Y.N.; Wu C.W.; Lin Y.D. & Lai Y.C. (2008), A Latency and Modulation Aware Bandwidth Allocation Algorithm for WiMAX Base Stations, *IEEE Wireless Communications and Networking Conference*, pp. 1408-1413, ISBN 978-1-4244-1997-5, Las Vegas, Nevada, US, March 31 - April 3 2008

Liu C.Y. & Chen Y.C. (2008), An Adaptive Bandwidth Request Scheme for QoS Support in WiMAX Polling Services, *Proceedings of The 28th International Conference on Distributed Computing Systems Workshops*, pp. 60-65, ISBN 978-0-7695-3173-1, Beijing , China, June. 17-20 2008

ns2, Available from http://www.isi.edu/nsnam/ns/

Park E.C. (2009), Efficient Uplink Bandwidth Request with Delay Regulation for Real-Time Service in Mobile WiMAX Networks, *IEEE Transaction on Mobile Computing*, Vol. 8, Iss. 9, pp. 1235-1249, ISSN 1536-1233, Sept. 2009

Peng Z.; Guangxi Z.; Haibin S. & Hongzhi L. (2007), A Novel Bandwidth Scheduling Strategy for IEEE 802.16 Broadband Wireless Networks, *Proceedings of International Conference on Wireless Communications, Networking and Mobile Computing*, pp. 2000-2003, ISBN 978-1-4244-1311-9, Shanghai, China, Sept. 21-25, 2007

Rong B.; Qian Y. & Chen H.H. (2007), Adaptive power allocation and call admission control in multiservice WiMAX access networks, *IEEE Wireless Communications*, Vol. 14, Iss. 1, pp. 14-19, ISSN 1536-1284, Feb 2007

Shu'aibu D. S.; Syed-Yusof S. K. & Fisal N. (2010), Partition-base bandwidth managements for mobile WiMAX IEEE802.16e, *International Review on Computers and Software*, vol. 5, no. 4, pp. 445–452

Shu'aibu D. S. (2011), Fuzzy Logic Partition-Based Call Admission Control for Mobile WiMAX, *ISRN Communications and Networking*, Article ID 171760, ISSN 2090-4355, April 11, 2011

Tao S. Z. & Gani A. (2009), Intelligent Uplink Bandwidth Allocation Based on PMP Mode for WiMAX, *Proceedings of the 2009 International Conference on Computer Technology and Development*, pp. 86-90, ISBN 978-0-7695-3892-1, Kuala Lumpur, Malaysia, Nov. 13-15 2009

Vaughan-Nichols, S. J. (2004), Achieving Wireless Broadband with WiMAX, *IEEE Computer*, Vol. 37 Iss. 6 , pp. 10-13, ISSN: 0018-9162

Wee K. K. & Lee S. W. (2009), Priority based bandwidth allocation scheme for WIMAX systems, *Proceedings of 2nd IEEE International Conference on Broadband Network & Multimedia Technology*, pp. 15-18, ISBN 978-1-4244-4590-5, Cyberjaya, Malaysia, Oct. 18-20 2009

7

Automatic Modulation Classification for Adaptive Wireless OFDM Systems

Lars Häring

Department of Communication Systems, University of Duisburg-Essen
Germany

1. Introduction

The flexible adaption of the transmission scheme to the current channel state becomes more and more a key issue in future communication systems. One efficient solution in multicarrier systems like Orthogonal Frequency Division Multiplexing (OFDM) has been proven to be adaptive modulation (AM) where the modulation scheme is selected on a subcarrier-basis or group of subcarriers. A lot of research has been carried out on AM or bit loading algorithms (Campello, 1998; Chow et al., 1995; Czylwik, 1996; Fischer & Huber, 1996; Hughes-Hartogs, 1987).

A basic disadvantage, however, is that the receiver requires the knowledge about the selection of the modulation schemes to decode the transmitted data. The conventional measure is to transmit the so-called bit allocation table (BAT) via a signaling channel.[1] This leads to a considerable reduction of the effective data rate. In contrast to wired communication links like the digital subscriber line (DSL) in which AM is already well-established, the time-variance of mobile radio channels usually necessiates a continuous and fast update of the BAT. Even sophisticated signaling schemes using state-dependent source coding of signaling bits reduce the throughput by 3 − 4% for short packets (Chen et al., 2009). If the channel statistics are not known, the signaling overhead is significantly larger.

In order to lower the amount of the signaling overhead and to obtain more flexibility, the BAT can be automatically detected at the receiver side. Such automatic modulation classification (AMC) algorithms have already been explored intensively since several decades, primarily for military applications but not for civil radio communication systems. The classifiers can be categorized into two types: likelihood-based (Boiteau & Martret, 1998; Long et al., 1994; Polydoros & Kim, 1990; Sills, 1999; Wei & Mendel, 2000) and feature-based methods (Dobre et al., 2004; Hsue & Soliman, 1989; Nandi & Azzouz, 1998; Swami & Sadler, 2000). While likelihood-based approaches arise from a defined optimality criterion, feature-based methods are usually heuristically motivated using e. g. higher-order moments. On the other hand, likelihood algorithms tend to require a higher complexity. A comprehensive overview of existing AMC algorithms is given in (Dobre et al., 2007).

In this book chapter, we will highlight the classification of digital quadrature amplitude modulation (QAM) schemes in wireless adaptive OFDM systems using the likelihood principle (Edinger et al., 2007; Huang et al., 2007; Lampe, 2004). We particularly focus on

---

[1] The BAT contains the information about the modulation schemes on each subcarrier.

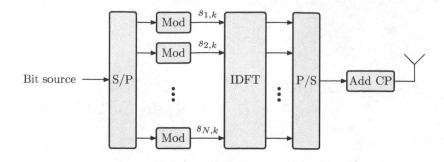

Fig. 1. Block diagram of an OFDM transmitter

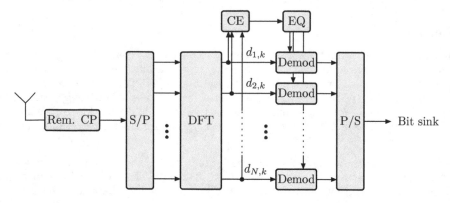

Fig. 2. Block diagram of an OFDM receiver

time-division duplex (TDD) systems in which the channel can be regarded as reciprocal. In contrast to other research work, a lot of new constraints are taken into account. Namely, many parameters are known by the receiver that can be utilized to enhance the classification reliability (Häring et al., 2010a; Häring et al., 2010b; 2011).

2. System model

2.1 Signal model

In Fig. 1 and 2, the baseband models of an OFDM transmitter and receiver are depicted. In OFDM, information data are transmitted blockwise. A sequence of bits is split into blocks, fed to different subcarriers and modulated. For the k-th block, an inverse discrete Fourier transform (IDFT) of length N on the symbols of all carriers is carried out. Subsequently, in order to combat interblock interference, a cyclic prefix of sufficient length N_g is preceded before transmission via the frequency-selective radio channel.

At the receiver side, the cyclic prefix is removed. In order to decode in OFDM, a discrete Fourier transform (DFT) is carried out. In a perfectly synchronized OFDM system, the received symbol $d_{n,k}$ on the n-th subcarrier ($1 \leq n \leq N$) of the k-th OFDM block ($1 \leq k \leq K$)

can be modeled by

$$d_{n,k} = H_n \cdot s_{n,k} + v_{n,k} \, ,\tag{1}$$

where $s_{n,k}$ and H_n denote the transmitted data symbol and the transfer function value on the n-th subcarrier of the k-th OFDM block, respectively. We consider a propagation scenario with slowly time-variant channels, typical for indoor communications. Thus the channel transfer function does not change significantly during one transmission frame, i. e. it holds: $H_{n,k} = H_n$. The additive white noise exhibits a complex Gaussian distribution: $v_{n,k} \sim \mathcal{CN}(0, \sigma_v^2)$. Due to the multicarrier principle, low-data rate signals are transmitted via flat-fading subchannels. This enables a simple frequency domain channel estimation (CE) and equalization (EQ) shown in Fig. 2.

In OFDM systems using adaptive modulation, symbols on different subcarriers can emanate from different symbol alphabets. Without loss of generality, we restrict ourselves to the digital modulation schemes with maximum bandwidth efficiencies 6 bit/symbol according to Table 1. In Fig. 3, the respective signal constellations are depicted.

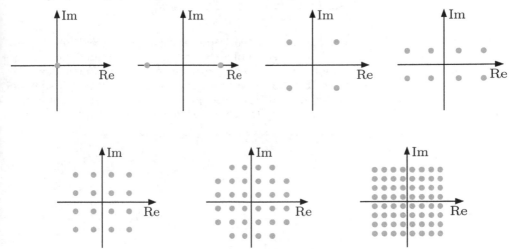

Fig. 3. QAM signal constellations: no modulation, BPSK, 4QAM, 8QAM, 16QAM, 32QAM, 64QAM

bandwidth efficiency [bit/symbol]	0	1	2	3	4	5	6
modulation type	null	BPSK	4QAM	8QAM	16QAM	32QAM	64QAM

Table 1. Considered digital modulation types

2.2 Adaptive modulation

Due to the frequency-selective nature of the radio propagation channel, some subcarriers exhibit good channel conditions whereas others suffer from a low signal-to-noise power ratio

(SNR). The overall system performance in terms of the raw bit-error ratio is dominated by the poor subcarriers.

The idea of adaptive modulation is to distribute the total amount of data bits among all subcarriers in an optimal way. If the subcarrier SNR is high, more bits than the average are loaded and higher-order modulation schemes are used. If the subcarrier SNR is low, less or even no bits are loaded such that the bit-error ratios on different subcarriers are evened out.

Using this principle, either the average bit-error ratio can be decreased at the same data rate or the data rate can be increased at the same target bit-error ratio. Since the knowledge about the data rate turns out to be an important feature of the AMC, the first approach with a fixed data rate is investigated here.

A huge amount of research on adaptive modulation algorithms has been carried out during the last twenty years (Campello, 1998; Chow et al., 1995; Czylwik, 1996; Fischer & Huber, 1996; Hughes-Hartogs, 1987). In the following, we focus on algorithms that utilize the bit metric:

$$b_n = \log_2\left(1 + \frac{\gamma_n}{k \cdot \gamma}\right) \quad \text{s.t.} \quad \sum_{n=1}^{N} b_n = N_b \,, \tag{2}$$

where γ and γ_n denote the average signal-to-noise power ratio and the SNR of the n-th subcarrier. This bit metric b_n is motivated by the channel capacity formula which takes the SNR gap (Starr et al., 1999) into account. As an example of adaptive modulation, the magnitude of the channel transfer function $|H_n|$ in a typical indoor propagation scenario (dashed line) and the corresponding bandwidth efficiencies (solid line) are shown in Fig. 4. There are two challenges involved in the application of AM in practical systems:

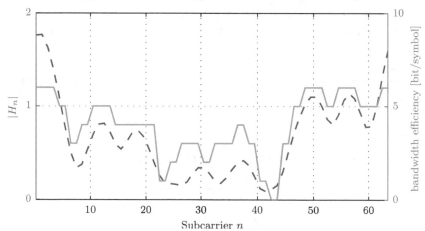

Fig. 4. Example of adaptive modulation

- *Channel knowledge at transmitter side*
  In order to be able to apply AM, the transmitter must know about the subcarrier SNRs. There are two ways to obtain this knowledge: 1) via feedback from the receiver or 2) using reciprocity in time-division duplex systems. Here, the focus is on TDD systems.

In our analysis, the channel transfer factors H_n are therefore obtained by a preamble-based channel estimation in the receive mode.

- *BAT knowledge at receiver side*
  In order to be able to decode the transmitted information, the receiver must know about the bit allocation table which includes the assignment of modulation schemes to subcarriers. Either this information is transmitted via a signaling channel or it is automatically classified.

2.3 Problem formulation

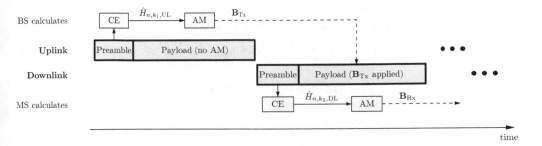

Fig. 5. TDD system structure

Fig. 5 shows the signal flow at the initiation of adaptive modulation in downlink transmission of the considered time-division duplex system model:

1a) In the uplink (UL), the mobile station (MS) transmits a frame consisting of training information and payload. When establishing the link, the MS does not know the channel state. Hence, for the first frame transmitted, no AM can be applied.

b) Based on the received preamble, the base station (BS) estimates the channel $\hat{H}_{n,k_1,\text{UL}}$ and calculates the optimal bit allocation table $\mathbf{B}_{\text{Tx}}$ using an AM algorithm.

2a) In the downlink (DL), the BS transmits a frame composed of training symbols and payload according to $\mathbf{B}_{\text{Tx}}$.

b) Based on the received preamble, the MS estimates the channel $\hat{H}_{n,k_2,\text{DL}}$ and calculates the optimal bit allocation table $\mathbf{B}_{\text{Rx}}$ using the same AM algorithm as the BS.

. . .

In order to decode the payload that has been sent by the base station, the mobile station requires the knowledge about $\mathbf{B}_{\text{Tx}}$. Since we assume that the BAT information is not signaled, the receiver must *automatically* classify the modulation schemes on each subcarrier solely based upon the received signal. In this analysis it is shown that utilization of side information that is typically available in wireless communication systems can significantly boost the classification reliability. More specifically, the AMC algorithms can exploit:

- channel correlation in frequency direction (e. g. subcarrier grouping)
- channel correlation in time direction (fixed modulation order on subcarriers and/or subgroups for entire frame)

- channel reciprocity in TDD mode
- knowledge about overall data rate (total number of loaded bits)

3. Automatic modulation classifier

In this section, automatic modulation classifiers that are based on different levels of knowledge are introduced. Denote the group of M possible digital QAM schemes by ($1 \leq m \leq M$)

$$\mathcal{I}^{(m)} = \{S_1^{(m)}, S_2^{(m)}, \ldots, S_{L^{(m)}}^{(m)}\} , \tag{3}$$

where $L^{(m)}$ is the constellation size (number of constellation points) of the m-th modulation scheme and $S_i^{(m)}$ denote the complex constellation symbols.

After collecting the received symbols $\mathbf{d}_n^{\mathrm{T}} = [d_{n,1}, d_{n,2}, \ldots, d_{n,K}]$, where K is the data frame length, the following M hypotheses are tested for each subcarrier n:

$$\mathcal{H}_n^{(m)} \cong \text{the used modulation scheme of the received data symbols}$$
$$\mathbf{d}_n \text{ was } \mathcal{I}^{(m)} .$$

Based upon these symbols, the underlying modulation scheme is to be detected.

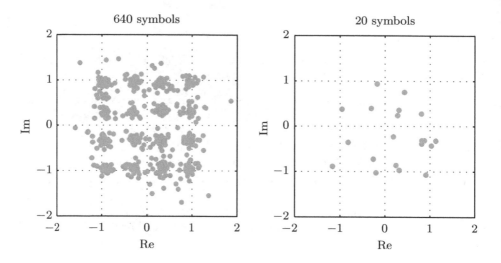

Fig. 6. Constellation diagrams of equalized 16QAM symbols at an average SNR of $\gamma = 20\,\mathrm{dB}$

As a motivating example, random signal constellations of received (and equalized) 16QAM symbols $d_{n,k}/\hat{H}_n$ at an average SNR of $\gamma = 20\,\mathrm{dB}$ are shown in Fig. 6. Whereas it seems obvious that the symbols emanate from a 16QAM scheme if a large number of symbols is available, it becomes much more difficult to classify the underlying modulation scheme if only a small number of symbols is available. Hence, the exploitation of additional information turns out to be a key aspect for a robust and reliable classification.

3.1 Maximum-likelihood (ML)

The maximum-likelihood (ML) approach chooses the hypothesis whose likelihood-function is maximum:

$$\hat{\mathcal{H}}_{n,\text{ML}} = \arg\max_{\mathcal{H}_n^{(m)}} p(\mathbf{d}_n|\mathcal{H}_n^{(m)}) . \tag{4}$$

The probability density function of the received symbols under the condition that the m-th modulation scheme was used (hypothesis $\mathcal{H}_n^{(m)}$) is

$$p(d_{n,k}|\mathcal{H}_n^{(m)}) = \sum_{i=1}^{L^{(m)}} p(d_{n,k}|S_i^{(m)}) \cdot p(S_i^{(m)}|\mathcal{I}^{(m)}) . \tag{5}$$

Each symbol within its constellation is equiprobable, i.e. $p(S_i^{(m)}|\mathcal{I}^{(m)}) = \frac{1}{L^{(m)}}$. Since v_n is assumed to be Gaussian distributed, it holds:

$$p(d_{n,k}|\mathcal{H}_n^{(m)}) = \frac{1}{L^{(m)}} \sum_{i=1}^{L^{(m)}} \frac{1}{\pi\sigma_v^2} \cdot \exp\left(-\frac{|d_{n,k} - H_n S_i^{(m)}|^2}{\sigma_v^2}\right) . \tag{6}$$

If symbols of different OFDM blocks $1 \le k \le K$ are statistically independent, the joint probability density function $p(\mathbf{d}_n|\mathcal{H}_n^{(m)})$ is given by:

$$p(\mathbf{d}_n|\mathcal{H}_n^{(m)}) = \prod_{k=1}^{K} p(d_{n,k}|\mathcal{H}_n^{(m)}) . \tag{7}$$

The log-likelihood function is:

$$\ln p(\mathbf{d}_n|\mathcal{H}_n^{(m)}) = \sum_{k=1}^{K} \ln p(d_{n,k}|\mathcal{H}_n^{(m)}) \tag{8}$$

$$= -K \cdot \ln L^{(m)} + c + \sum_{k=1}^{K} \ln \left(\sum_{i=1}^{L^{(m)}} \exp\left(-\gamma \cdot |d_{n,k} - H_n S_i^{(m)}|^2\right)\right) \tag{9}$$

with the average SNR $\gamma = E\{|S_i^{(m)}|^2\}/\sigma_v^2$ (= average symbol energy to noise spectral density E_S/N_0), $E_S = E\{|S_i^{(m)}|^2\} = 1$ and $E\{|H_n|^2\} = 1$. Neglecting irrelevant terms for the maximization, the ML-based classifier can be formulated as:

$$\hat{\mathcal{H}}_{n,\text{ML}} = \arg\max_{\mathcal{H}_n^{(m)}} J_{\text{ML}}(\mathbf{d}_n, \mathcal{H}_n^{(m)}) \quad \text{with}$$

$$J_{\text{ML}}(\mathbf{d}_n, \mathcal{H}_n^{(m)}) = \sum_{k=1}^{K} \ln \left(\sum_{i=1}^{L^{(m)}} \exp\left(-\gamma \cdot |d_{n,k} - H_n S_i^{(m)}|^2\right)\right) - K \cdot \ln L^{(m)} . \tag{10}$$

An example for the probability of correct classifications is given in Fig. 7. At the transmitter, the bandwidth efficiencies from 0 to 6 bit/symbol have been loaded by the AM algorithm in (Chow et al., 1995).

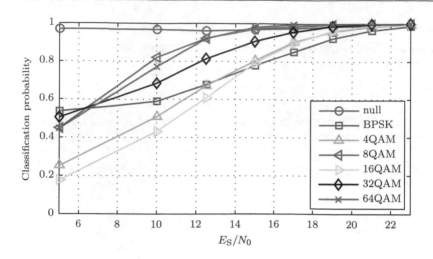

Fig. 7. Classification probability of the ML algorithm versus E_S/N_0, frame length $K = 10$

3.2 Maximum-a-posteriori (MAP)

One major drawback of the ML algorithm is that it cannot take into account that the hypotheses under test are not equally likely. Depending on the current channel status, however, some modulation schemes will be used more frequently than others. The ML algorithm is therefore not suitable for OFDM-based systems with adaptive modulation.

A first step to improve the performance is to maximize the a-posteriori probability $p(\mathcal{H}_n^{(m)}|\mathbf{d}_n)$ instead of $p(\mathbf{d}_n|\mathcal{H}_n^{(m)})$. The main difference can be seen by using the Bayes theorem:

$$p(\mathcal{H}_n^{(m)}|\mathbf{d}_n) = \frac{p(\mathbf{d}_n|\mathcal{H}_n^{(m)}) \cdot p(\mathcal{H}_n^{(m)})}{p(\mathbf{d}_n)} \ . \tag{11}$$

Since $p(\mathbf{d}_n)$ is irrelevant for the maximization of $p(\mathcal{H}_n^{(m)}|\mathbf{d}_n)$, the MAP classifier can be formulated as:

$$\hat{\mathcal{H}}_{n,\text{MAP}} = \arg\max_{\mathcal{H}_n^{(m)}} J_{\text{MAP}}(\mathbf{d}_n, \mathcal{H}_n^{(m)}) \quad \text{with} \tag{12}$$

$$J_{\text{MAP}}(\mathbf{d}_n, \mathcal{H}_n^{(m)}) = \sum_{k=1}^{K} \ln\left(\sum_{i=1}^{L^{(m)}} \exp\left(-\gamma \cdot |d_{n,k} - H_n S_i^{(m)}|^2\right)\right) - K \cdot \ln L^{(m)} + \ln p(\mathcal{H}_n^{(m)}) \ .$$

The a-priori information about the occurence probabilities $p(\mathcal{H}_n^{(m)})$ is utilized.[2] But still, the MAP algorithm in its current form is not able to sufficiently consider the specific characteristics of adaptive modulation.

---

[2] If these probabilities are equal, i.e. it holds: $p(\mathcal{H}_n^{(m)}) = 1/M$, then the MAP approach reduces to the ML algorithm.

3.3 MAP algorithms exploiting channel reciprocity (MAP-R)

One key feature to increase the classification reliability is the utilization of channel reciprocity in TDD systems. Under ideal conditions (perfect channel reciprocity, channel time-invariance and channel state information (CSI)), the receiver can perfectly reconstruct the transmit BAT by applying the same AM algorithm based on the propagation channel in the receive direction. In that case, the bit allocation table $\mathbf{B}_{Rx}$ equals $\mathbf{B}_{Tx}$. To be more realistic, we assume that the channel is time-variant and CSI is taken from a data-aided channel estimation. This causes $\mathbf{B}_{Tx}$ and $\mathbf{B}_{Rx}$ to be different but still correlated.

Fig. 8 shows an illustrating example of the magnitudes of the channel transfer function at transmitter and receiver side and the corresponding BATs $\mathbf{B}_{Tx}$ and $\mathbf{B}_{Rx}$ at an SNR of $\gamma = 10\,\mathrm{dB}$, a frame duration of $T_{fr} = 0.1\,\mathrm{ms}$ and a Doppler frequency of $f_{dop} = 15\,\mathrm{Hz}$. A typical preamble-based zero-forcing method to estimate the channels has been applied.

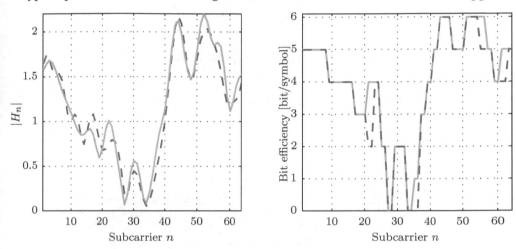

Fig. 8. Example channel transfer functions (left) and corresponding BATs (right) at transmitter (dashed lines) and receiver side (solid lines)

Both the channel transfer functions as well as the BATs differ for the transmitter and receiver, respectively. However, they are very similar which an AMC algorithm can take advantage of.

Two different concepts to benefit from this channel reciprocity are discussed now.

3.3.1 Receive bit allocation table (MAP-RQ)

Let $\hat{\mathcal{H}}_{n,Rx}$ be the modulation scheme for the n-th subcarrier in $\mathbf{B}_{Rx}$ which was computed by the AM algorithm. The method in (11) is extended by the knowledge about $\hat{\mathcal{H}}_{n,Rx}$. The modified approach which exploits channel reciprocity in terms of the quantized information $\hat{\mathcal{H}}_{n,Rx}$ can be written as:

$$\hat{\mathcal{H}}_{n,\mathrm{MAP-RQ}} = \arg\max_{\mathcal{H}_n^{(m)}} p(\mathcal{H}_n^{(m)}|\mathbf{d}_n, \hat{\mathcal{H}}_{n,Rx}) . \tag{13}$$

By applying the Bayes theorem and then neglecting irrelevant terms, we obtain:

$$\hat{\mathcal{H}}_{n,\text{MAP--RQ}} = \arg\max_{\mathcal{H}_n^{(m)}} \frac{p(\mathbf{d}_n, \hat{\mathcal{H}}_{n,\text{Rx}}|\mathcal{H}_n^{(m)}) \, p(\mathcal{H}_n^{(m)})}{p(\mathbf{d}_n, \hat{\mathcal{H}}_{n,\text{Rx}})} \tag{14}$$

$$= \arg\max_{\mathcal{H}_n^{(m)}} p(\mathbf{d}_n, \hat{\mathcal{H}}_{n,\text{Rx}}|\mathcal{H}_n^{(m)}) \, p(\mathcal{H}_n^{(m)}) \,. \tag{15}$$

Since $p((A,B)|C) = p(A|(B,C)) \cdot p(B|C)$, it holds:

$$p((\mathbf{d}_n, \hat{\mathcal{H}}_{n,\text{Rx}})|\mathcal{H}_n^{(m)}) = p(\mathbf{d}_n|(\hat{\mathcal{H}}_{n,\text{Rx}}, \mathcal{H}_n^{(m)})) \, p(\hat{\mathcal{H}}_{n,\text{Rx}}|\mathcal{H}_n^{(m)}) \,. \tag{16}$$

The simplification $p(\mathbf{d}_n|(\hat{\mathcal{H}}_{n,\text{Rx}}, \mathcal{H}_n^{(m)})) \approx p(\mathbf{d}_n|\mathcal{H}_n^{(m)})$ turns out to be reasonable for increasing correlation between $\mathbf{B}_{\text{Tx}}$ and $\mathbf{B}_{\text{Rx}}$. We set:

$$\hat{\mathcal{H}}_{n,\text{MAP--RQ}} = \arg\max_{\mathcal{H}_n^{(m)}} \left\{ p(\mathbf{d}_n|\mathcal{H}_n^{(m)}) \, p(\hat{\mathcal{H}}_{n,\text{Rx}}|\mathcal{H}_n^{(m)}) \, p(\mathcal{H}_n^{(m)}) \right\} \,. \tag{17}$$

Taking the logarithm and using $p(\hat{\mathcal{H}}_{n,\text{Rx}}, \mathcal{H}_n^{(m)}) = p(\hat{\mathcal{H}}_{n,\text{Rx}}|\mathcal{H}_n^{(m)})p(\mathcal{H}_n^{(m)})$, the MAP classifier based on the receive BAT is:

$$\hat{\mathcal{H}}_{n,\text{MAP--RQ}} = \arg\max_{\mathcal{H}_n^{(m)}} J_{\text{MAP--RQ}}(\mathbf{d}_n, \mathcal{H}_n^{(m)}) \quad \text{with}$$

$$J_{\text{MAP--RQ}}(\mathbf{d}_n, \mathcal{H}_n^{(m)}) = \sum_{k=1}^{K} \ln\left(\sum_{i=1}^{L^{(m)}} \exp\left(-\gamma \cdot |d_{n,k} - H_n S_i^{(m)}|^2\right) \right) - K \cdot \ln L^{(m)}$$

$$+ \ln p(\hat{\mathcal{H}}_{n,\text{Rx}}, \mathcal{H}_n^{(m)}) \,. \tag{18}$$

The joint probability $p(\hat{\mathcal{H}}_{n,\text{Rx}}, \mathcal{H}_n^{(m)})$ could be numerically determined in advance and stored in look-up tables. Simulation results have shown that a coarse knowledge of these values is sufficient to achieve a significant performance improvement compared to the conventional ML algorithm. However, the probabilities $p(\hat{\mathcal{H}}_{n,\text{Rx}}, \mathcal{H}_n^{(m)})$ are usually not available in practice and, moreover, strongly depend on the transmission system and the propagation scenario. In order to overcome this disadvantage we present a heuristic approach to obtain these values. The classification performance is similar to the optimal one.

Approximation of joint probabilities

Suppose that $\hat{\mathcal{H}}_{n,\text{Rx}} = \mathcal{I}^{(\mu)}$ is the modulation scheme for the n-th subcarrier in $\mathbf{B}_{\text{Rx}}$. Due to the correlation between transmit and receive BAT, $\mathcal{I}^{(\mu)}$ is said to be more likely than the other $M - 1$ possible modulation schemes. Consequently, we set

$$p(\hat{\mathcal{H}}_{n,\text{Rx}} = \mathcal{I}^{(\mu)}, \mathcal{H}_n^{(m)}) = \alpha \cdot \begin{cases} w & , \; m = \mu \\ \frac{1-w}{M-1} & , \; m \neq \mu \end{cases} \tag{19}$$

with $\frac{1}{M} \leq w \leq 1$. The proportional factor α is irrelevant for the metric maximization. The probabilities of all other hypotheses $m \neq \mu$ have been set equally to $\frac{1-w}{M-1}$. Further numerical investigations have shown that distributing the "residual" probability $1 - w$ in a more sophisticated way does not lead to a significant advantage.

The optimal values of the weighting factors w depend on a multitude of effects: channel quality, channel estimation method, adaptive modulation algorithm etc. The more correlated

$\mathbf{B}_{\text{Tx}}$ and $\mathbf{B}_{\text{Rx}}$ are, the higher the value of w should be chosen. Unfortunately, the analytical search for the optimum seems to be intractable.

The simplicity of this heuristic approach appears attractive for a practical implementation. The receiver needs to calculate $\mathbf{B}_{\text{Rx}}$ anyway for the application of adaptive modulation in the next transmission frame.

A general drawback of using the BAT in order to exploit channel reciprocity is the quantization of bandwidth efficiencies.

3.3.2 Receive channel state information (MAP-RS)

An even better way to incorporate the channel correlation in transmit and receive direction into the AMC algorithm is described now. We assume that the AM algorithm at transmitter and receiver side is based upon bit loading according to the widely used criterion (Chow et al., 1995) for the estimated bandwidth efficiency:

$$\hat{b}_{n,\text{Rx}} = \log_2\left(1 + \frac{\gamma_n}{k \cdot \gamma}\right) = \log_2\left(1 + \frac{|H_n|^2}{k}\right) , \tag{20}$$

in which k is adapted such that the target bit rate is achieved. The classification method that utilizes the *soft* channel information $\hat{b}_{n,\text{Rx}}$ instead of the *hard* BAT information $\hat{\mathcal{H}}_{n,\text{Rx}}$ can be formulated as:

$$\hat{\mathcal{H}}_{n,\text{MAP}-\text{RS}} = \arg\max_{\mathcal{H}_n^{(m)}}\left\{ p(\mathcal{H}_n^{(m)}|\mathbf{d}_n, \hat{b}_{n,\text{Rx}})\right\} . \tag{21}$$

Following the same steps as in the previous section, the solution of (21) is:

$$\hat{\mathcal{H}}_{n,\text{MAP}-\text{RS}} = \arg\max_{\mathcal{H}_n^{(m)}} J_{\text{MAP}-\text{RS}}(\mathbf{d}_n, \mathcal{H}_n^{(m)}, \hat{b}_{n,\text{Rx}}) \quad \text{with}$$

$$J_{\text{MAP}-\text{RS}} = \sum_{k=1}^{K} \ln\left(\sum_{i=1}^{L^{(m)}} \exp\left(-\gamma \cdot |d_{n,k} - H_n S_i^{(m)}|^2\right)\right) - K \cdot \ln L^{(m)}$$

$$+ \ln p(\hat{b}_{n,\text{Rx}}, \mathcal{H}_n^{(m)}) . \tag{22}$$

Whereas the AMC algorithm based on the receive BAT in section 3.3.1 requires the knowledge about $p(\hat{\mathcal{H}}_{n,\text{Rx}}, \mathcal{H}_n^{(m)})$, here the function $p(\hat{b}_{n,\text{Rx}}, \mathcal{H}_n^{(m)})$ must be known. Fig. 9 depicts simulation examples of the joint probability density function $p(\hat{b}_{n,\text{Rx}}, \mathcal{H}_n^{(m)})$ and the probabilities $p(\hat{\mathcal{H}}_{n,\text{Rx}}, \mathcal{H}_n^{(m)})$.

Approximation of joint probability density functions

It is unrealistic to assume that the joint probabilities $p(\hat{b}_{n,\text{Rx}}, \mathcal{H}_n^{(m)})$ are available in practical systems. Again, we find approximations based on a heuristic approach: Suppose that the AM algorithm at the receiver has computed $\hat{b}_{n,\text{Rx}} = b_0$ for subcarrier n. Then it is obvious that those hypotheses which are "closer" to b_0 are more likely than others. We use the heuristic measure

$$p(\hat{b}_{n,\text{Rx}} = b_0, \mathcal{H}_n^{(m)}) = \beta \cdot \exp\left(-\left(\frac{b(\mathcal{H}_n^{(m)}) - b_0}{\sqrt{2}\sigma}\right)^2\right) \tag{23}$$

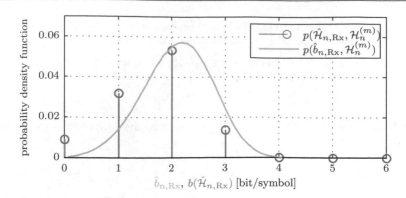

Fig. 9. Example of simulated probability density functions $p(\hat{b}_{n,\mathrm{Rx}}, \mathcal{H}_n^{(m)})$ and $p(\hat{\mathcal{H}}_{n,\mathrm{Rx}}, \mathcal{H}_n^{(m)})$ for $b(\mathcal{H}_n^{(m)}) = 2\,\mathrm{bit/symbol}$ and $\gamma = 10\,\mathrm{dB}$; the average bandwidth efficiency is $4\,\mathrm{bit/symbol}$

parameter	value
sampling period T	50 ns (20 MHz bandwidth)
FFT length N	64
cyclic prefix length N_{cp}	16
Channel model	(Medbo & Schramm, 1998) (indoor)
Delay spread τ_{ds}	100 ns
Doppler frequency f_{dop}	15 Hz (Jakes spectrum)

Table 2. Simulation set-up

thus assuming a Gaussian distributed deviation with the design parameter σ. Here, the operator $b(\mathcal{H})$ denotes the number of bits transmitted for hypothesis $\mathcal{H}$; β is a proportional factor that is irrelevant for the maximization.

The more correlated the channels at transmitter and receiver are, the smaller σ should be chosen. Due to the complex influence of the AM algorithm, it seems to be intractable to find the optimal value σ analytically.

However, compared to the quantized information that is used in the MAP-RQ algorithm, the MAP-RS method benefits from the more reliable soft information leading to a higher classification performance.

3.3.3 Discussion

In Fig. 10, the classification error probability $P_{\mathrm{ce}} = \Pr\{\hat{\mathbf{B}}_{\mathrm{Tx}} \neq \mathbf{B}_{\mathrm{Tx}}\}$ versus E_S/N_0 for various AMC methods is shown. If already one modulation scheme in $\mathbf{B}_{\mathrm{Tx}}$ is incorrectly classified, the whole packet would get lost. Throughout the entire contribution, the performance discussion is based upon a common simulation scenario: The main parameters concordant with a WLAN IEEE 802.11a/n system (IEEE, 2005) are summarized in Table 2. The bandwidth efficiencies, however, vary between $0 - 6\,\mathrm{bit/symbol}$. In average, $4\,\mathrm{bit/symbol}$ are loaded by the AM algorithm proposed in (Chow et al., 1995). Due to the fact that the subcarrier spacing is small compared to the channel coherence bandwidth, 2 neighbouring subcarriers are grouped without sacrificing significant system performance. The AMC algorithms utilize this

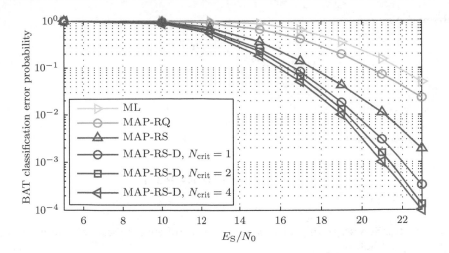

Fig. 10. BAT classification error probability P_{ce} of AMC algorithms versus E_S/N_0, frame length $K = 10$

subcarrier grouping by additively combining the corresponding metric values. A zero-forcing channel estimation is carried out on 2 training OFDM blocks with a total length of 160 samples. It is assumed that the receiver has perfect knowledge about the total data rate since this single information can be protected against errors well by suitable channel coding. The design parameters for the reciprocity-based classifiers are set to $w = 0.8$ and $\sigma = 0.3$.

First, we refer only to the ML algorithm (section 3.1) and the reciprocity-based methods MAP-RQ and MAP-RS: Due to the additional reciprocity information used, the reciprocity-based MAP algorithms outperform the classical ML algorithm significantly. With the soft information used in MAP-RS, the number of modulation classification errors can even be further reduced compared to MAP-RQ. Another step towards a higher reliability is to include even more side information in the modulation classification.

3.4 MAP-R algorithms exploiting the knowledge about the data rate (MAP-R-D)

In communication systems, the overall data rate is typically signaled to the receiver via a control channel. Therefore, we can incorporate the available information about the total number of transmitted bits N_b per OFDM block into the MAP-R algorithms.[3]

As an example, we describe the extension for the algorithm MAP-RS. The corresponding modifications in MAP-RQ are analogous. The scheme that jointly classifies the bandwidth efficiencies on all subcarriers can be formulated as:

$$\hat{\mathcal{H}}_{n,\text{joint}} = \arg\max_{\mathcal{H}_n^{(m)}} \left\{ p(\mathcal{H}_n^{(m)} | \mathbf{d}_n, \hat{b}_{n,\text{Rx}}) \right\} \quad, 1 \leq n \leq N,$$

$$\text{s.t.} \sum_{n=1}^{N} b(\mathcal{H}_n^{(m)}) = N_b \ . \tag{24}$$

---

[3] Note that the BAT is fixed for the entire transmission burst.

With the abbreviations

$$\vec{\mathcal{H}} = [\mathcal{H}_1^{(m)}, \ldots, \mathcal{H}_N^{(m)}] \tag{25}$$

$$\hat{\mathbf{b}}_{\mathrm{Rx}} = [\hat{b}_{1,\mathrm{Rx}}, \ldots, \hat{b}_{N,\mathrm{Rx}}] \tag{26}$$

$$\hat{\vec{\mathcal{H}}}_{\mathrm{joint}} = [\hat{\mathcal{H}}_{1,\mathrm{joint}}, \ldots, \hat{\mathcal{H}}_{N,\mathrm{joint}}] \tag{27}$$

$$\vec{\mathbf{d}} = [\mathbf{d}_1, \ldots, \mathbf{d}_N] \tag{28}$$

for the hypotheses under test, the bandwidth efficiencies computed at the receiver, the classified hypotheses and the collected received data symbols on all subcarriers, we reformulate (24):

$$\hat{\vec{\mathcal{H}}}_{\mathrm{joint}} = \arg\max_{\vec{\mathcal{H}}} \left\{ p(\vec{\mathcal{H}} | \vec{\mathbf{d}}, \hat{\mathbf{b}}_{\mathrm{Rx}}) \right\}$$

$$\text{s.t. } b(\vec{\mathcal{H}}) = N_\mathrm{b} . \tag{29}$$

Since the modulation schemes and data symbols on different subcarriers are independent from each other, it holds:

$$p(\vec{\mathcal{H}} | \vec{\mathbf{d}}, \hat{\mathbf{b}}_{\mathrm{Rx}}) \approx \prod_{n=1}^{N} p(\mathcal{H}_n^{(m)} | \mathbf{d}_n, \hat{b}_{n,\mathrm{Rx}}) , \tag{30}$$

and hence:

$$\hat{\vec{\mathcal{H}}}_{\mathrm{joint}} = \arg\max_{\vec{\mathcal{H}}} J_{\mathrm{joint}}(\vec{\mathbf{d}}, \vec{\mathcal{H}}) \quad \text{s.t. } b(\vec{\mathcal{H}}) = N_\mathrm{b} \quad \text{with} \tag{31}$$

$$J_{\mathrm{joint}}(\vec{\mathbf{d}}, \vec{\mathcal{H}}) = \sum_{n=1}^{N} \left[\sum_{k=1}^{K} \ln \left(\sum_{i=1}^{L^{(m)}} \exp\left(-\gamma \cdot |d_{n,k} - H_n S_i^{(m)}|^2 \right) \right) - K \cdot \ln L^{(m)} \right.$$

$$\left. + \ln p(\hat{b}_{n,\mathrm{Rx}}, \mathcal{H}_n^{(m)}) \right] . \tag{32}$$

The maximum search must be carried out over a set of all possible hypothesis *candidate combinations* $\vec{\mathcal{H}}$. Due to its high complexity, such a joint search is not feasible in practice. We investigate a trade-off between this joint algorithm and the subcarrier-independent methods instead.

First, we split the set of all subcarriers $\mathcal{S} = \{1, \ldots, N\}$ into two subsets:

- $\mathcal{S}_{\mathrm{rel}} := \{n \mid \text{reliable decision}\}$ including subcarriers for which *reliable* decisions are possible, and

- $\mathcal{S}_{\mathrm{crit}} := \{n \mid \text{critical decision}\} = \mathcal{S} \setminus \mathcal{S}_{\mathrm{rel}}$ including subcarriers with *critical* decisions.

The number of elements in $\mathcal{S}_{\mathrm{crit}}$ is denoted as N_{crit}. The distinction is based upon an ordering procedure according to the absolute distance between the largest and second largest metric value of the subcarrier-independent metrics (18) or (22), respectively. Subcarriers with the largest absolute distances are inserted in $\mathcal{S}_{\mathrm{rel}}$; the remaining N_{crit} subcarriers are included in $\mathcal{S}_{\mathrm{crit}}$. In other words, decisions are said to be critical if the two largest metric values are similar.

Surprisingly, ordering by subcarrer SNR values turns out to be unfavourable. Due to adaptive modulation in the considered OFDM systems, channel gains and modulation orders

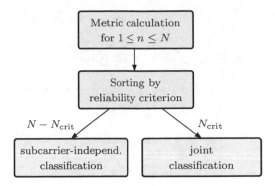

Fig. 11. Signal flow of joint algorithm

are mutually coupled – bandwidth efficiencies increase with increasing SNR. High-order modulation schemes like 64QAM are, however, more difficult to classify than low-order modulation schemes like BPSK.

The automatic modulation classification is performed for the two subsets in a different way:

- For set $\mathcal{S}_{\mathrm{rel}}$, we apply the subcarrier-wise MAP-R algorithms described in section 3.3, e. g.:

$$\hat{\mathcal{H}}_{n,\mathrm{MAP-RS-D}} = \arg\max_{\mathcal{H}_n^{(m)}} J_{\mathrm{MAP-RS}}(\mathbf{d}_n, \mathcal{H}_n^{(m)}) \; \forall n \in \mathcal{S}_{\mathrm{rel}} \; . \tag{33}$$

- For set $\mathcal{S}_{\mathrm{crit}}$, we apply the joint MAP algorithm following (31):

$$\vec{\hat{\mathcal{H}}}_{\mathrm{MAP-RS-D}} = \arg\max_{\vec{\mathcal{H}}_{\mathcal{S}_{\mathrm{crit}}}} J_{\mathrm{joint}}(\vec{\mathbf{d}}, \vec{\mathcal{H}}_{\mathcal{S}_{\mathrm{crit}}}) \quad \mathrm{s.t.} \;\; b(\vec{\mathcal{H}}_{\mathcal{S}_{\mathrm{crit}}}) = N_{\mathrm{b}} - b(\vec{\mathcal{H}}_{\mathcal{S}_{\mathrm{rel}}}) \tag{34}$$

with $\vec{\mathcal{H}}_{\mathcal{S}_{\mathrm{crit}}}$ and $\vec{\mathcal{H}}_{\mathcal{S}_{\mathrm{rel}}}$ denoting the vectors that contain the hypotheses of all subcarriers in set $\mathcal{S}_{\mathrm{crit}}$ and $\mathcal{S}_{\mathrm{rel}}$, respectively.

The choice of the design parameter N_{crit} balances performance and complexity.

3.4.1 Discussion

We refer to Fig. 10 again. It shows, among others, the classification reliability of the MAP-RS-D algorithm for different values of N_{crit}. A significant increase of the reliability can be seen, independent of the value of N_{crit}. For $N_{\mathrm{crit}} > 2$, the performance improvement saturates whereas the complexity grows rapidly. In numerous cases at high SNR, only very few decisions are ambiguous. By incorporating additional information, the reliability of these vague decisions can be considerably improved. Against the background of a practical design, small values of N_{crit} are an appropriate choice.

3.5 MAP algorithm with reduced complexity (MinMAP)

The complexity of all presented MAP methods is still rather high and may be prohibitive for real-time applications. In order to reduce the complexity, we take a closer look e. g. at the classification metric (22) illustrated in Fig. 12 which is based on the expression

$$\sum_{k=1}^{K} \ln \left(\sum_{i=1}^{L^{(m)}} \exp\left(-\gamma \cdot |d_{n,k} - H_n S_i^{(m)}|^2\right) \right) - K \cdot \ln L^{(m)} + \ln p(\hat{b}_{n,\mathrm{Rx}}, \mathcal{H}_n^{(m)}) . \tag{35}$$

Both logarithmic and exponential functions in the metric are intensive operations which are

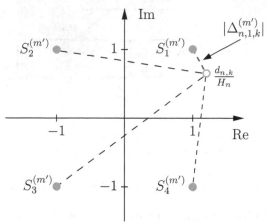

Fig. 12. Example of metric calculation for 4QAM transmission: Dominant contribution term $\exp(-\gamma_n |\Delta_{n,i,k}^{(m)}|^2)$ for fixed n and k arises from closest constellation point

not suitable for implementations in hardware structures. The metric evaluates weighted and squared distances $\Delta_{n,i,k}^{(m)}$

$$\gamma \cdot |d_{n,k} - H_n S_i^{(m)}|^2 = \gamma_n \cdot \left| \frac{d_{n,k}}{H_n} - S_i^{(m)} \right|^2 = \gamma_n \cdot |\Delta_{n,i,k}^{(m)}|^2 \tag{36}$$

between the received equalized symbol and all possible constellation symbols of the modulation scheme under test.

Let us consider the high SNR region obtaining a received equalized symbol $d_{n,k}/H_n$ located close to a possible constellation point. Thanks to the fast decrease of the exponential function for decreasing arguments, only the term $\exp(-\gamma_n \cdot \min\{|\Delta_{n,i,k}^{(m)}|^2\})$ will significantly contribute to the inner sum in (22):

$$\sum_{i=1}^{L^{(m)}} \exp\left(-\gamma_n |\Delta_{n,i,k}^{(m)}|^2\right) \approx \max_{S_i^{(m)} \in \mathcal{I}^{(m)}} \left\{ \exp\left(-\gamma_n |\Delta_{n,i,k}^{(m)}|^2\right) \right\} \tag{37}$$

$$= \exp\left(-\min_{S_i^{(m)} \in \mathcal{I}^{(m)}} \left\{\gamma_n |\Delta_{n,i,k}^{(m)}|^2\right\}\right) . \tag{38}$$

The resulting simplified metric is denoted by MinMAP as the operations $\log(\cdot)$ and $\exp(\cdot)$ are replaced by a simple minimum search. As an example, the MinMAP-RS metric is:

$$\hat{\mathcal{H}}_{n,\text{MinMAP}-\text{RS}}^{(m)} = \arg\max_{\mathcal{H}_n^{(m)}} J_{\text{MinMAP}-\text{RS}}(\mathbf{d}_n, \mathcal{H}_n^{(m)}) \quad \text{with}$$

$$J_{\text{MinMAP}-\text{RS}} = \sum_{k=1}^{K} \ln\left(\exp\left(-\min\left\{\gamma_n |\Delta_{n,i,k}^{(m)}|^2\right\}\right)\right) - K \ln L^{(m)} + \ln p(\hat{b}_{n,\text{Rx}}, \mathcal{H}_n^{(m)})$$

$$= -\gamma_n \sum_{k=1}^{K} \min_{S_i^{(m)} \in \mathcal{I}^{(m)}} \left\{|\Delta_{n,i,k}^{(m)}|^2\right\} - K \ln L^{(m)} + \ln p(\hat{b}_{n,\text{Rx}}, \mathcal{H}_n^{(m)}). \tag{39}$$

No exponential and almost no logarithm operations are needed in the classifier. This leads to an essentially decreased complexity. The algorithms MAP-RS-D, MAP-RQ and MAP-RQ-D are modified alike, i. e. denoted as MinMAP-RS-D, MinMAP-RQ and MinMAP-RQ-D.

Metric analysis for high SNR:

It will be shown that the error caused by the simplification (37) tends to zero (under mild conditions) if $\gamma \to \infty$. For simplicity reasons, we neglect all subcarrier, modulation and block indices. First, we analyze the sum:

$$\ln\left(\sum_{i=1}^{L} e^{-\gamma|\Delta_i|^2}\right) = \ln\left(e^{-\gamma|\Delta_1|^2} + e^{-\gamma|\Delta_2|^2} + \ldots + e^{-\gamma|\Delta_L|^2}\right). \tag{40}$$

Here, we have sorted $|\Delta_i|$ in a ascending order with $|\Delta_1|$ being the minimum and $|\Delta_L|$ being the maximum distance. Now, let us define the error ε between the optimal MAP and the simplified MAP metrics ($K = 1$):

$$\varepsilon = \ln\left(\sum_{i=1}^{L} e^{-\gamma|\Delta_i|^2}\right) - \ln\left(e^{-\gamma|\Delta_1|^2}\right) \tag{41}$$

$$= \ln\left(\frac{\sum_{i=1}^{L} e^{-\gamma|\Delta_i|^2}}{e^{-\gamma|\Delta_1|^2}}\right) \tag{42}$$

$$= \ln\left(\sum_{i=1}^{L} e^{-\gamma(|\Delta_i|^2 - |\Delta_1|^2)}\right) \tag{43}$$

$$= \ln\left(1 + \sum_{i=2}^{L} e^{-\gamma(|\Delta_i|^2 - |\Delta_1|^2)}\right) \geq 0. \tag{44}$$

Under the condition $|\Delta_i| \neq |\Delta_1|$ for $i \neq 1$, the error in the metric function for $\gamma \to \infty$ is:

$$\lim_{\gamma \to \infty} \varepsilon = \ln(1 + 0) = 0, \tag{45}$$

since $|\Delta_i|^2 - |\Delta_1|^2 > 0 \ \forall \ 2 \leq i \leq L$. Numerical results have shown moderate deviations between the simplified and the optimal metrics already for practical SNR ranges.

Assume that the received equalized symbols converge to their constellation points for high SNR. Then e. g. the special case $|\Delta_2| = |\Delta_1|$ or $|\Delta_2| \approx |\Delta_1|$ that we excluded in our consideration so far can only occur if the hypothesis under test is incorrect. Since the neglect of terms in (37) lowers the metric value of this incorrect hypothesis, this condition can even have a favourable effect on the discrimination of the modulation schemes.

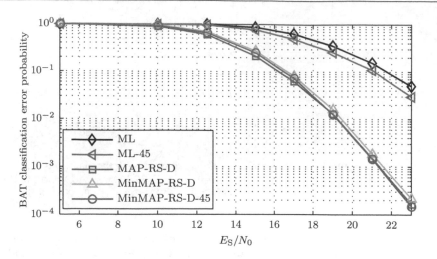

Fig. 13. BAT classification error probability P_{ce} of AMC algorithms versus E_S/N_0, frame length $K = 10$, $N_{crit} = 2$

An overview of the presented metrics is appended at the chapter end in Table 3.

3.5.1 Discussion

Fig. 13 shows the classification performance of the algorithm MAP-RS-D with and without the previously described metric simplification. It indicates that the influence of the metric simplification on the classification performance is minor. Only a small performance degradation is visible which will even decrease with increasing SNR. Obviously, the MinMAP-RS-D approach seems to be an proper tradeoff between performance and complexity.

3.6 QAM symbol rotation

Apart from the algorithm design of the receiver, also the transmitter can be factored into the modulation classification for performance improvements. A simple example is given here: Similar to pattern recognition, automatic classification becomes effective if the objects to discriminate are *as different as possible* and can therefore be easily separated.

However, especially the constellation sets of the higher-order modulation schemes are very *similar*. A large number of received symbols must be observed to judge safely from which set the symbols have been generated. As a first measure to achieve a better separation of the constellation sets we rotate all 16QAM symbols by 45° as shown in Fig. 14 (right). Note that the optimal constellation modifications would take the rotation of all modulation schemes into account.

3.6.1 Discussion

The algorithms using the constellation modification at the transmitter side are denoted as ML-45 and MinMAP-RS-D-45, respectively. A slight performance improvement due to the

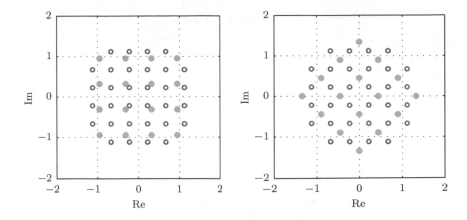

Fig. 14. Constellation diagrams of 32QAM and 16QAM (left) and 32QAM and rotated 16QAM (right)

symbol rotation by 45° can be seen in Fig. 13, especially in case of the ML algorithm. For the advanced techniques, the effect of the symbol rotation becomes less significant.

However, we expect further potential in jointly optimizing the adaptive modulation at the transmitter and the automatic modulation classification at the receiver side.

4. Overall system performance

Finally we analyze the overall system performance of a typical adaptive OFDM-based transmission system which applies AMC. We are primarily interested in the influence of errors caused by imperfect AMC on the packet error ratio (PER). A packet error is observed here if either the BAT or the payload data is detected erroneously. For these PER simulations, a hard-decision Viterbi algorithm decodes the information bits that have been encoded with a convolutional code of rate $R_C = 1/2$. Since each frame consists of 10 data OFDM blocks, the payload size amounts to 10 blocks · 64 subcarriers/block · 4 bit/subcarrier · $R_C = 1280$ bit $= 160$ bytes.

Fig. 15 depicts the PER versus E_S/N_0 for the following four scenarios:

- non-adaptive: The transmitter uses the same modulation scheme on all subcarriers.

- adaptive, ML: The transmitter applies AM; the receiver detects the transmit BAT automatically using the ML algorithm.

- adaptive, MinMAP-RS-D: The transmitter applies AM; the receiver detects the transmit BAT automatically using the MinMAP-RS-D algorithm.

- adaptive, BAT known: The transmitter applies AM; the receiver has perfect knowledge of $\mathbf{B}_{Tx}$ (reference).

For the frequency-selective indoor propagation channel considered here, the transmission system benefits significantly from adaptive modulation; the number of packet errors is lowest in the adaptive case with perfect knowledge of $\mathbf{B}_{Tx}$. Clearly, AMC degrades the system performance due to BAT classification errors. Whereas the degradation in case of the ML

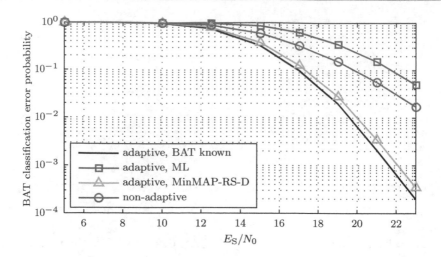

Fig. 15. BAT classification error probability P_{ce} of the MinMAP algorithm versus E_S/N_0, frame length $K = 10$, $N_{crit} = 2$

algorithm is considerable and even overcompensates the benefit from adaptive modulation, the overall PER with perfect knowledge and classified BAT according to MinMAP-RS-D is very similar. The longer the frames, the smaller the influence classification errors on the PER will be. On the one hand, AMC benefits from more symbols to average. On the other hand, payload detection errors will occur more frequently with increasing frame lengths.

By using the presented AMC approach, we can fully gain from adaptive modulation (at costs of a higher complexity) without loss in the effective data rate due to signaling of the BAT, also in wireless communication scenarios.

5. Summary

In this contribution, a framework of likelihood-based automatic modulation classification algorithms for wireless orthogonal frequency division duplex systems with adaptive modulation has been presented. Instead of signaling the bit allocation table to the receiver, the bit allocation table can be efficiently detected solely based upon the received signal and side information which is available in time-division duplex communication systems.

It has been shown that the well-known maximum-likelihood algorithm does not offer a sufficiently high classification reliability in typical wireless communication scenarios. Therefore, an improved maximum-a-posteriori technique has been presented that utilizes additional information, i.e. a fixed bit allocation table per frame, channel reciprocity in time-division duplex systems and the information about the overall data rate. A metric simplification is possible which reduces the computational burden considerably without sacrificing much performance.

By using these advanced classifiers, there is almost no performance loss in terms of the packet error ratio compared to the case with perfect knowledge of the bit allocation table. Due to the moderate computational complexity and high classification reliability even for short packets, the application of automatic modulation classification can be an attractive alternative to conventional signaling schemes.

The automatic modulation classification could be even further improved if the transmitter and receiver signaling processing is considered jointly. A brief example has been given by rotating the symbols of the conventional 16QAM scheme. However, this sophisticated topic of a joint transmitter and receiver design will be part of future research.

Algorithm	Metric
ML	$\sum_{k=1}^{K} \ln\left(\sum_{i=1}^{L^{(m)}} \exp\left(-\gamma_n \cdot |\Delta_{n,i,k}^{(m)}|^2\right)\right) - K \cdot \ln L^{(m)}$
MAP	$J_{\mathrm{ML}}(\mathbf{d}_n, \mathcal{H}_n^{(m)}) + \ln p(\mathcal{H}_n^{(m)})$
MAP-RQ	$J_{\mathrm{ML}}(\mathbf{d}_n, \mathcal{H}_n^{(m)}) + \ln p(\hat{\mathcal{H}}_{n,\mathrm{Rx}}, \mathcal{H}_n^{(m)})$
MAP-RS	$J_{\mathrm{ML}}(\mathbf{d}_n, \mathcal{H}_n^{(m)}) + \ln p(\hat{b}_{n,\mathrm{Rx}}, \mathcal{H}_n^{(m)})$
MAP-RQ-D	two-step procedure in section 3.4 based on MAP-RQ
MAP-RS-D	two-step procedure in section 3.4 based on MAP-RS
MinMAP-RQ	$-\gamma_n \cdot \sum_{k=1}^{K} \min_{S_i^{(m)} \in \mathcal{I}^{(m)}} \left\{|\Delta_{n,i,k}^{(m)}|^2\right\} - K \ln L^{(m)} + \ln p(\hat{\mathcal{H}}_{n,\mathrm{Rx}}, \mathcal{H}_n^{(m)})$
MinMAP-RS	$-\gamma_n \cdot \sum_{k=1}^{K} \min_{S_i^{(m)} \in \mathcal{I}^{(m)}} \left\{|\Delta_{n,i,k}^{(m)}|^2\right\} - K \ln L^{(m)} + \ln p(\hat{b}_{n,\mathrm{Rx}}, \mathcal{H}_n^{(m)})$
MinMAP-RQ-D	two-step procedure in section 3.4 based on MinMAP-RQ
MinMAP-RS-D	two-step procedure in section 3.4 based on MinMAP-RS

Table 3. Metrics overview

6. References

Boiteau, D. & Martret, C. L. (1998). A generalized maximum likelihood framework for modulation classification, *Proc. of IEEE International Conference on Acoustics, Speech and Signal Processing (ICASSP)*, pp. 2165–2168.

Campello, J. (1998). Optimal discrete bit loading for multicarrier modulation systems, *Proc. IEEE International Symposium on Information Theory*, p. 193.

Chen, Y., Häring, L. & Czylwik, A. (2009). Reduction of AM-induced signaling overhead in WLAN-based OFDM systems, *Proc. of the 14th International OFDM-Workshop (InOWo)*, Hamburg, Germany, pp. 30–34.

Chow, P., Cioffi, J. & Bingham, J. (1995). A practical discrete multitone transceiver loading algorithm for data transmission over spectrally shaped channels, *IEEE Transactions on Communications* 43(2/3/4): 773–775.

Czylwik, A. (1996). Adaptive OFDM for wideband radio channels, *Proc. of Global Telecommunications Conference GLOBECOM '96*, pp. 713–718.

Dobre, O. A., Abdi, A., Bar-Ness, Y. & Su, W. (2007). Survey of automatic modulation classification techniques: classical approaches and new trends, *IET Communications* 1(2): 137–156.

Dobre, O., Bar-Ness, Y. & Su, W. (2004). Robust QAM modulation classification algorithm using cyclic cumulants, *Proc. of IEEE Wireless Communication and Networking Conference (WCNC)*, Vol. 2, pp. 745–748 Vol.2.

Edinger, S., Gaida, M. & Fliege, N. J. (2007). Classification of QAM signals for multicarrier systems, *Proc. of the European Signal Processing Conference (EUSIPCO)*, pp. 227–230.

Fischer, R. F. H. & Huber, J. B. (1996). A new loading algorithm for discrete multitone transmission, *Proc. of IEEE Global Telecommunications Conference GLOBECOM'96*, pp. 724–728.

Häring, L., Chen, Y. & Czylwik, A. (2010a). Automatic modulation classification methods for wireless OFDM systems in TDD mode, *IEEE Trans. on Communications* (9): 2480 – 2485.

Häring, L., Chen, Y. & Czylwik, A. (2010b). Efficient modulation classification for adaptive wireless OFDM systems in TDD mode, *Proc. of the Wireless Communications and Networking Conference*, Sydney, Australia, pp. 1–6.

Häring, L., Chen, Y. & Czylwik, A. (2011). Utilizing side information in modulation classification for wireless OFDM systeme with adaptive modulation, *Proc. of the IEEE Vehicular Technology Conference 2011-Fall*, San Francisco, USA.

Hsue, S. Z. & Soliman, S. S. (1989). Automatic modulation recognition of digitally modulated signals, *Proc. of IEEE MILCOM*, pp. 645–649.

Huang, Q.-S., Peng, Q.-C. & Shao, H.-Z. (2007). Blind modulation classification algorithm for adaptive OFDM systems, *IEICE Trans. Commun.* E.90-B No. 2: 296–301.

Hughes-Hartogs, D. (1987). Ensemble modem structure for imperfect transmission media, *U.S. Patent 4,679,227* .

IEEE (2005). IEEE 802.11n, *Technical report*, http://grouper.ieee.org/groups/802/11.

Lampe, M. (2004). *Adaptive Techniques for Modulation and Channel Coding in OFDM Communication Systems*, PhD thesis.

Long, C., Chugg, K. & Polydoros, A. (1994). Further results in likelihood classification of QAM signals, *Proc. of IEEE MILCOM*, pp. 57–61.

Medbo, J. & Schramm, P. (1998). Channel models for HiperLAN/2 in different indoor scenarios, ETSI/BRAN document no. 3ERI085B.

Nandi, A. & Azzouz, E. (1998). Algorithms for automatic modulation recognition of communication signals, *IEEE Trans. on Communications* 46(4): 431–436.

Polydoros, A. & Kim, K. (1990). On the detection and classification of quadrature digital modulations in broad-band noise, *IEEE Trans. on Communications* 38(8): 1199–1211.

Sills, J. A. (1999). Maximum-likelihood modulation classification for PSK/QAM, *Proc. of IEEE MILCOM*, pp. 57–61.

Starr, T., Cioffi, J. M. & Silverman, P. J. (1999). *Understanding Digital Subscriber Line Technology*, Prentice Hall.

Swami, A. & Sadler, B. M. (2000). Hierarchical digital modulation classification using cumulants, *IEEE Trans. on Communications* 48: 416–429.

Wei, W. & Mendel, J. (2000). Maximum-likelihood classification for digital amplitude-phase modulations, *IEEE Trans. on Communications* 48(2): 189–193.

8

Super-Broadband Wireless Access Network

Seyed Reza Abdollahi, H.S. Al-Raweshidy and T.J. Owens
WNCC, School of Eng. and Design,
Brunel University, Uxbridge, London
UK

1. Introduction

Today's communication network deployment is driven by the requirement to send, receive, hand off, and deliver voice, video, and data communications from one end-user to another. Current deployment strategies result in end-to-end networks composed of the interconnection of networks each of which can be classified as falling into one of three main categories of network: core, metropolitan and access network. Each component network of the end-to-end communication network performs different roles. Nowadays, the increase in the number and size of access networks is the biggest contributor to the rapid expansion of communication networks that transport information such as voice, video and data from one end-user to another one via wired, wireless, or converged wired and wireless technologies. Such services are commonly marketed collectively as a triple play service, a term which typically refers to the provision of high-speed Internet access, cable television, and telephone services over a single broadband connection. The metropolitan networks perform a key role in tripleplay service provision in delivering the service traffic to a multiplicity of access networks that provide service coverage across a clearly defined geographical area such as a city over fiber or wireless technologies infrastructure. The core networks or long haul networks are those parts of the end-to-end communication network that interconnect the metropolitan area networks. The core network infrastructure includes optical routers, switches, multiplexers and demultiplexers, used to deliver triple play service traffic to the metropolitan networks and route traffic from one metropolitan network to another.

Fig. 1, shows a simplified diagram of network connecting tripleplay service providers to end-users of the service. In this network, the uplink traffic from the end-users is input to the network via wireless or wired access network connections in the user's home. The packets associated with this traffic are multiplexed together and forwarded to the local metropolitan network for delivery to a long haul network for transporting to the service providers' access network and hence to the service provider. The downlink traffic from different service providers which is typically traffic corresponding to requested services is input to the network via local access network connections in the service provider premises. The downlink traffic from a particular access network is multiplexed together and delivered to the local metropolitan network for forwarding to a core network (or in some cases another metropolitan network) and hence to the end-users access networks for delivery to the end users. As many access networks are connected to a metropolitan network the traffic data rates throughout a metropolitan network are significantly higher than those throughout an access network. As many metropolitan networks feed traffic into a core network the traffic handling capabilities of a core network are significantly higher than those of a metropolitan network. The network traffic on core networks is expected to reach the order of hundreds exabytes in the near future, (Laskar et al., 2007). The rapidly changing face of networked communications has seen a continued growth in the need to transfer enormous amounts of information across large distances. A consequence of this is that technologies that are used extensively for transferring information such as coaxial cable, satellite, and microwave radio are rapidly running out of spare capacity, (Mcdonough, 2007).

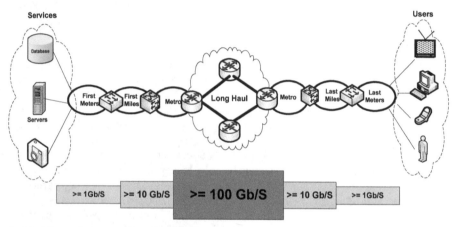

Fig. 1. Near term future network capacity requirements.

Therefore, transportation of the traffic volumes that will be demanded by users in the near future will require significantly greater network transmission bandwidth than that provided by the current infrastructure. Consequently, in the near term each category of component network of existing end-to-end networks will face different and increasingly difficult challenges with respect to transmission speed, cost, interference, reliability, and delivery of the demanded traffic to or from end-users. Currently, super-broadband penetration and the on-going growth in the internet traffic to and from business and home users are placing a huge bandwidth demand on the existing infrastructure.

Broadband wireless sits at the confluence of two of the most remarkable growth stories of the telecommunication industry in recent years. Wireless and broadband have each enjoyed rapid mass-market adoption. Wireless mobile services grew from 11 million subscribers worldwide in 1990 to more than 5 billion by the end of 2010. The world's largest manufacturer of mobile phones has forecast that the number of mobile users accessing the internet via mobile broadband will grow to over 2 billion globally by the end 2014. Fixed broadband subscribers numbered only 57,000 in 1998 and rapidly increased to 555 million subscribers by the end of 2009. The number of fixed broadband subscribers is projected to exceed 720 million by 2015 despite the current economic situation, (OASE, 2010; ITU, 2011). The growth in the numbers of mobile telephone subscribers, broadband and internet users over the last decade and the projections for the growth in these numbers are depicted in Fig. 2.

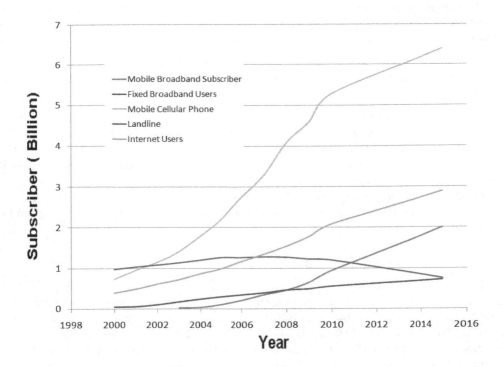

Fig. 2. Worldwide subscriber growth in the numbers of mobile telephony, internet, and broadband access users.

It follows that the demand for use of the available radio spectrum is very high, with terrestrial mobile phone and broadband internet systems being just one of many types of access technology vying for bandwidth. Mobile telephony and internet applications require the systems that support them to operate reliably in non-line-of-sight environments with a propagation distance of 0.5-30 km, and at velocities up to 100 km/h or higher. These operating environment constraints limit the maximum radio frequency the systems can use as operating at very high frequencies, i.e. approaching microwave frequencies, results in excessive channel path loss, and excessive Doppler spread at high velocity. This limits the spectrum suitable for mobile applications making the value of the radio spectrum extremely high. As an example, in Europe auctions of 3G licenses for the use of radio spectrum began in 1999. In the United Kingdom, 90 MHz of bandwidth was auctioned off for £22.5 billion (GBP). In Germany, the result was similar, with 100 MHz of bandwidth raising $46 billion (US). This represents a value of around $ 450 million (US) per MHz. The duration of these license agreements is 20 years. Therefore, it is vitally important that the spectral efficiency of the communication system should be maximized, as this one of the main limitations to providing low cost high data rate services, (OMEGA ICT Project, 2011; Yuen et al., 2004). By deploying converged fiber and wireless communication (Fi-Wi) technologies, network operators and service providers can meet the challenges of providing low cost high data rate services to wireless users. Only the relatively huge bandwidth of a fiber-optic access network can currently support low cost high data rate services for wired and wireless users.

This chapter makes the case for radio over fiber (RoF) networks as a future proof solution for supporting super-broadband services in a reliable, cost-effective, and environmentally friendly way.

This chapter is organized as follows: In Section II, the evolution of Internet traffic driven by the growth in wired and wireless subscribers worldwide is discussed. In Section III, solutions for cost effective transportation of traffic volumes in line with the demand expected as a result of anticipated growth in interactive video, voice communication and data services are presented. In Section IV, the radio over fiber (RoF) network as a future proof solution for supporting super-broadband services is described as a reliable, cost-effective and environmentally friendly technology. Finally, concluding remarks are given in Section V.

2. Evolution of data traffic and future demand

Globally, mobile communication data traffic is expected to increase 26-fold between 2010 and 2015 and reach 6.3 exabytes per month by 2015. Furthermore, the compound annual growth rate (CAGR) of mobile data traffic is expected to reach 92 percent over the period 2010 to 2015. Moreover, during 7 years from 2005 to 2012 mobile data traffic will have increased a thousand-fold. In 2010, about 49.8% of mobile data traffic was video traffic. By deploying a converged fiber and wireless communication (Fi-Wi) technologies, the operators and service providers can meet the challenges they face from the continued dramatic growth in mobile data traffic volumes.

By the end of 2011, video traffic over mobile networks reached about 52.8% of the total traffic on mobile networks. It is expected that almost 67% of the world's total mobile traffic will be video by 2015 and that the volume of video traffic on mobile networks will have doubled every year over the period 2010 to 2015, (FP7, 2010, Cisco Visual Networking Index, 2011). In Fig. 3, the worldwide growth in data traffic rates per month are compared for mobile terminals and other devices. Fig. 3 (a) shows the anticipated growth of data traffic by user terminal type for the following terminal types: tablets, machine-to-machine (M2M), home gateways, smartphones, laptops, non-smartphones, and other portable devices. It is predicted that in 2015 82.4% of all network data traffic, about 5.768 exabbytes per month, will be being transported to and from just by two types of portable wireless devices. Specifically, it is predicted that 55.8% and 26.6% of all network data traffic will relate to laptop and smartphone users, respectively. As shown in Fig. 3 (b), the expectation is that the data traffic rate relating to mobile devices will be about 6.3 exabaytes per month by the end of 2015, (Cisco Visual Networking Index, 2011).

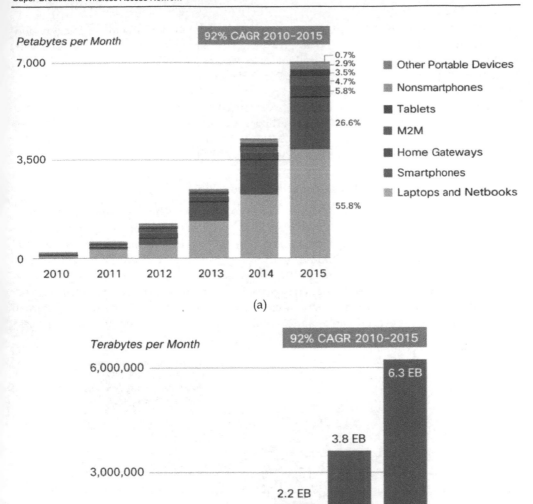

Fig. 3. The anticipated growth of data traffic (a): by user terminal type , (b) forecast of mobile data traffic growth by 2015, (Cisco Visual Networking Index, 2011).

High-Definition Television (HDTV) can now be provided in many countries throughout the world while Ultra High Definition Television (UHDTV) is now being studied in Japan as the most promising candidate for next-generation television beyond HDTV, and Super-High-

Definition Television (SHDTV). UHDTV consists of extremely high-resolution imagery and multi-channel 3D video and sound to give viewers a stronger sensation of presence. The UHDTV project's commercializing outlook is to become available in domestic homes over the period 2016 to 2020. For example, in 2005, NHK demonstrated a live relay of a UHDTV program using dense wavelength division multiplexing (DWDM) with 24 Gbit/s speed over a distance of 260 km on a fiber optic network. In 2006 NHK demonstrated a solution for bandwidth efficient delivery of UHDTV, utilizing a codec developed by NHK the video was compressed from 24 Gbit/s to 180–600 Mbit/s and the audio was compressed from 28 Mbit/s to 7–28 Mbit/s, (Sugawara et al., 2007; Kudo, 2005).

3. Deployment of super-broadband services

Globally the evolution of internet video services will be in the three following phases: 1) experiencing a growth of internet video as viewed on the PC, 2) internet delivery of video to the TV, and 3) interactive video communications, Fig. 4. Considering the future ultra high, super high and high definition resolution of end-user demanded and generated data traffic, each phase will impact on a different aspect of the end-to-end delivery network such as bandwidth, spectral efficiency, cost, power consumption, architecture, and technology. In addition to internet video, there is very high growth in the internet protocol (IP) transport of cable and mobile IPTV, and video on-demand services, (OASE, 2010).

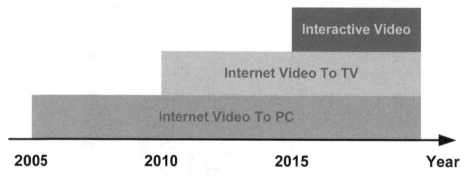

Fig. 4. Three waves of consumer Internet traffic growth.

Mobile voice services are already considered a necessity by many end-users, and mobile data, video, and TV are now becoming an essential part of some end-users' lives. The number of mobile subscribers' is growing rapidly and is expected to reach over 6.2 billion subscribers by 2015. Mobile users' bandwidth demand due to video services is increasing. Therefore, there is an essential need to increase the capacity of delivery networks for mobile broadband, data access, and video services to retain subscribers as well as keep cost in check.

Major considerations in planning the deployment of next-generation mobile networks are an increasing need for service portability and interoperability driven by the proliferation of mobile and portable digital devices and an accompanying need for the networks to enable such devices, including smartphones, tablets, laptops, and non-smartphones, to connect to them seamlessly. The expansion of wireless ubiquity will result in increasing numbers of consumers depending on mobile networks creating a need for increasing economies of scale to deliver lower cost per-bit. According to a prediction of future combined consumer and advertiser spend on mobile media and associated data, which includes handset browsing, mobile applications, mobile games, mobile music, mobile TV, ringtones, wall papers and alerts, spend will rise from just under $75 billion at the end of 2010 to $138 billion by 2015, at a 13.17 CAGR, (MacQueen, 2010). Moreover, it is predicted (RNCOS Industry Research Solution, 2011) that the number of mobile TV subscribers worldwide will grow at a CAGR of around 43% during 2011-2014 to reach about 792.5 million by the end 2014.

In response to this remarkable development, core and metro networks have experienced a tremendous growth in bandwidth and capacity with the widespread deployment of fibre-optic technology over the past decade, (OASE, 2010). Fiber optic transmission has become one of the most exciting and rapidly changing fields in telecommunication engineering. Fiber optic communication systems have many advantages over more conventional transmission systems. They are less affected by noise, are completely unaffected by electromagnetic interference (EMI) and radio frequency interference (RFI), do not conduct electricity and therefore, provide electrical isolation, are completely unaffected by lightning and high voltage switching, and carry extremely high data transmission rates over very long distances, (Guo et al., 2007). As shown in Fig. 1, data speeds in metro and long-haul systems are evolving from 10 Gbps to 40 Gbps transmission. A 100 Gbps per wavelength channel system is taking shape as a next step for core and metro networks, (FP7, 2010). Wavelength division multiplexing (WDM) techniques, such as: dense WDM (DWDM), and highly DWDM (HDWDM) offer the potential for huge bandwidth fiber optic networks with all-optical switching and routing in the future.

In the recent years wireless services have been taking a steadily increasing share of the telecommunications market. End users not only benefit from their main virtue, mobility, but are also demanding ever larger bandwidth. Larger wireless capacity per user requires the reduction of the wireless cell size, i.e. establishing pico-cells. These can be realised using Wi-Fi systems based on the wireless Local Area Network (LAN) IEEE 802.11n standard which offers data rates of up to 600 Mbit/s. Furthermore, the Wi-Fi Alliance and the Wireless Gigabit Alliance (WiGig) announced that they will cooperate on multi-gigabit wireless schemes that are likely to bring robust wireless networking from the 60 GHz frequency band to consumers whose devices are equipped with Wi-Fi. The partnership will pave the way for new wireless devices that will operate in the 2.4, 5 and 60 GHz bands. It is anticipated that data transfer rates up to 7 Gbps can be achieved, although the highest data rates are is likely to be available only over short distances within living room-sized areas. Nevertheless, the highest rates will be more than 10 times faster than 802.11n (Anthony, 2011). Furthermore, Worldwide interoperability for Microwave Access second generation (WiMAX 2), the marketing name for systems based on the IEEE 802.16m standard, is expected to expand capacity to 300 Mbps peak rates via advances in antennas, channel stacking and frequency re-use over the period 2012 to 2013, (Schwarz, 2011). Looking further ahead the recently ratified IEEE 802.15.3c standard has been defined for the frequency band of 57.0–66.0 GHz, allocated by regulatory agencies in Europe, Japan, Canada, and the United

States. According to this standard, single carrier mode in millimeter wave PHY supports a variety of modulation and coding schemes (MCSs) that support up to 5 Gb/s, (Guo and Kuo, 2007).

Super-broadband access not only provides faster web surfing and quicker file download, but also enables several multimedia applications such as real-time high definition audio and video streaming, multimedia conferencing, and interactive gaming. Broadband connections are currently being used for voice telephony using Voice-over-Internet-Protocol (VoIP) technology. More advanced broadband access systems, such as fiber to the home (FTTH) and very high data rate digital subscriber line (VDSL), enable applications such as entertainment–quality video, including HDTV, and Video on Demand (VoD) to be provided, but for SHDTV and UHDTV services a super-broadband network is essential. As the broadband market continues to grow, several new applications are likely to emerge and it is difficult to predict which ones will succeed in the future.

Broadband wireless is about bringing the broadband experience to a wireless context, which offers users certain unique benefits and convenience. There are two fundamentally different types of broadband wireless services. The first attempts to provide a set of services similar to that of the traditional fixed-line broadband but using wireless as the medium of transmission. This type, called fixed wireless broadband, can be thought of as a competitive alternative to DSL or cable modem. The second type of broadband wireless, called mobile broadband, offers the additional functionality of connectivity in mobility. Mobile broadband attempts to bring broadband applications to new user experience scenarios and hence can offer the end-user a very different value proposition.

Long Term Evolution (LTE) is a new radio platform technology that will allow operators to achieve even higher peak throughputs than High Speed Packet Access evolution (HSPA+) in higher spectrum bandwidth. Furthermore, the overall objective for LTE is to provide an extremely high performance radio-access technology that offers full vehicular mobility and can readily coexist with HSPA and earlier networks. Because of scalable bandwidth, operators will be able to migrate their networks and users from HSPA to LTE easily over time. LTE assumes a full IP network architecture, (Rysavy Research, 2007).

Fig. 5 shows the evolution of the 3GPP family of standards towards LTE Advanced (Chang et al., 2007; Rodrigo et al., 2009). LTE uses OFDMA (Orthogonal Frequency Division Multiplexing Access) on the downlink and FDMA (Frequency Division Multiple Access) on the uplink for better power performance of the end-user's handset, which is well suited to achieving high peak data rates in high spectrum bandwidth, achieving peak rates in the 1 Gbps range with wider radio channels. However, wider channels would result in highly complex terminals and is not simply achievable with the conventional communication infrastructure. Moreover, access bandwidth requirements for delivering multi-channel HDTV, SHDTV, and UHDTV signals and online gaming services are expected to grow beyond several Gbps in the near future and the current subscriber access networks have not been scaled up commensurately. To avoid being the bottleneck in the last miles and last meters, and exploit the benefits of both wired and wireless technologies, mobile and wireless communication service providers and operators are actively seeking convergent network architecture to deliver multiple super-broadband services to serve both fixed and mobile users, (Nokia, 2009; PIANO+, 2010).

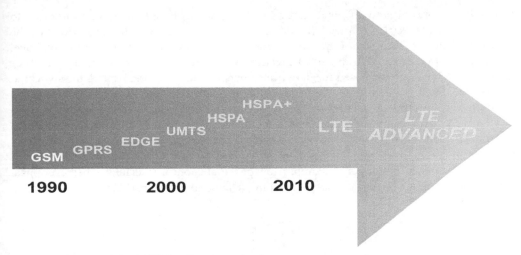

Fig. 5. Evolution of the 3GPP family of standards.

In this regard, optical-wireless access technologies have been considered the most promising solution to increase the capacity, coverage, bandwidth, and mobility in environments such as conference centers, airports, hotels, shopping malls, and ultimately to homes and small offices. As a result, research activity in the field of optical networks and converged optical and wireless communication technologies has grown rapidly and steadily over the last several years. This is because optical communication is a promising choice to fulfill the ever-increasing demand on bandwidth via the vast available capacity of optical fiber and its economic cost. Wireless communication technology on the other hand can provide mobility during communication periods and it is entering a new phase where the focus is shifting from voice to high definition multimedia services. Present consumers are no longer interested in the underlying technology; they simply need reliable and cost effective communication systems that can support end-users' demanded services anytime, anywhere, any media, that they want.

3.1 Core and metro networks

The two main categories of network to be considered from the point of view of establishing super-broadband access networks are core and metro networks. In this subsection, the two main challenges facing core and metro networks are discussed. These challenges are realising the bandwidth potential of fiber optic core networks by appropriate wavelength allocation and switching strategies. Therefore in this subsection, the discussion focussed on optical switching paradigms and dynamic wavelength allocation.

The main barrier to the use of most existing core and metro networks for future traffic transportation arises from their active electrical switching and routing systems which delay packets when processing them for switching in the electrical domain. It takes time to convert a signal from the optical to the electrical domain and vice versa. In addition the synchronization and data retiming processing takes time. Indeed, a great part of the research into optical networks is dedicated to transparency in optical networks in order to bypass Optical/Electrical/Optical (O/E/O) conversions in the intermediate nodes of the network. Thus, a number of network protocols such as MPLS (Multi Protocol Label Switching, GMPLS, etc. (Larkin, 2005) together with switching strategies (circuit- burst- or packet-switching) are proposed for data transparency in the network. Among the switching strategies, burst switching is the most compatible with the current optoelectronic technologies in terms of data transparency and switching speed. Packet switching is more efficient for data communication, but due to the limited speed of electrical networks compared to the current optical networks and the insufficient evolution of all-optical signal processing alternatives, packet based optical networks are not a practical solution for transparent optical networks. A comparison of the all optical switching schemes, optical circuit switching (OCS), optical burst switching (OBS) and optical packet switching (OPS) is shown in Fig. 6.

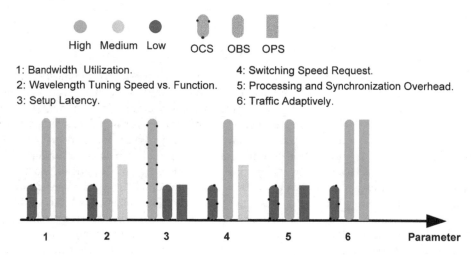

Fig. 6. Comparison of all-optical switching technologies in terms of relative magnitudes of performance measures.

Optical packet switching (OPS) is a viable candidate switching scheme for future networks because it is a purely-connectionless networking solution that is fully compatible with IP-centric data traffic and offers the finest network granularity, the best bandwidth utilization, flexibility, high-speed, and the ability to use the resources available economically.

OPS places more demanding prerequisites on the network than OBS because it processes packets on the fly. The most feasible approach to implementing of OPS involves processing synchronously transmitted packets with fixed lengths. However, in this case the hardware overhead is on the implementation of the packet synchronizer at the input to the switch. Despite their feasibility limitations, OPS demonstrators assisted the development of numerous ultra-fast switching and processing techniques regarding wavelength conversion, header encoding/decoding and processing, label swapping, fast clock extraction, and regeneration.

The main challenges in OPS are the implementation of the optical header processing mechanism, the development of an intelligent switch controller, the realization of ultra fast switching at a nanosecond timescale, and the exploitation on buffering mechanisms to reduce packet blocking (Rodrigo et al, 2009; Raffaelli et al. 2008; Le Rouzic et al., 2005).

Furthermore, the channel allocation and spectral efficiency are other key points for super-broadband network deployment. There are different schemes for channel allocation and multiplexing techniques such as wavelength division multiplexing (WDM), Dense-WDM(DWDM), Highly DWDM, Orthogonal WDM (Goldfarb et al., 2007; Llorente et al., 2005) that are suitable for super-broadband network deployment. The WDM multiplexing based schemes are in addition to multiplexing schemes including time, frequency, and code division multiplexing techniques, which are used in current wired and wireless communication networks and perform well on them. Moreover, cognitive channel and spectrum allocation improves the network's throughput and reduces the cost-over head significantly.

3.2 Access network

Ultra-fast and super-broadband are recognized as becoming increasingly important as demands for bandwidth multiply. Investment in the development of next-generation optical-wireless converged access technologies will enable a future network to be deployed that will radically reduce Fiber-Wireless (FiWi) infrastructure costs by removing local exchanges and potentially much of the metro network. To integrate fiber and wireless technologies, there are important challenges. First, it will be crucial to have mechanism in place to control system load, which will translate into the physical characteristics of the different radio access technologies of wireless systems, the variability of users' requirements and the data rate of on-going wireless connections, complicating the resource management/sharing in FiWi access networks. This raises technical issues such the required protocol interfaces between the resource management entities of tightly coupled networks, and calls for the design of very flexible and effective protocols to allow enhanced routing and link adaptation that makes the best usage of the available resources while dynamically accommodating the users' traffic properties and quality of services requirements.

3.2.1 Passive optical network

There are different topologies for deploying the fiber network from a central exchange station to end-user's premises such as: 1) point-to-point (P-to-P): where individual fibers run from the central station to end-users, 2) point-to-multi-point (P-to-MP) active star architecture: where a single feeder fiber carries all traffic to an remote active node close to the end-users, and from there individual short branching fibers run to the end-users. In this architecture, the fiber network implementation cost is less than that of a point-to-point topology but the main disadvantages of this architecture are a) the bandwidth of the feeder fiber is shared between several end-users and the allocated dedicated bandwidth for each end-user is less than in the point-to-point architecture. b) the requirement for active equipment in a remote node will impose some restrictions on network deployment such as the availability of a reliable and uninterruptable power supply, proper space for installation of active equipment, air conditioning and ventilation, and maintenance costs, 3) point-to-multi-point passive star architecture: in which the active node of the active star topology is replaced by a passive optical power splitter/combiner that feeds the individual short range fibers to end-users. This topology has become a very popular and is known as the passive optical network (PON). In this topology, in addition to the reduction in installation cost, the active equipment is completely replaced by passive equipment avoiding the powering and related maintenance costs, (Koonen, 2006).

Besides the technical issues of implementation, the maintenance and operation cost overhead should be accounted for as it plays a key role in choosing a particular architecture. In the P-to-P architecture, for each end-user, two dedicated optical line terminations (OLT) are needed, while, in the P-to-MP scheme, for each end-user one dedicated OLT is required at the end-user side, another shared OLT at the central station is interfaced between several end-users. When the number of customers increases, the system costs of the P-to-P architecture grow faster than those of the P-to-MP architecture, as more fibers and more line terminating modules are needed. Therefore, sharing the implemented infrastructures between several operators, service providers, technologies, and end users is an essential solution to reduce the infrastructure network cost overhead. . As shown in Fig. 7, the initial cost of P-to-P topology ($Cost_{P-to-P}(N_1)$) for N_1 users is lower than initial cost of P-to-MP topology ($Cost_{P-to-MP}(N_1)$), while by increasing the duct length at point L_0, the $Cost_{P-to-P}(N_1)$ crosses the $Cost_{P-to-MP}(N_1)$ graph and will be greater than it for fibre lengths greater than L_0. Furthermore, the initial cost of P-to-MP topology for N_2 users, where $N_2 > N_1$, is more cost effective than the initial cost of P-to-P topology for N_2 users.

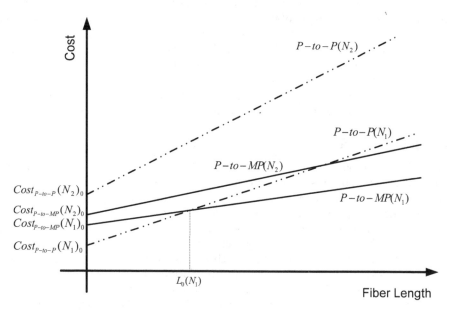

Fig. 7. The comparison of systems cost of FTTH different topology networks versus duct length to end-users premises.

In the P-to-P and P-to-MP active star architectures, each fiber link are only carries a data stream between two electro-optic converters, and the traffic streams of the end-users are multiplexed electrically at these terminals. Therefore, there is no risk of collision of optical data streams. Whereas, the traffic multiplexing is done optically in a Passive Optical Network (PON) topology by integration of the data streams at the passive optical power combiner; to avoid collisions between individual data streams it is necessary to implement a well-designed multiplexing technique. A model of WDM PON network is shown in Fig. 8.

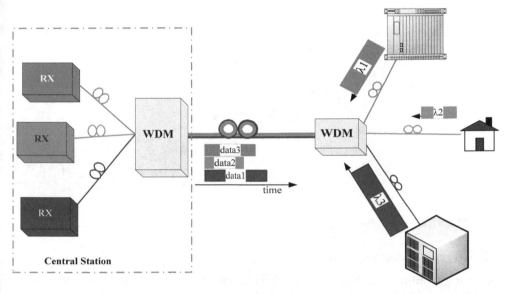

Fig. 8. A model of a point-to-multi-point passive optical network topology.

Several multiplexing techniques are used in PON networks, such as time division multiple access (TDMA), subcarrier multiple access (SCMA), wavelength division multiple access (WDMA), and optical code division multiple access (OCDMA). Excluding, the wavelength division multiplexing technique, these multiplexing techniques are available in wireless or wired telecommunication systems. As shown in Fig. 9, in a WDM PON, each optical network unit (ONU) uses a different wavelength channel to send its packets to an OLT in a central office. The wavelength channels can be routed from the OLT to the appropriate ONUs and vice versa by a wavelength demultiplexing/multiplexing device located at the PON splitting point. This wavelength multiplexing technique constitutes independent communication channels and the network could be able to transport different signal formats; even if the channels use different multiplexing techniques no time synchronization between the channels is needed.

Currently Fiber to the home (FTTH) access technologies provide huge bandwidth to users, but are not flexible enough to allow roaming connections. On the other hand, wireless networks offer mobility to users, but do not possess sufficient bandwidth to meet the ultimate demand for multi-channel video services with high definition quality. Therefore, seamless integration of wired and wireless services for future-proof access networks will lead to a convergence to high bandwidth provision for both fixed and mobile users in a single, low-cost transport platform, This can be accomplished by using the developed hybrid optical and wireless networks, which not only can transmit signals received wirelessly over fiber at the BS, but also simultaneously provide services received over fiber to wireless the end users.

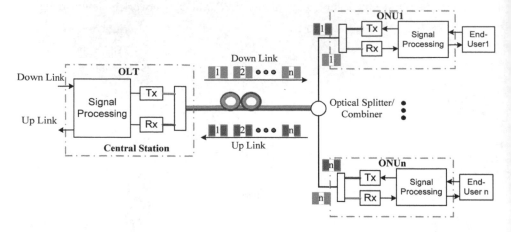

Fig. 9. WDM over a passive optical network.

3.2.2 Dynamic wavelength allocation

By creating multiple wavelengths in a common fiber infrastructure, the capabilities of this infrastructure can be extended into an additional dimension. This wavelength dimension can implement independent communication planes between nodes. For example, interconnections in this plane can be asynchronous, have different quality-of-service requirements, and can transport signals with widely differing characteristics. By using the WDM technique, the access network can: 1) separate services; 2) separate service providers; 3) enable traffic routing; 4) provide higher capacity; 5) improve scalability. For assignment of wavelengths to channels the system may follow different scenarios such as: a) static allocation; b) semi-static allocation; c) dynamic allocation, (Urban et al., 2009).

The static wavelength multiplexing scheme sets a virtual P-to-P topology up between two nodes of the network. However, the rapid growth in access network traffic requires flexible and adaptive planning of the wavelength allocation to each different channel or wired and wireless service to avoid congestion resulting from variable data rates demanded or to guaranteed data traffic transportation or services to/from the end-users. By using adaptive wavelength allocation deliverable services will be more cost-effective on the same network and the vast potential bandwidth of fiber optical networks will be more fully exploited. By assigning the wavelength dynamically at the Optical Network Unit (ONU), with flexible wavelength routing, the access network capabilities can be considerably enhanced. This configuration allows setting up a new wavelength channel before breaking down the old one. Alternatively, it may use wavelength tuneable transmitters and receivers, which can in principle, address any wavelength in a certain range. The network management and control system commands to which downstream and to which upstream wavelength channel each ONU transceiver is switched. By issuing these commands from a central station, the network operator actually controls the virtual topology of the network, and thus is able to allocate the networks resources in response to the traffic at the various ONU sites. By changing the wavelength selection at the ONUs, the network operator can adjust the system's capacity allocation in order to meet the local traffic demands at the ONU sites.

In this scenario, as soon as the traffic to be sent upstream by an ONU grows and does not fit anymore within its wavelength channel, the network management system can command the ONU to be allocated another wavelength channel, in which sufficient free capacity is available. Obviously, this dynamic wavelength reallocation process reduces the system's blocking probability, i.e. it allows the system to handle more traffic without blocking and thus it can increase the revenue of the operator from a given pool of communication resources at the central station.

4. Radio over fiber network

The deployment of optical and wireless access network infrastructure is starting to proliferate throughout the world. When these heterogeneous access networks converge to a highly integrated network via a common optical feeder network, network operators can reap the benefits of lowering the operating costs of their access networks and meeting the capital costs of future upgrades more easily. In addition, the converged access network will facilitate greater sharing of common network infrastructure between multiple network operators. Signals received wirelessly and transported over optical fiber (RoF) links will be a possible technology for simplifying the architecture of remote base stations (BSs). By relocating key functions of a conventional BS to a central location, BSs could be simplified into remote antenna units that could be inter-connected with the central office (CO) via a high performance optical fiber feeder network.

Wireless networks typically show considerable dynamics in the traffic loads of their radio access points (RAPs) due to the fluctuations in the number and nature of mobile and wireless services demanded by the networks users. Using the traditional RAPs approach this requires all the wireless nodes to be equipped to cater for the highest capacity likely to be demanded of them which results in the inefficient use of network resources. The design of dynamic reconfigurable micro/pico or femo wireless cells increases network complexity but can significantly increase network efficiency. Similarly, within the optical access network layer WDM PONs allow an extra level of reconfiguration as wavelengths can be assigned either by static or dynamic routing.

The numbers of wireless subscribers are increasing and these subscribers are demanding more capacity for ultra-high data rate transfer at speeds of 1Gbps and up while the radio spectrum is limited. This requirement of more bandwidth allocation places a heavy burden on the current operating radio spectrum and causes spectral congestion at lower microwave frequencies. Millimetre Wave (mm-Wave) communication systems offer a unique way to resolve these problems (Ji, et al. 2009). Radio over fiber (RoF) technology is currently receiving a lot of attention due to its ability to provide simple antenna front ends, increased capacity, and multi wireless access coverage.

An analog RoF (ARoF) also known as RoF is the technique of modulating a radio frequency (RF) sub-carrier onto an optical carrier for distribution over a fiber network. An ARoF link includes optical source, modulator, optical amplifier and filters, optical channel and a photodiode as a receiver, electronic amplifiers and filters; a simple ARoF architecture is shown in Fig. 10. In this system, for a downlink at a central station, a signal received wirelessly is modulated onto an optical carrier generated by a laser diode (LD) and the modulated optical signal is transported over a fiber optic cable. The transported optical signal is detected at base station using a photo diode (PD). The received signal, recovered after performing analog signal processing, is fed to an antenna for wireless transmission. For uplink signal transmission from a base station to the central station, the signal received at an antenna is directed to a low noise amplifier (LNA) and modulated onto an optical carrier that is generated by another LD. The generated optical signal is sent back to the central station for any signal processing and detection. In some cases the RF signal is directly modulated by optical source, but as the laser is usually a significant source of noise and distortion in a radio over fiber link, the laser diode normally exhibits nonlinear behavior. When the LD is driven well above its threshold current, its input/output relationship can be modeled by Volterra series of order 3. Therefore, in high data rate links indirect modulation has better performance. However, an ARoF link suffers from the nonlinearity of both microwave and optical components that constitute the optical link (Al-Raweshidy & Komaki, 2002; Cox, 2004; Li & Yu, 2003). Fig. 11, shows an ARoF link architecture with indirect intensity modulation that uses an electro-optical modulator for modulating an electrical signal representing the information in a wireless signal onto a continuous wave laser source.

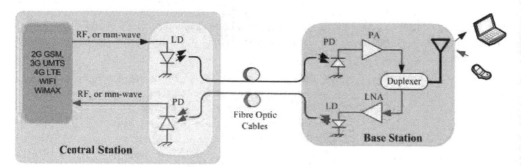

Fig. 10. A direct intensity modulation and detection full-duplex ARoF architecture.

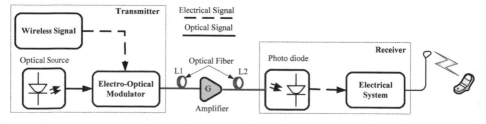

Fig. 11. Downlink architecture of a ARoF link with indirect intensity modulation.

The ARoF technique has been considered a cost-effective and reliable solution for the distribution of future services to wireless devices by using optical fiber with vast transmission bandwidth capacity. An ARoF link is used in remote antenna applications to distribute signals to a Microcell or Picocell base station (BS). The downlink RF signals are distributed from a central station (CS) to a BS known as a Radio Access Point (RAP) through optical fibers. The uplink signals received at RAPs are sent back to the CS for any signal processing. RoF has the following main features: (1) it is transparent to bandwidth or modulation techniques; (2) it only needs simple and small BSs; (3) centralized operation is possible; (4) it supports multiple wired and wireless standards, simultaneously. (5) its power consumption is relatively low. Furthermore, the implementation of the RoF technique faces the following challenges: fiber optic network implementation cost, optical communication components nonlinearity and fiber dispersion. Consequently, in last decade several research projects have sought to develop and discover new solutions to overcome these challenges and broaden the benefits of RoF.

4.1 Radio over fiber's link architecture

The signal that is transmitted over the optical fiber can either be originally an RF, intermediate frequency (IF) or baseband (BB) signal. For the IF and baseband (BB) transmission cases, additional hardware for up converting the signal to the RF band is required at the BS. At the optical transmitter, the RF/IF/BB signal can be modulated onto the optical carrier by using direct or external modulation of the laser light. In an ideal case, the output signal from the optical link will be a copy of the input signal. However, there are some limitations because of non-linearity and frequency response limits in the laser and modulation devices as well as dispersion in the fiber. The transmission of analog signals puts certain requirements on the linearity and dynamic range of the optical link. These demands are different and more exacting than requirements placed on digital transmission systems.

In Fig. 12, typical RoF link configurations are shown, which are classified based on the kinds of frequency bands transmitted over an optical fiber link. In the downlink from the CS to the BS, the information signal from a public switched telephone network (PSTN), an Internet Service Provider (ISP), a mobile telecommunications operator, an Intelligent Transportation System (ITSs) or another CS is fed into the optical network at the CS. The signal that is either RF, IF or BB band modulates an optical signal from a LD. As described earlier, if the RF band is low, it's possible to modulate the LD signal using the RF band signal directly. If the RF band is high, such as the mm-wave band, it's better to use electro- optical modulators (EOMs), like Mach-Zehnder Modulators. The modulated optical signal is transmitted to the BS via optical fibers. At the BS, the RF/IF/BB band signal is recovered by detecting the modulated optical signal by using a PD. The recovered signal, which needs to be upconverted to RF band if an IF or BB signal is to be transmitted to a mobile handset (MHs) via the antenna of the BS.

In the configuration shown in Fig. 12 (a), the modulated signal is generated at the CS in an RF band and directly transmitted to a BS by an EOM, which is called "RF-over-Fiber". At the BS, the modulated signal is recovered by detecting the modulated optical signal with a PD and directly transmitted to a MH. Signal distribution using RF-over-Fiber has the advantage of a simplified BS design but is susceptible to fiber chromatic dispersion that severely limits the transmission distance (Gliese et al., 1996). In the configuration shown in Fig. 12 (b), the modulated signal is generated at the CS in an IF band and transmitted to a BS by an EOM, which is called "IF-over-Fiber". At each BS, the modulated signal is recovered by detecting the modulated optical signal with a PD, up converted to an RF band, and transmitted to a MH. In this scheme, the effect of fiber chromatic dispersion on the distribution of IF signals is much reduced. However, the antennas of the BSs a RoF system incorporating IF-over-Fiber transport require additional electronic hardware such as an mm-wave frequency LO for frequency up- and down conversion.

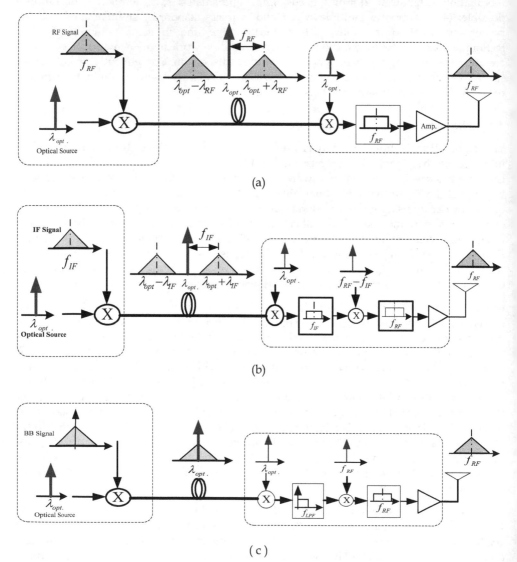

Fig. 12. Different schemes for signal modulation onto an optical carrier for distribution: (a) Radio over Fiber; (b) IF over Fiber; (c) BB over Fiber.

In Fig. 12 (c), the modulated signal is generated at the CS in baseband and transmitted to a BS, which is referred to as "BB-over-Fiber". At the BS, the modulated signal is recovered by detecting the modulated optical signal with a PD, up converted to an RF band through an IF band or directly, and transmitted to a MH. In baseband transmission, the impact of fiber dispersion effects is negligible, but the BS configuration is the most complex. This is especially important when RoF in mm-wave bands is combined with dense wavelength division multiplexing (DWDM). This increases the amount of equipment at the BSs because an up converter for the downlink and a down converter for the uplink are required. In the RF subcarrier transmission, the BS configuration can be simplified only if an mm-wave optical external modulator and a high-frequency PD are implemented in the electric-to-optic (E/O) convertor and the optic-to-electric (O/E) converter, respectively.

Optical links are mainly transmitting microwave and mm-wave signals by applying an intensity modulation technique onto an optical carrier (Al-Raweshidy & Komaki, 2002). Fundamentally, two methods exist for transmission of the microwave/mm-wave signals over optical links with intensity modulation: (1) direct intensity modulation, (2) external modulation.

In direct intensity modulation an electrical parameter of the light source is modulated by the information RF signal. In practical links, this is the current of the laser diode, serving as the optical transmitter. In Fig. 10, the simplest and most cost-effective architecture of intensity-modulation direct-detection (IMDD) is depicted. In this architecture, the detection is performed using a photo diode (PD). In the direct-modulation process a semiconductor laser directly converts an electrical small-signal modulation (around a bias point set by a dc current) into a corresponding optical small-signal modulation of the intensity of the photons emitted (around the average intensity at the bias point). Thus, a single device serves as the optical source and the RF/optical modulator. An important limitation in this architecture for super broadband access are the restrictions placed on the modulation bandwidth by the laser and the mm-wave band while a simple laser's linewidth can be modulated to frequencies of several Gigahertzes. Furthermore, it is reported that direct intensity modulation lasers can operate at up to 40 GHz or even higher, but, these are expensive and are not cost-effective in the commercial market. Therefore, at frequencies above 10 GHz, external modulation rather than direct modulation is applied.

In the external modulation technique, Fig. 11, an unmodulated light source is modulated with an information RF signal using an electro-optical intensity modulator. Because the number of BSs is high in RoF networks, simple and cost-effective components must be utilized. Therefore, in the uplink of a RoF network system, it is convenient to use direct intensity modulation with cheap lasers; this may require down conversion of the uplink RF signal received at the BS. In the downlink either lasers or external modulators can be used.

4.2 Application of WDM in a radio over fiber system

The application of WDM in RoF networks has many advantages including simplification of the network topology by allocating different wavelengths to individual BSs, enabling easier network and service upgrades and providing simpler network management. Thus, WDM in combination with optical mm-wave transport has been widely studied (Grifin et al., 1999; Toda et al., 2003).

The implementation of WDM in a RoF network is illustrated in Fig. 13, where for simplicity, only downlink transmission is depicted. Optical mm-wave signals from multiple sources are multiplexed and the generated signal is optically amplified, transported over a single fiber and demultiplexed to address each BS concerned separately. A challenging issue is that the optical spectral width of a single optical mm-wave source may approach or exceed WDM channel spacing. Therefore, there have been several reports on dense WDM (DWDM) applied to RoF networks (Grifin et al., 1999; Grifin, 2000; Toda et al., 2003); by utilizing the large number of available wavelengths in the DWDM technique, the lack of free transmission channels for the deployment of more BSs in mm-wave bands can be overcome. Another issue is related to the number of wavelengths required per BS. It is desirable to use one wavelength to support full-duplex operation. In (Nirmalathas et al., 2001), a wavelength reuse technique has been proposed, which is based on recovering the optical carrier used in downstream signal transmission and reusing the same wavelength for upstream signal transmission.

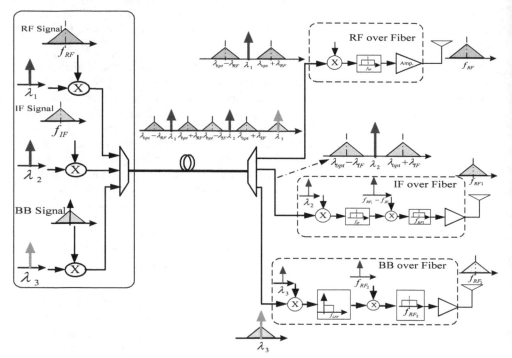

Fig. 13. Schematic illustration of the implementation of WDM in a RoF network.

4.3 Digital radio over fiber

Digital systems are more flexible, more conveniently interface with other systems, are more reliable and robust against additive noise from devices and channels, and achieve a better dynamic range than analog systems. Analog to digital and digital to analog converters (ADC and DAC, respectively) (Walden R. H., 1999) are the link between the analog world and the digital world of signal processing and data handling. In an analog system the bandwidth is limited by devices performance and parasitic components are introduced.

In a Digital RoF (DRoF) system, an electrical RF signal is digitized by using an Electronic ADC (EADC) (Vaughan et al., 1991). Then, the generated digital data is modulated with a continuous coherent optical carrier wave either using a direct modulation technique or by using an external electro-optical modulator as shown in Fig. 14. The modulated optical carrier is transmitted through the fiber. At the base station, after detecting the optical signal using a photo diode, the detected digital data is converted back to the analog domain using an EDAC. Finally, the analog electrical signal is fed to an antenna (Li et al., 2009; Kuwano, 2006, 2008; Lim et al., 2010). Current EDAC systems experience problems such as jitter in the sampling clock (Stephens, 2004; Hancock, 2004), the settling time of the sample and hold circuit, the speed of the comparator, mismatches in the transistor thresholds and passive component values. The limitations imposed by all of these factors become more severe at higher frequencies. Wideband analog to digital conversion is a critical problem encountered in broadband communication and radar systems (Valley, 2007; Kim et al., 2008). For the future beyond Gigabit/s mobile and wireless end-user traffic rates (Abdollahi et al., 2010) due to the limitations of electronic technology for implementing ultra high-speed, high performance EADC, and the resolution of existing EDAC, the deployment of conventional DRoF links (Li et al., 2009; Kuwano, 2006, 2008; Lim et al., 2010) is not simply achievable.

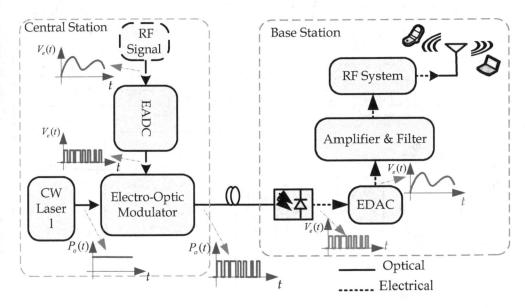

Fig. 14. Conventional DRoF architecture using EADC (downlink) (Li et al., 2009).

Moreover, if a conventional DRoF link could be achieved for Gigabit/s traffic rates the generated digital traffic creates a new challenge, namely, for this architecture to use more electro-optical modulators and photo diodes to implement the wavelength division multiplexing (WDM) technique to diminish the chromatic dispersion caused by the restrictions on the modulation bandwidth for super broadband access by RoF.

4.4 All-photonic digital radio over fiber

An all-photonic DRoF architecture has been proposed (Abdollahi et al. 2011) and is depicted in Fig. 15. This architecture uses an electro-optical modulator, which is simultaneously shared as an optical sampling and modulating device at the CS. A photonic ADC (PADC) by using a mode-locked laser (MLL) and an electro-optical modulator is able to scale the timing jitter of the laser sources to the femtosecond level, (Kim et al. 2007; Bartels et al. 2003), which allows designers to push the resolution bandwidth by many orders of magnitude beyond what electronic sampling systems can currently achieve. The proposed system includes an all-photonic signal processing block for optical quantization and wavelength conversion of the sampled and symmetrically split signal's power. By using the WDM technique to distribute the generated traffic over different wavelengths exceeding the modulation bandwidth of the fiber on a particular wavelength is prevented.

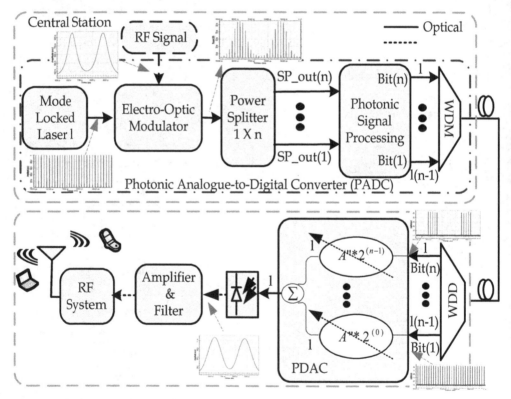

Fig. 15. All-photonic DRoF architecture, (downlink), (Abdollahi et al., 2011).

In Fig. 15, at the CS, the RF signal is sampled and modulated by optical train pulses that are generated using a passive mode-locked laser. The optical power of the sampled pulses is split into n levels using a symmetrical optical splitter, where n denotes the number of quantization bits. Finally, the split signals are fed to a photonic signal processing block for quantization and wavelength conversion operations.

The quantization procedure is performed by the process of Fig. 16 in which A and A' are constant parameters. At the first stage of this process, the stage number is equal to '1' (S=1). In this process entire stages are equal to number of quantization bits, i.e., for each output bit there is a corresponding quantization stage. For quantization of the most significant bit (MSB) the received signal from output number 'n' of the symmetrical splitter SP_out(M) that is defined by the generic number 'M' which is equal to 'n' in this stage. This output optical signal is compared with a reference quantization level equal to '2(M-S) *A'. If the signal power square is greater than or equal to '2(M- S) *A', the output quantization bit is '1'. Otherwise, it is '0'. In this scheme, for performing the pipeline architecture, the quantized bits are converted back into analog domain. Therefore, in stage number '(M-S)', the converted back analog signals from stages 'n' to '(M-S+1)' of the process , are subtracted from the input of the split output signal SP_out(M-S). Then, the given signal is compared with '2(M-S) *A'. The quantization process is repeated in parallel 'n' times for quantizing each sampled optical signal into 'n' bits.

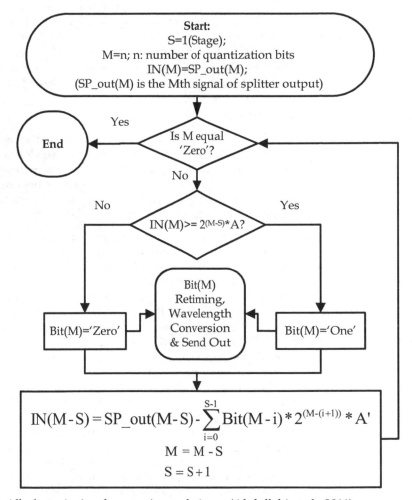

Fig. 16. All-photonic signal processing technique, (Abdollahi et al., 2011).

Subsequent to the wavelength conversion, the digital photonic signals are multiplexed in the wavelength domain by using a WDM and transmitted over a fiber. At the BS, the received signal is demultiplexed by wavelength division demultiplexer (WDD) and fed to the photonic digital-to-analog converter (PDAC). The PDAC subsystem, receives digital optical signals on different wavelengths, and converts them back to the equivalent analog signal at wavelength λ by using a passive PDAC and all-optical wavelength conversion. In the following of this stage, by using a photo diode (PD), the RF signal is recovered and after some RF signal processing it is passed to a multi-band distributed antenna system.

According to results provided (Abdollahi et al., 2011), it is demonstrated that ARoF is more dependent on fiber network impairments and length than DRoF. However, very low phase noise photonic sampling pulses and high speed signal conversion rates can be achieved in an all-photonic DRoF system compared with high-speed electronic circuits generated sampling pulses, signal conversion and processing. Consequently, an all-photonic DRoF system can support a digitized RF signal transmission system for providing super-broadband access to remote distributed wired and wireless access networks. It follows that, compared to the present digital optical communication infrastructures the number of CS would decrease with the introduction of all-photonic DRoF systems and as a result the service providers and network operators cost overheads per bit would be reduced.

5. Conclusions and chapter summery

Mobile and wireless networks generated traffic rates are growing very fast and are expected to double each year. The expectation is for delivering at least 1 Gbps multi-services traffic to each end-user in the near future for personal and multimedia communication services. Therefore, deploying super-broadband networks will be essential for service providers and operators. In this chapter, the convergence of wireless and optical communication technology for deploying future super broadband networks has been discussed.

Fiber optic transmission is rapidly becoming the dominant infrastructure medium for the transportation of fixed and mobile video on the internet. By replacing electronic switching with ultra fast photonic switching fiber optic transmission is expected to meet the need for super-broadband capacity. Radio-over-Fiber is a potential solution for deploying wireless access to broadband and super-broadband seamlessly. It can provide dynamic allocation of resources and can be realised with simple and small BSs with centralized operations. The requirement for more bandwidth allocation places a heavy burden on the current operating radio frequency (RF) spectrum and causes spectral congestion at lower microwave frequencies. Millimeter wave (mm-Wave) communication systems offer a unique way to resolve the bandwidth problems. When heterogeneous access networks converge to a highly integrated network via a common optical feeder network, network operators can reap the benefits of lowering the operating cost of access networks and meeting the capital costs of future upgrades easily. In addition, the converged access network will facilitate greater sharing of common network infrastructure between multiple network operators. Radio signals transportation over optical fiber (RoF) links will be a possible technology for simplifying the architecture of remote base stations (BSs). By relocating key functions of a conventional BS to a central location, BSs could be simplified into remote antenna units that could be inter-connected with a central office (CO) via high performance optical fiber feeder network. Wireless networks typically show considerable dynamics in traffic load of the radio access points (RAPs), due to fluctuations in the number mobile and wireless service users using them and the services they demand.

On the other hand, using the traditional RAPs approach requires equipping all of the wireless nodes for the highest capacity demanded which results in the inefficient use of recourses. The design of dynamic reconfigurable micro/pico or femo wireless cells increases network complexity but also greatly increases network efficiency. Within the optical access network layer WDM PONs allow an extra level of reconfiguration as wavelengths can be assigned to channels as part of static or dynamic routing. Therefore, integrating dynamic wavelength routing with RoF technology facilitates future flexible, low cost and reconfigurable super-broadband wired and wireless access network.

DRoF links are more independent of fiber network impairments and length than ARoF links. By using very low phase noise photonic sampling pulses and high speed signal conversion rates in place of high-speed electronic circuits generated sampling pulses, signal conversion and processing in an all-photonic DRoF system, digitized RF signal transmission for delivering future super-broadband remote distributed wired and wireless access networks traffic can be realised. Consequently, compared to the present digital optical communication infrastructure the number of CS will decrease in an all-photonic DRoF infrastructure and as a result, service providers and operators cost overhead per bit will be significantly reduced.

6. References

Abdollahi, S. R.; Al-Raweshidy, H.S.; Nilavalan, R. (2011). Fully-Photonic Analogue-to-Digital Conversion Technique for Super-Broadband Digitized-Radio over Fibre Link, Proceedings of 16th European Conference on Networks and Optical Communication, (July 2011, pp. 72-75), UK.

Abdollahi, S. R.; Al-Raweshidy, H.S.; Nilavalan, R.; Darzi A. (2010).Future Broadband Access Network Challenges, IEEE WOCN , (Sep. 2010), pp. 1-5.

Al-Raweshidy, H.; Komaki, S. (2002). Radio over Fiber Technology for Mobile Communication Networks, Artech House, 685 Canton Street, MA 02062, (2002), pp. 136-138.

Anthony, S., (2011). WiGig: 7Gbps Unified Data/Audiovisual Wi-Fi Coming in 2012, Available from http://www.extremetech.com/computing/89904-wigig-7gbps-data-display-and-audio-mid-range-networking-coming-in-2012, (July 2011).

Bartels, A.; Diddams, S. A.; Ramond, T. M.; Holberg, L. (2003). Mode-locked Laser Pulse Trains with Subfemtosecond Timing Jitter Synchronized to an Optical Reference Oscillator, Optic Letters, Vol.28, (2003), pp. 663-665.

Chang, G.-K.; Chowdhury, A.; Yu, J.; Jia, Z.; Younce, R. (2007). Next generation 100Gbit/s Ethernet Technologies," APOC 2007, Invited Paper, (November 2007) , Wuhan China.

Cisco Visual Networking Index (2011). Global Mobile Data Traffic Forecast Update, (Sep 2011).

FP7 (2010). A Converged Copper-Optical-Radio OFDMA-Based Access Network with High Capacity and Flexibility, ICT Objective 1.1 The Network of the Future, (Jan. 2010).

Cox, C. H. (2004). Analog Optical Links, Cambridge University Press, (2004), Cambridge UK.

Gliese, U; . Norskow, S.; Nielsen, . T. N. (1996). Chromatic Dispersion in Fiber-Optic Microwave and Millimeter-Wave Links, IEEE Transaction of Microwave Theory Technology, Vol. 44, No. 10, (Oct. 1996), pp. 1716.1724.

Goldfarb, G.; Li, G.; Taylor, M. G. (2007). Orthogonal Wavelength-Division Multiplexing Using Coherent Detection, IEEE Photonics Technology Letters, Vol. 19, No. 24, (Dec., 2007), pp. 2015-2017.

Grifin, R. A. (2000). DWDM Aspects of Radio-over-Fiber, Proceedings of LEOS 2000 Annual Meeting, Vol.1, (Nov. 2000), pp. 76-77.

Grifin, R. A.; Lane, P. M.; O'Reilly, J. J.; (1999). Radio-Over-Fiber Distribution Using an Optical Millimeter-Wave/DWDM Overlay, Proceeding of OFC/IOOC 99, Vol.2, (Feb. 1999), pp. 70-72.

Guo,Y.F.; Kuo, G.S. A novel QoS-guaranteed power-efficient management scheme for IEEE 802.15.3 HR-WPAN, 2007 4th Annual IEEE Consumer Communications and Networking Conference CCNC 2007 (2007) Pages: 634-638

Hancock, J. (2004), Jitter-Understanding it, Measuring it, Eliminating it, Part 1: Jitter Fundamentals, High Frequency Electronics, Summit Technical Media, LLC, (April, 2004) pp. 44-50.

ITU (2009). The World in, ICT Facts and Figs, Available from http://www.itu.int, (2009).

Ji, H. C.; Kim, H.; Chung, Y. C. (2009). Full-Duplex Radio-Over-Fiber System Using Phase-Modulated Downlink and Intensity-Modulated Uplink, IEEE Photonics Technology Letters, Vol.21, No.1, (Jan. 2009), pp 9-11.

Kim, J.; Chen, J.; Cox, J.; Kartnr, F. X. (2007). Attosecond-Resolution Timing Jitter Characterization of Free-Running Mode-Locked lasers, Optics Letters, Vol.32, (2007), pp. 3519-3521.

Kim, J.; Park, M. J.; Perrott, M. H.; Kartner, F. (2008). Photonic Sub-Sampling Analog-to-Digital Conversion of Microwave Signals at 40-GHz with Higher than 7-ENOB Resolution, Optics Express, Vol.16, No.21, (2008), pp. 16509-16515.

Koonen, T. (2006), Fiber to the Home/Fiber to the Premises: What, Where, and When? Proceedings of the IEEE , Vol. 94, No. 5, (May 2006), pp. 911-934.

Kudo, K. (2005). Introduction of High-Definition TV System in NHK News Center, ABU Technical Committee 2005 Annual Meeting, (20-24 November 2005), Hanoi.

Kuwano, S.; Suzuki, Y.; Yamada, Y.; Fujinio, Y.; Fujiti, T.; Uchida, D.; Watanabe, K. (2008). Diversity Techniques Employing Digitized Radio over Fiber Technology for Wide-Area Ubiquitous Network, IEEE Global Telecommunication Conference, (Globecom) , (Dec. 2008), pp. 1-5.

Kuwano, S.; Suzuki, Y.; Yamada, Y.; Watanabe, K. (2006). Digitized Radio-over-Fiber (DRoF) System for Wide-Area Ubiquitous Wireless Network", IEEE International Topical Meeting on Microwave Photonics, (2006), pp. 1-4.

Larkin, N. ASON & GMPLS; The Battle for the Optical Control Plane; Available from http://www.dataconnection.com/network/download/whitepapers/asongmpls.pdf.

Laskar, J.; Pinel, S.; Dawn, D.; Sarkar, S.; Perumana, B.; Sen, P.; (2007). The Next Wireless Wave is a Millimeter Wave. Microwave Journal, Vol.90, No. 8, (August 2007), pp 22-35.

Le Rouzic, E.; Gosselin, S. (2005). 160 Gb/s Optical Networking: A Prospective Techno-Economic Analysis. Journal of Lightwave Technology, Vol.23, No.10, (Oct. 2005), pp 3024-3033.

Li., G. L.; Yu, P. K. L.; (2003). Optical Intensity Modulators for Digital and Analog Applications, Journal of Lightwave Technology. Vol. 21, (2003), pp. 2010-2030.

Li, T.; Crisp, M.; Penty, R. V.; White, I. H. (2009). Low Bit Rate Digital Radio over Fiber system, IEEE International Topical Meeting on Microwave Photonic, (2009), pp. 1-4.

Lim, C.; Nirmalathas, A.; Bakaul, M.; Gamage, P.; Lee, K. L.; Yang, Y.; Novak, D.; Waterhouse, R. (2010). Fiber-Wireless Networks and Subsystem Technologies, Journal of Lightwave Technology, Vol. 28, No. 4, (2010) , pp. 390-405.

Llorente, R.; Lee, J. H.; Clavero, R. (2005). Orthogonal Wavelength-Division-Multiplexing Technique Feasibility Evaluation, Journal of Lightwave Technology, Vol.23, No.3, (March 2005), pp. 1145-1151.

MacQueen, D., (2010). Global Mobile Forecast 2001-2015, Mobile Media Strategies, (March 2010), pp. 10.

Mcdonough, J. (2007). Moving Standards to 100 Gbps and Beyond, IEEE Communication Magazine, Vol. 45, No.11, (Nov. 2007), pp. 6-9.

Nirmalathas, A.; Novak, D.; Lim, C.; Waterhouse, R. B. (2001). Wavelength Reuse in the WDM Optical Interface of a Millimeter-Wave Fiber-Wireless Antenna Base Station, IEEE Transaction of Microwave Theory Technology, Vol.49, No.10, (Oct. 2001), pp. 2006-2012.

Nokia. (2009). LTE-Delivering the Optimal Upgrade Path for 3G Networks, Available from www.nokia.com/NOKIA.../Nokia.../LTE_Press_Backgrounder.pdf.

OASE (2010). Optical Access Seamless Evolution. Available from http://cordis.europa.eu/fetch?CALLER=PROJ_ICT&ACTION=D&CAT=PROJ&R CN=93075, (Feb. 2010).

OMEGA ICT Project. (2011). Gigabit Home Networks, Seven Framework Programme, Available from http://www.ict-omega.eu, (2008-2011).

PIANO+ (2010). Photonic-based Internet Access Networks of the Future. Available from http://www.trdf.co.il/eng/kolkoreinfo.php?id=448, (Feb., 2010).

Raffaelli, C.; Vlachos, K.; Andriolli, N.; Apostolopulus, D. (2008). Photonic in Switching: Architectures, Systems and enabling technologies. Computer Networks Volume 52, Issue 10, (July 2008), pp. 1873-1890.

Rodrigo, M. de V.; Latouche, G., Remiche, M. –A. (2009). Modeling Bufferless Packet-Switching Networks with Packet Dependencies. Computer Networks 53 (Feb. 2009), pp. 1450–1466.

RNCOS Industry Research Solution, (2011). Global Mobile TV Forecasting to 2013, (Aug. 2011).

Rysavy Research, (2007). Edge, HSPA and LTE the Mobile Broadband Advantage, 3G Americas, (Sep., 2007).

Schwarz, Y. (2011). WiMAX 2: the future Super Broadband 4G Network, Avaiable from http://www.goingwimax.com/ (July 5, 2011).

Stephens, R. (2004). Analyzing Jitter at High Data Rates, IEEE Optical Communication, Feb. 2004, pp. 6-10.

Stephens R. Jitter Analysis: The Dual-Dirac Model, RJ/DJ, and Q-sacle, Agilent Technical Note, Dec. 2004.

Sugawara, M.; Masaoka, K.; Emoto, M.; Matsuo, Y.; Nojiri, Y. (2007). Research on Human Factors in Ultra-high-definition Television to Determine its Specifications, SMPTE Technical Conference, (October 2007).

Toda, H.; Yamashita, T.; Kuri, T.; Kitayama, K. (2003).Demultiplexing Using an Arrayed-Waveguide Grating for Frequency-Interleaved DWDM Millimeter-Wave Radio-on-Fiber Systems, Journal of Lightwave Technology, Vol. 21, No. 8, (Aug. 2003), pp. 1735-1741.

Urban, P. J.; Huiszoon, B.; Roy, R.; de Laat, M. M.; Huijskens, F. M.; Klein, E. J.; Khoe, G. D.; Koonen, A. M. J.; Waardt, H. D. (2009), High-Bit-Rate Dynamically Reconfigurable WDM–TDM Access Network, Journal of Optical Commuincation Network Vol.1, No. 2, (July 2009), pp. 143-159.

Valley, G. C. (2007). Photonic analog-to-digital converters, Optics Express. Vol.15, No.5, (2007), pp. 1955-1982.

Vaughan, R. G.; Scott, N. L.; White, D. R., (1991). The Theory of Bandpass Sampling, IEEE Transaction on Signal Processing, Vol.39, No.9, (Sep. 1991), pp.1973-1984.

Walden R. H. (1999).Analog-to-Digital Converter Survey and Analysis, IEEE Journal of Selected Areas in Communications, Vol.17, No.4, (1999), pp. 539-550.

Yuen, R.; Fernando, X. N.; Krishnan, S. (2004). Radio Over Multimode Fiber for Wireless Access, IEEE Canadian Conference on Electrical and Computer Engineering, Vol.3, (May 2004) pp. 1715-1718.

9

Wireless Technologies in the Railway: Train-to-Earth Wireless Communications

Itziar Salaberria, Roberto Carballedo and Asier Perallos
Deusto Institute of Technology (DeustoTech), University of Deusto
Spain

1. Introduction

Since the origins of the railway in the XIX century most of the innovation and deployment efforts have been focused on aspects related to traffic management, driving support and monitoring of the train state (Shafiullah et al., 2007). The aim has been to ensure the safety of people and trains and to meet schedules, in other words, to ensure the railway service under secure conditions. To achieve this it has been necessary to establish a communication channel between the mobile elements (trains, infrastructure repair machinery, towing or emergency vehicle, and so on) and the earth fixed elements (command posts and stations, signals, tracks, etc.) (Berrios, 2007).

Nowadays, safety is a priority too, but new requirements have arisen, mainly concerning the quality improvement of the transport service provided to the passengers (Aguado et al., 2005). Moreover, the current European railway regulation by establishing that railway services be managed by railway operators independent of railway infrastructure managers makes it necessary for infrastructure fixed elements to share information with mobile elements or trains (handled by railway operators). This new policy results in additional requirements on the exchange of information between different companies. How to fulfil these requirements is a new technological challenge in terms of railway communications (Shafiullah et al., 2007) that is explored in this chapter.

The use of wireless technologies and Internet is growing in the railway industry which allows the deployment of new services that need to exchange information between the trains and terrestrial control centres (Shafiullah et al., 2007). In this sense, there are suitable solutions for other environments which allow to manage the bandwidth in terms of data rates. Therefore, these solutions are not designed for railway needs, and do not cover all the requirements that the railway industry has (California Software Labs, 2008; Marrero et al., 2008).

This chapter describes a specific wireless communications architecture developed taking into account railway communications needs and the restrictions that have to be considered in terms of broadband network features. It is based on standard communication technologies and protocols to establish a bidirectional communication channel between trains and railway control centres.

The second section of this chapter includes a brief description of the state of art in railway communications. The third one describes a specific train-to-earth wireless communication architecture. The fourth section describes the main challenges concerning with the management of the quality of service in train-to-earth communications. The fifth identifies some services that are arising as result of using this connectivity architecture and the way in which they interoperate. The sixth section shows the future lines of work oriented to improve the proposed communication channel. Finally, the seventh section of the chapter establishes the main conclusions of this work.

2. State of the art in railway communications

Railway communications emerged almost exclusively from the communication between fixed elements to carry out traffic management and circulation regulation. The technologies that communicate fixed elements with mobile elements (trains) are relatively recent, and they have contributed to improve and simplify the work required for rail service exploitation. Therefore, focusing on the network topology, two categories can be identified within the field of railway communications: a first one involving only fixed elements, and a second one involving both, fixed and mobile elements (called "train-to-earth" communications) (Salaberria et al., 2009). For the former, the most efficient solutions are based on wired systems. The latter has undergone great change in recent years, requiring wireless and mobile communications (Laplante & Woolsey, 2003).

Traditionally, the communication between fixed elements and trains has been established using analogical communication systems, such as the traditional telephone or PMR (Private Mobile Radio) based on radio systems (ETSI, 2008). These analogical systems are still used for voice communications and issues related with signalling. However, their important limitations in terms of bandwidth are causing the migration to digital systems, which offer a higher bandwidth.

Among the technologies of communication "train-to-earth", one of the most important advances of the last decade has been the GSM-R (Global System for Mobile Communications - Railway) (International Union of Railways, 2011). This system is based on the GSM telephony, but has been adapted to the field of railways. GSM-R is designed to exchange information between trains and control centres, and has as key advantages its low cost, and worldwide support.

Another technology that provides a wide circulation in the rail sector is the radio system TETRA (Terrestrial Trunked Radio) (ETSI, 2011). TETRA is a standard for digital mobile voice communications and data communication for closed user groups. The system includes a series of mobile terminals, similar to walkie-talkies, which allow establishing direct communication between control centres, train drivers and maintenance personnel, in addition to being able to establish communications with earthlines and mobile phones. Being a private mobile telephone system, its deployment in the rail sector is very simple, because it is based on the placement of a series of antennas at stations or control centres along the route.

In addition, the special-purpose technologies mentioned so far include the growing use of wireless communication technologies based on conventional mobile telephony (GSM, GPRS and UMTS) and broadband solutions such as WiFi (IEEE 802.11, 2007) or WiMax (IEEE 802.16.2, 2004). The wireless local area networks WiFi enable the exchange of information, at much higher speeds and bandwidths than with other technologies. The cost of deployment of such networks is very low, but these are limited in terms of coverage or distance they cover. To address this limitation, the WiMax technology has emerged extending the reach of WiFi, and is a very suitable technology to establish radio links, given its potential and high-capacity at a very competitive cost when compared with other alternatives (Aguado et al, 2008).

All technologies discussed so far aim to establish a wireless communication channel between fixed elements and mobile elements of the railway field, but what happens with the services offered by means of this communication channel?, how can they have access to the channel?, how can they share it?. To address these questions, a categorization of railway services is necessary. Traditional applications or services of the railway can be classified into two major groups: (1) services related with signalling and traffic control; and (2) services oriented to train state monitoring.

The first group of services is based on the exchange of information between infrastructure elements (tracks, signals, level crossings, and so on) and control centres, all of them fixed elements. Additionally, it uses voice communication between train drivers and operators in the control centres. Therefore, for this type of service, traditional communication systems based on analogical technology remain significant.

The second group of services requires the exchange of information in the form of "data" between the trains and the control centres. In this case, the new services use any of the wireless technologies mentioned so far, but on an exclusive basis, which means that each application deployed on the train must be equipped with its own wireless communications hardware. This leads to have an excessive number of communications devices, often underused. In addition, there are still many applications that require a physical connection "through a wire" between the train devices and a computer for information retrieval and updating tasks.

On the other hand, a new set of services around the end user (passenger, or companies who need to transport some goods) is emerging. These services are oriented to providing a transport service of higher quality that not only is safe, but provides additional benefits such as: detailed information about the location of trains and schedules, contextual advertising services, video on demand, and so on. All these services are characterized by their need of a wireless communication channel with high bandwidth and extensive coverage (Garstenauer & Pocuca, 2011). As a result, the following needs are identified: (1) to standardize the way to exchange train state information between the trains and the control centres; and (2) to define a wireless communications architecture suitable for the new 'end user'-oriented services (Aguado et al., 2005).

In this chapter, a specific communications architecture based on standard technologies and protocols; that is designed to manage train-to-earth connectivity at application layer, will be presented in order to fulfil such needs.

3. Managing train-to-earth wireless communications

This section describes a general purpose wireless communications architecture to address the needs for high bandwidth and wide coverage. This solution is based on the management of a wireless communications channel at the application layer. The architecture proposed is currently being deployed in some railway companies from Spain (Euskotren and ETS from Basque Country, and Renfe from Spain), will be presented (Gutiérrez et al., 2010). It is an innovative general purpose wireless communication channel which allows the train to communicate with the railway control centres in such a way that the applications or services are unaware of communication issues such as: establishment and closure of the communication, management of the state of connectivity, prioritization of information and so on.

This new wireless communication architecture has to respond to the demand for communication and transmission of information from any application, so it will have to take into account the nature of the information to be sent. The information exchanged between two applications (one on earth and the other on a train) may have different urgency degrees depending on their purpose or treatment with respect to the exchanged information. In fact, there is information that needs to be transmitted at the time that it is generated, for example in case of positioning information or alarms in some critical train operation elements. On the other hand, there may be less urgent information whose transmission can be postponed, such as train CCTV images, or audio files used by the background music. In addition, the urgent or priority information is usually smaller than the non-priority information.

3.1 Towards a train-to-earth wireless communications architecture

In this section the core components of the mentioned wireless communications architecture are described. This architecture allows a full-duplex transmission of information between applications and devices deployed in the trains, and applications that are in the railway control centres. The description of the architecture will be made at two levels: conceptual and physical level. The first level defines the basic concepts of the architecture, and the second one illustrates the technologies used to implement the architecture in two real scenarios.

3.1.1 Conceptual level

From a conceptual point of view, two issues are especially important: the elements that manage architecture's behaviour and the ways in which the different applications (terrestrial and on-board) transmit the information.

Our architecture hosts both terrestrial and train-side applications, so in order to manage its behaviour two main entities are defined: Terrestrial Communication Manager (TCM) and On-board Communication Manager (OCM). The former manages terrestrial aspects of the architecture and the latter train-side issues. Although the managers have a different physical location, both of them have nearly the same responsibilities:

- Delivery and reception of the information,
- Dynamic train addressing,
- Medium access control,
- Security and Encryption, and
- Communication error management.

Due to the information transmission needs and for a correct and optimized used of the communication architecture, two types of communications are distinguishes: "slight" and "heavy" communications. These two types take into account characteristics of both information and communication technologies, such us: the volume and the priority of the information, the existence of coverage, and the cost of the communication.

- **Slight communications:** This type of communication is for the transmission of small volumes of information (few kB.) and with high priority. In general, information that has low latency (milliseconds or a pair of seconds) and needs to be transmitted exactly when it is generated or acquired (for instance, the GNSS location of a train, or a driving order to the train driver). For example, in the first case, if the information about positions is not sent immediately after its generation (real-time transmission), it loses all relevance.
- **Heavy communications:** This type of communication is tied to the transmission of large volumes of information (in the order of MB) and with low priority. The importance of this information is not affected by the passage of time, so it doesn't need to be transmitted at the exact time it is generated (no real-time transmission).

3.1.2 Physical level

In this section the technological aspects of the wireless architecture are described. They refer to the protocols and the communication technologies used for the development of the train-to-earth architecture (showed in Fig. 1).

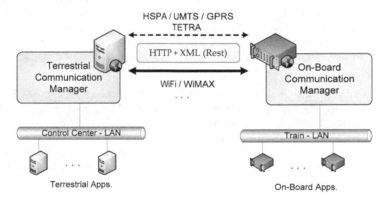

Fig. 1. Train-to-earth wireless communications architecture.

It is important to point out that the protocols and technologies for the development of the new architecture have been selected with regard to: standardization, robustness, security, scalability and compatibility with existing and potential applications and systems. The major aim has been the ease of integration of any application or system into the new communication architecture. Concerning with this objective, Web Services constitute the transport technology for the communication between final applications and the "local" Communication Managers. All the information is interchanged in XML format, in order to allow future extensions.

On the other hand, the communication between the Terrestrial (TCM) and the On-board Communication Managers (OCM) is based on REST (Representational State Transfer) technology. This communication technology uses the HTTP (HyperText Transfer Protocol) protocol and XML formatted messages. This solution is similar to traditional XML Web Services but with the benefit of a low overload and computational resources consumption.

Although the information interchanged between the TCM and the OCM is not encrypted, using the HTTP protocol allows the easy migration to HTTPS (HyperText Transfer Protocol Secure) that offers encryption and secure identification. It can be seen that every communication has to go through two core elements that can result in the loose of channel availability in case of failure. This problem is tackled by means of the use of web services because this solution deploys support web services in a way similar to traditional web architectures. It can be said that the selected technologies and architectures are well known and broadly used in different application areas or contexts, but they are novel in the railway 'train-to-earth' communication field.

In order to establish a wireless communications channel between the trains and the railway control centres, mobile and radio technologies have been selected (Yaipairoj et al., 2005). In this case, slight and heavy communications use different technologies due to different transmission characteristics.

Due to de necessity of delivery of information in real-time, mobile technologies such as GPRS/UMTS/HSPA (Gatti, 2002) are used for the *slight communications*. These technologies do not offer a great bandwidth nor a 100% coverage and they have a cost associated to the information transmission. Despite this, these technologies are a good choice for the delivery of high-priority and small sized information. The selection of the specific technology (GPRS/UMTS/HSPA) depends on whether the service is provided or not, (by a telecommunications service provider), and the coverage in a specific area.). To increase coverage availability, the hardware installed in each train has two phone cards belonging to different telephone providers. This allows switching from one to the other depending on coverage availability. Therefore, the idea is to have a predetermined operator, and only switch to the second when the former is unable to send.

On the other hand, for the *heavy communications*, WiFi radio technology has been chosen. This technology allows the transmission of large volumes of information, does not have any costs associate to the transmission and its deployment cost is not very expensive. In this case, a private net of access points is needed. This net does not need to cover the complete train route because the heavy communications are thought for the transmission of big amounts of information at the end of train service (for example the video recorded by the security cameras).

Although each separate technology can't achieve 100% coverage of the train route, the combination of both comes very close to complete coverage (Pinto et al., 2004). As the application layer protocols are standard, other radio technologies such as TETRA or WiMAX (Aguado et al, 2008) can easily substitute the ones selected now. These technologies can achieve a 100% coverage and neither one has a transmission cost. However, there are certain limitations such as the cost of deploying a private TETRA network, and the cost and the stage of maturity of the WiMAX technology

3.2 How to manage wifi based broadband communications

As it was explained previously, there are some railway applications that need a high bandwidth to interchange large amount of information without time restrictions (real time communication is not needed). This kind of communications will enable 'train-to-earth' information exchange for train side systems update/maintenance and multimedia information download/upload such as videos or pictures.

With the purpose of providing an innovative broadband communications architecture suitable for the railway, a number of WiFi networks have to be settled in places where the trains are stopped long enough to ensure the discharge of a certain amount of information. This is: stations in the header that starts or ends a tour and garages. In this way, WiFi coverage is not complete, but broadband communications are designed to update large amount of information, which, usually do not need to take place in real time.

Therefore, at this point it has to be taken into account aspects such as bandwidth, coverage or communications priorities. The existing broadband management systems, which are used in other (non mobile) environments, do not satisfy all the needs of the railway applications (California Software Labs, 2008; Marrero et al., 2008). Furthermore, some additional problems have to be solved on this environment. In one hand, it is necessary to find a mechanism to locate the trains because they don't have a known IP address all the time. A dynamic IP assignment is used for every WiFi network so a train obtains a different IP address every time it is connected to a network, and a certain IP address could be assigned to different trains in different moments. On the other hand, there are several applications that want to transmit information to/from the trains at the same time. This implies the existence of a bandwidth monopolization problem.

To tackle these challenges, it is necessary a smart intermediate element which manages when the applications (both terrestrial and train-side) can communicate with each other. That is to say: a Broadband 'train-to-earth' Communications Manager, whose design and functional architecture is described below.

3.2.1 Design of the broadband communications manager

The Broadband Communications Manager (BCM) (Carballedo et al., 2010a) is a system that arbitrates and distributes shifts to communicate terrestrial applications and train-side systems (see Fig. 2); in this way, the terrestrial applications request a turn when they want to establish a communication with a train. This distribution shift is managed on the basis of the state of the train connection to a WiFi network (known at all times) and a system of priorities, which are allocated according to the terrestrial application that wants to communicate with a specific train.

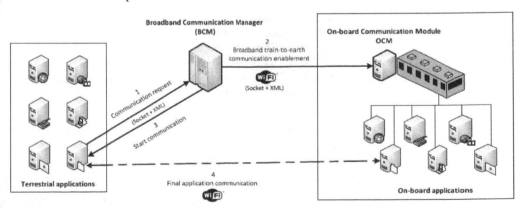

Fig. 2. The Broadband Communications Manager (BCM) arbitrating the communications between terrestrial applications and train-side systems.

1. **Communication establishment protocol.** When the BCM decides to give a shift to communicate a terrestrial application and a train-side system, it sends an authorization to both (application and train system). To do this, the manager establishes a communication with each entity through TCP Sockets. Within these TCP Sockets a series of XML messages, that define the communication protocol, are used.

To explain in a simple way the operation of the BCM, here is the description of a typical scenario:

-   Firstly, a terrestrial application designed to communicate with a train system is connected to the manager through a TCP Socket.
-   The terrestrial application will make communication request, and will give it a certain priority. By the time the manager receives the request, it orders the request in the queue of the destination train's requests. This queue is always sorted by different criteria.
-   When a train arrives at a station, it connects to the WiFi network and it gets an IP address. This address is supplied to the BCM. If the train has pending communication requests, the terrestrial application is notified so that it can start the communication.
-   At this moment there is a direct communication between the terrestrial application and the train-side system, through the WiFi network. The responsibility for starting the communication relies on the terrestrial application because it knows the IP address of the train.
-   When the communication ends, the terrestrial application informs the BCM, which is ready to serve the next communication request.

It is important to emphasize that the BCM does not set any limitation or condition in the communication between the terrestrial application and the train-side system. The manager's work focuses only in defining the time at which this communication must be carried out, and warns of this fact to the entities involved in the information interchange. It does not define any structure or format of the information being exchanged; it only establishes a mechanism to know the IP address of the destination train (because it is dynamic), and manages the transmission shifts to prevent the monopolization of the communications channel.

2. **Multithreading management.** To carry out its work, the BCM must establish connections with multiple applications and train-side systems at the same time. To manage all these communications efficiently a multithreaded design has been chosen for the management of the connections. Every communication that the BCM performs with any external element (terrestrial applications and train-side systems) is carried out independently and concurrently, using a dedicated thread in each case.

Both the Train-Side Systems Handler and the Terrestrial Applications Handler (see below, Fig. 4) are separate threads that are responsible for receiving connections from external agents. Upon receiving the connection message specified in the protocol, they generate a separate communication thread with the element which has sent the message.

3. **XML based protocol for data transmission.** All the communications are done through an architecture based on TCP sockets (one for the terrestrial applications and another one for the train-side systems) and XML messages exchange. A message will be defined for each requests/responses exchanged between the three elements that form the

architecture: terrestrial applications, train-side systems and Broadband Communication Manager. Fig. 3 shows a XML message of the communication protocol.

```
<?xml version="1.0" encoding="UTF-8"?>
 <request>
     <application name="CCTV" ip="130.88.10.56" />
     <train name="UT204" />
     <port number="3556" priority="1" />
 </request>
```

Fig. 3. An XML message with a request from terrestrial application to the BCM asking a communication with an on-board system of the train UT204.

In order to communicate terrestrial applications, train-side systems and BCM a XML messages base protocol has been defined. The choice of the TCP Socket schema and XML messages was taken due to the flexibility to add new functionality, and the simplicity of implementation (independent of platform and programming language).

Moreover, all the data handled by the BCM is stored in a relational database. These data contain information about the communication requests, trains, train-side systems, terrestrial applications, and the available communication ports between them. The BCM's design contains a data layer that abstracts the data source of the business layer, so that the changes of this data by another for a different data source does not create any problems in the proper functioning of the BCM.

4. **Port to IP address translation schema.** To finish, we will make a brief description of the management of the applications installed on the trains (train-side systems), which are the target of the communication from terrestrial applications. These train-side systems are implemented on a computer that will have a private IP address (within the on-board Local Area Network) and is not accessible from outside the train. Therefore, it has been defined an addressing scheme to allow access from the IP address of the terrestrial application to the IP address of the train-side system. This is achieved through PAT filtering, associating each private IP address to a port number. Thus, whenever a train acquires an IP address from a WiFi network, the port number becomes the way to access the train-side systems. PAT filtering schema also ensures the security of communications and the information transmitted.

In each train there is a communications module which is responsible for performing this filtering of port numbers to IP addresses. This module is also responsible for communicating with the BCM, and manages the opening and closing of the ports that are associated to each train-side system.

3.2.2 Functional architecture of the broadband communications manager

Functional architecture of the BCM is based on message exchange between the manager itself and two types of external entities such as terrestrial applications and train-side systems. The BCM is divided into 5 modules (Fig. 4) that handle processing and deployment of all the functionality.

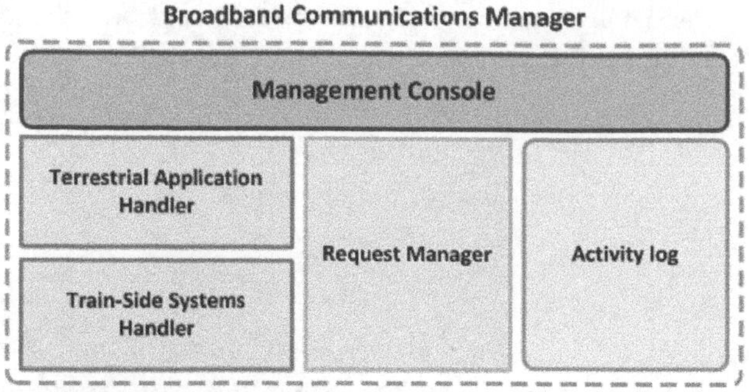

Broadband Communications Manager

Fig. 4. Functional architecture of BCM, composed of five modules: Terrestrial Application Handler, Train-Side Systems Handler, Request Manager, Activity Log and Management Console.

To have a global vision of the performance of the BCM, it is necessary to focus on three modules which carry out the most important functionality:

1. **Terrestrial Applications Handler.** It will be responsible for managing all the messages exchanged between each terrestrial application and the BCM. Its basic functionality is to receive the XML messages coming from terrestrial applications and generate an appropriate response. This communication is bidirectional, and it is the responsibility of the terrestrial application to start and finish it.

To streamline the management of communications between terrestrial applications and the BCM, the connections are managed independently (through a dedicated thread). The main functionality offered by this module would be the next one:

- Establish and close the connection between terrestrial applications and the BCM.
- Receive communication requests.
- Send messages to a terrestrial application in order to start communication requests.
- Receive communication completed messages from terrestrial applications.

2. **Train-Side Systems Handler.** It will be responsible for managing all the messages exchanged between the communication module of each train and the BCM.

This module is similar to the Terrestrial Applications Handler. It receives XML messages from the communication module of each train and generates the responses. In this case, the primary goal of the module is to indicate when a train is connected to a WiFi network and its IP address. This data is very important for terrestrial applications to communicate with train-side systems.

There is a very important task that Train Communication Module manages, it is: the disconnection or closure of the connection between the train and the BCM. When a train reaches a station with WiFi connectivity, it connects to the WiFi network and establishes a communication with the BCM. After that, two scenarios can occur: in the first one, the train has no pending communication request from terrestrial applications. In this case, the manager sends a connection ending message to the train and the connection is closed. In the second scenario, a train is disconnected from the WiFi network because of its movement or a communication failure. The manager is constantly checking if the connection with the train is lost so this situation is detected as soon as it happens. There is a problem when the connection fails in the middle of a communication between a terrestrial application and a train-side system because the communication request has not finished correctly. To solve this problem, the next time the train connects to the BCM it sends back the start message of the broken communication to the terrestrial application in order to regain restart the communication. This pattern is repeated until the communication request is completed correctly, or is discarded because it exceeded the threshold of retries.

3. **Request Manager.** It will be responsible for managing communication requests between terrestrial applications and train-side systems, and to control when and under what circumstances the requests need to be attended.

As discussed above, the BCM splits communication shifts to terrestrial applications based on requests that they have performed. These requests are grouped by train, so the manager handles requests addressed to each train independently. The communication request for each train is sorted by the following criteria: (1) priority, which represents the 'urgency' by which a request must be addressed; (2) retries, it is taken into account the number of attempts to start a communication, to avoid the monopolization of the communication channel; and (3) parallelism, the manager can handle communications from multiple applications simultaneously with several trains.

The priorities associated with the communication requests are managed centrally and the BCM assigns these priorities to each terrestrial application. In addition, the manager also controls the train-side systems that can communicate with each single terrestrial application, identifying the ports that can be accesses by each of those terrestrial applications.

To complete the communication shifts service and management algorithm, it has prepared a final criteria, variable in this case (Noh-sam & Gil-Haeng, 2005). This approach takes into account two factors that are related directly with the communications that have been carried out previously. (1) The first factor is based on the calculation of average duration that takes the communications of a particular application. (2) The second factor takes into account the average duration of trains stopping in a particular station. Thus, the manager calculates a numeric value that represents the fitness of serving a request, knowing that the lower average duration of both factors will be most appropriate, since the risk of communication to be split because the train leaves the station will be less. Once calculated this criteria, it is used to discern which communication request is served, if the criteria explained a few paragraphs above is not sufficient

4. **Management Console (MC) and Activity Log (AL).** As for the remaining two modules, the MC contains a small management utility for monitoring the status of existing communication requests, and cancelling or changing the priority of unfinished requests. Through this interface it is possible to configure parameter and information settings of the BCM such as terrestrial applications, train-side systems, communication ports and priorities. The MC utility is based on standard design patterns like Model View Controller (MVC) so that in the future this presentation layer can be replaced by a more suitable one. The other supporting module is the AL. It stores each of the activities undertaken by the BCM.

To validate the improvement in the management of broadband communications produced by the BCM, there were a series of laboratory tests, which have subsequently been carried out in a real scenario. At first, the Broadband Communications Manager was tested in devising single communications between train and CCTV application. But to prove the performance improvement of the available bandwidth use, it has been necessary to include other terrestrial applications such as a document updating tool, and two other fictitious applications that simulate communications with the train.

The performance tests have taken into account two key parameters: 'train-to-earth' data transfer average time; and average waiting time between each communication. Table 1, shows the results obtained in the management of communications between four terrestrial applications (with different volume of data) and a train at the same station without the Broadband Communications Manager, while the second table shows the same scenario with the Broadband Communications Manager.

Data Volume (MB)	Data Transfer Time (seconds)	Waiting time (seconds)
< 1	1.10	0
1-10	11.30	0
11-50	58.84	0
51-100	184.62	0

Table 1. Results without the Broadband Communications Manager.

In the first table we can see that the absence of a communications manager allows communications to be made in parallel sharing the bandwidth. This greatly slows down the transfer rate, increasing the transfer time as the volume of information grows.

Data Volume (MB)	Data Transfer Time (seconds)	Waiting time (seconds)
< 1	0.76	0
1-10	7.69	0.76
11-50	38.46	8.45
51-100	115.38	49.91

Table 2. Results with the Broadband Communications Manager.

The second table shows how communications are conducted from smaller to larger amounts of data transferred thanks to the algorithm developed for the communication request service. The average time of transfer is lower than in Table 1, and the fact that communications are conducted one-by-one implies that there is a timeout that does not exist if they were carried out all at same time. At the conclusion of the tests it was determined that communications are carried out about 30% faster with the Broadband Communications Manager than without it, although there are wait times.

4. Considering the quality of service

In previous sections there have been described a train-to-earth communication architecture that enables two kinds of communications schemes: (1) slight communications and (2) heavy communications. Slight communications aims to respond priority and real-time application communication needs that no requires broadband communications bandwidth capabilities, whereas heavy communications were designed to lower priority large information volumes transmission management with no real-time requirements.

Based on this previous work, future work aims to go a step further by combining both schemes mentioned before to enable a real-time broadband communication platform which responds to train-to-earth applications communication needs. Thus, the objective is to enable several physical network communication links between train and ground system, choosing the network link considered as the best at every moment according with the bandwidth availability. Not having final applications to get involved in the network management. So, the system should respond to several requirements:

- **High availability:** each train should be enabled with one or more physical network communication links (3G, WiFi, etc.). Providing continuous train-to-earth connectivity in order to respond to the final applications communications demand in real-time.
- **The best bandwidth:** the purpose of this platform is to enable real-time train-to-earth broadband communications, using the best possible bandwidth. Thus, the system will always select the physical link considered as the best in order to respond to final application communication requirements.
- **Quality of Service (QoS):** this solution aims to make a service quality management too. Therefore it is necessary to know the bandwidth availability offered by the network link which is active at every moment, as well as the bandwidth offered by the rest of communications links (although they are not being used). At this point it is essential to establish a set of connection procedures which permit to reserve a certain bandwidth for a particular communication.

Hence this broadband train-to-earth communication platform has three principal functions:

1. Multiple physical device management, considering the dynamic selection of the best one (best bandwidth) and their abstraction into a single virtual device.
2. QoS implementation enabling the reservation and release of channels (virtual links) with a given bandwidth.
3. Message routing.

So, to carry out the train-to-earth communications management and arbitration, presented solution manages a set of criteria for prioritizing final applications communication requirements which will focus primarily on the criticality of the information transmitted and the required bandwidth, as well as their chronological arrival order.

4.1 Capabilities of the Real-time broadband communications platform

The main capabilities of this communication platform are related to (1) communication prioritization, (2) selection of the physical communication network link and (3) train-to-earth information exchange management.

4.1.1 Communication prioritization

The objective is to prioritize train-to-earth communications based on several criteria so that the transmission of critical information have more priority over other information that need less "immediacy" when being transmitted.

Therefore, this platform proposes a set of communication requests prioritization criteria in order to respond final applications communication demand (both on ground and onboard train). So, these criteria are applied to establish the order of the communication requests in train-to-earth communication prioritization queues.

4.1.2 Selection of the physical communication network link

The communication platform is based on different physical communication link existence so that the combination of these independent links offers a continuous train-to-earth connectivity in order to respond to the final applications communications demand in real-time. Thus, depending on the status of each of these media and their characteristics and restrictions, the platform must be designed to utilize the link that offers better performance in order to provide a high availability.

The system is designed to select at every moment the physical link that is most favorable for communications. Therefore, taking into account the availability of enabled different physical links, the system selects always as active link one that offers the best bandwidth (based on the features and coverage of the physical link).

At this point it should be emphasized that the basis is that the system always defines a single train-to-earth network link as active for communications (most favorable). So, all communications will always be generated by the channel set as active (WiFi, GSM / GPRS, Tetra, etc.) regardless of the availability of other physical channels simultaneously.

4.1.3 Train-to-earth Information exchange management

The main feature of the system is to manage the transmission of real-time broadband train-to-earth information. Therefore, the platform has to offer:

- Real-time bidirectional communication between train and terrestrial applications allowing generating new digital services with quality of service (QoS) guarantees. Thus, different kind of information exchange between final applications (multimedia, text, bytes, etc.) has to be supported.
- Train-to-earth communication management without requiring the participation of the final applications. However, transmission retries are delegated to the application logic. When an application information transmission is cut by the platform (because there is another higher priority request), and then it is re-established, it is responsibility of this application to decide if it continues transmitting from the point where it had left, or if the transmission is restarted from the beginning.
- Changing the requests bandwidth allocation in cases where the data traffic on the physical environment allows platform to assign applications' communications a greater bandwidth than initially requested from them.

4.2 Design of the Real-time broadband communications platform

This platform defines two main entities (Fig. 5): Terrestrial Communications Manager (TCM) and On-board Communications Manager (OCM). The former manages terrestrial aspects of the architecture and the latter train-side issues.

TCM interact with the OCM installed on each train in order to (1) select the best network link available at each time and then (2) manage applications communication requests serving those that are considered most priority first.

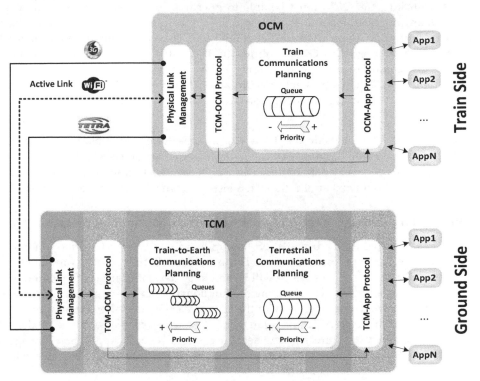

Fig. 5. Train and ground side components which conform the architecture of the real-time broadband communications platform.

4.2.1 Network active link selection

To establish train-to-earth communications, OCM and TCM can communicate through different communications network physical links. These two entities communicate each other to select the active link considered most favorable for communications. Therefore, OCM and TCM are continuously monitoring all enabled network link status, and switch from one to other in two cases: (1) when active link connectivity is lost and (2) when OCM and TCM select another link to be the new active link. In these two cases, the active communication link change is transparent for final applications that do not detect connection interruptions if these link changes occur while they are transmitting.

4.2.2 Priority application communication requests management

The broadband communication platform enables train and terrestrial railway applications to communicate each other. So when an application attempts to start a new communication makes a communication request to the platform. Then the system make a decision about what priority requests can be served concurrently by the system taking into account active link bandwidth limitations and requests QoS requirements.

On the train side, the OCM must be able to prioritize communication requests made by the on board applications. So, the OCM queues train applications' requests in base of established prioritization criteria. Then taking into account communication active link bandwidth properties, the OCM notifies to TCM about the on board most priority requests that could be served concurrently by the system respecting these requests QoS requirements. TCM will ultimately decide and notify the OCM which communications can be addressed at every moment, considering the rest of the terrestrial applications' request.

Therefore, on the ground side the TCM will manage terrestrial applications' requests as OCM do in the train. Besides, TCM will have a queue for each train on the system containing that train's requests (notified by its OCM) and terrestrial requests in order to make decisions about what applications' requests can communicate at every moment.

5. Developing railways services over train-to-earth communications

Now we will explore the benefits of having a train-to-earth wireless communication technology like the one presented before. These benefits will be justified by mean of the new valued added railway services which will be able to be developed using this communication architecture. In this section we will show the functionality of two specific services as well as the way in which they interoperate with the train-to-earth wireless communication channel. The first one is a Backup Traffic Management Service (BTMS) which uses the slight communication infrastructure and the second one is a Remote Application Management Service (RAMS) which uses the heavy communication model and integrates with the broadband communication manager.

5.1 Backup Traffic Management Service (BTMS)

Security in railway industry is a critical issue. Intelligent Transportation Systems are becoming a very valuable way to fulfill these critical security requirements. In fact, today, rail traffic management is performed automatically using Centralized Traffic Control systems (CTC) (Ambegoda et al., 2008). These systems are based on sensors and different elements fixed on the tracks. They allow real-time traffic management: (a) location of trains, (b) states of the signals, (c) status of level crossings and (d) orientation of the needles. Most of the infrastructure management entities have a CTC that handles centralized all these issues. The applications and systems that handle these tasks are very robust and have a performance index near 100%. Problems occur when these systems fail. In those situations, traffic management has to be performed manually and through voice communications between traffic operators and railway drivers (Sciutto et al., 2007).

In this section web described a support system to assist traffic operators in emergency situations in which CTC systems fail. The main objective of this system is to reduce human error caused by the situations in which priority systems do not work properly.

5.1.1 Functional requirements

CTC traditional systems are centralized and rely on wired communications. When CTC system or communications fail, no one knows the location of trains, thus increasing the chances of an accident. In these situations, the railway companies put into operation its security procedures that transfer the responsibility of traffic management to traffic operators, who are people that monitor traffic in the terrestrial control centres. These people should manage the traffic manually communicating through analogical radio systems to the drivers of the trains. As people get nervous in emergency situations and that leads to mistakes, the new service aims to reduce these errors by creating a new tool to help traffic operators in emergency situations. This new tool must be based on different technologies to those used by traditional CTC systems so that failure in the former does not cause failure in the latter.

Taking into account these motivations and requirements, a Backup Traffic Management Service (henceforth BTMS) has been developed (Carballedo et al., 2010b). This service will assist traffic operators when the primary system fails. The main functions of this new system are:

- **Traffic situation representation for the track stretches where the main system do not provide information.** The new service represents the affected line stretches situation (train locations, track section occupation states, etc.) from information received from train-side systems through real-time wireless 'train-to-earth' communications (see Fig. 6).
- **Traffic management environment.** The objective is to provide a traffic assistance application in order to assist operators in tasks related to traffic control when the main system fails partial or totally.
- **Statistical analysis.** About aspects related to the system performance and reliability.
- **Control message sending from control centre to trains.** This functionality will allow traffic operators to send messages to the train drivers in order to manage and control the traffic.

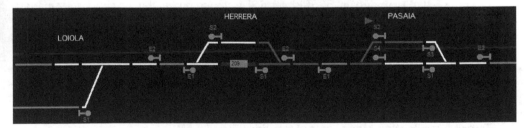

Fig. 6. Traffic situation representation.

The BTMS provides a traffic assistance application that works independently of the main CTC system. Thus, the new service is based on an application that informs about the position of the trains on track and permits to make tasks related to traffic management and control in an easier way. Moreover, this system permits a new way of communication between the traffic operators and trains drivers: exchanging control messages.

5.1.2 Architectural and design issues

In this section, we describe the most important technical considerations about the BTMS. The main issues are those related to (1) trains positioning, (2) wireless 'train-to-earth' communications and (3) added value services. These three issues are described below.

1. **Train positioning system.** The BTMS permits a new way of train positioning which works independently of the main system operation. In order to achieve this target, a new hardware/software module is boarded on each train. This module combines the positioning data provided by some hardware devices (accelerometers, gyroscope, odometer, etc.) with the coordinates given by a GPS module. To generate the most accurate positioning information, this system parts from a railway lines different tabulation ways. In this case, the tabulation is related to lines lengths (in kilometres) and the traffic signals positions. Based on this information, and the data extracted from the hardware and software modules boarded on trains (including GPS), this system translates this information to kilometric points (Shang-Guan et al., 2009). Then this positioning information is sent to the control centre in real-time.
    Besides, the BTMS communicates with an external positioning information system which permits the reception of train positioning information generated by the main CTC system.
2. **Real-time train-to-earth wireless communications.** The BTMS permits real-time train traffic management, so it is necessary to enable a real-time wireless communication channel between the BTMS installed on the control centre and the trains. For this reason, this service needs to use the previously explained train-to-earth wireless communications architecture, which enables slight communications and is based on mobile technologies (Aguado et al., 2005).
3. **Additional Services.** Using the mentioned train-to-earth communication architecture and the information provided by the on board positioning system positioning, two services have been developed related with the functionality of the BTMS.

The first one is a *Statistical Analysis Service*, which using the information stored by the positioning system on a data base, the BTMS can make statistical analysis related to the system's reliability level, GPS and GPRS coverage, and other system functionality aspects. Thus, one of the main goals of this service is to compare the received information, determining if the positioning provided by the train-side systems is according to the information generated by the primary CTC system. And the second is a *Control Message Exchanged Service*, which allows the procedural alarms transmission to the train-side systems. These kinds of alarms indicate anomalous situations to the train drivers: primary system failure, signal exceeds authorization to a certain point as a consequence of a failure of any electro-mechanical track component, etc.

5.2 Remote Application Management Service (Remote-AMS)

One of the main objectives of applications running inside the train is to provide information in order to facilitate the work of the train driver. Usually these applications need to use information generated in the ground centre. If this information changes, it needs to be updated remotely. In addition, there are terrestrial applications that use information generated by some on board applications. Therefore, it is necessary to be able to download that information from trains.

In order to resolve these issues, a new service that allows the remote management of on board applications (upgrade, download and deletion of information) will be proposed. This new service is composed by a terrestrial software module and an on board one that communicate each other to permit this remote management. So, this system would be integrated with the communication architecture described before based in heavy communications scheme (voluminous information with no real-time requirements).

5.2.1 Functionality

The main functionality of this service is to control the updating of the information used by applications running on the train terminal (for example track flat information or supporting documentation for the driver generated by the ground information systems) as well as downloading and deleting information generated by some on board applications (for example log files) remotely from the ground centre.

The solution consists of two software applications, one for the "ground" (control centre) and the other to be deployed in all train terminals. These applications are integrated with the previously described connectivity architecture via heavy communication since it involves the exchange of large volumes of information that do not require real-time communications.

Therefore, in this case, the Broadband Communication Manager (BCM) will be responsible for the arbitration of the train-to-earth heavy communication requests which are made by the software application running on the ground centre. This terrestrial application aims to communicate with those applications running inside the train to carry out all tasks related to the described service.

Thus, the terrestrial Remote-AMS application is installed on ground centre, and it will be responsible for managing the status of all applications in each terminal. On the other hand, on board Remote-AMS application will handle update, download and deletion requests made by the terrestrial Remote-AMS application. So, **terrestrial and on board Remote-AMS functionality** involves these issues:

- Knowledge about the configuration information for each application installed on each train terminal at any time (version, creation date, last updated date, etc).
- Knowledge about files and/or documents used by each application that can be updated, downloaded and/or deleted. This information will include: version, creation and last update date, update status (pending or not), etc. This management is done through a repository of information in a database.
- Management of information (files) of all current and future applications installed in the train terminals. For such management the Remote-AMS service will use FTP (File Transfer Protocol) configured to work locally (inside the train). So, the technology used is well known and standardized for remote management information.
- Management of the updating of configuration information of the applications installed on the train terminals.
- Management of the download of the information generated by the applications running on the train terminals.
- Management of the deletion of obsolete information in train terminals.

- Manage queries about the status information of the on board applications.
- Integration with the train-to-earth wireless communication architecture via heavy communications scheme.

5.2.2 Architecture

As mentioned before, this service is integrated with the previously described connectivity architecture via heavy communications scheme. So, when the terrestrial Remote-AMS generate tasks which involve downloading or uploading information from and to trains, it have to communicate with Broadband Communications Manager (BCM), because this is the entity who arbitrates heavy communications between ground and train applications. In this case, BCM arbitrates communications between terrestrial and on board Remote-AMS application. For proper integration with BCM, Remote-AMS (more specifically the terrestrial one) shall be compliant with the protocol of communication established by this management entity.

At this point it is important to remember that BCM does not interfere between final applications communication. The Fig. 7 shows Remote-AMS service architecture and its integration with the train-to-earth wireless communication technology.

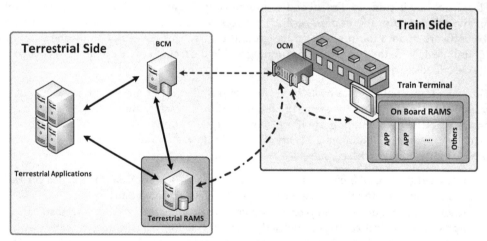

Fig. 7. Remote Application Management Service (Remote-AMS) and how it interoperates with the Broadband Communication Manager (BCM).

So, whenever terrestrial Remote-AMS schedules a task it has to send to the BCM a connection message. Once connected, there will be many communication requests as required. When BCM determines that a Remote-AMS request should be addressed, sends a notification to terrestrial Remote-AMS indicating to perform the service task corresponding to this request. Once the communication is completed, terrestrial Remote-AMS sends a notification to BCM which sets this request as completed and removes it from the corresponding communication prioritization queue. This same pattern is followed for all terrestrial Remote-AMS communication requests.

6. Future work

As future lines of work, the major efforts in train-to-earth wireless communication are focused on the improvement of the capabilities of the communication channel. Concerning with these improvements the followings are two of the hot topic to deal with:

1. **Communication network virtualization.** Where the technology and the terrestrial platform which takes part in the digital contents interchange are selected transparently to the front-end back-end applications depending on which technology suits better the communication features (information volume, nature and priority, communications cost; coverage, and so on).
   Nowadays, the wireless technologies available for this kind of communication between the railway and the terrestrial platform are: WiMax, WiFi, GPRS, UMTS, TETRA or GSM-R. Nevertheless, the proposed solutions have to be compatible with any other future communication technology.
2. **Hybrid self-managed and shared communication channel.** The challenge is to design a shared communications channel to be used for all applications, regardless of current or future functionality of those applications. With this new communications scenario, applications will only provide the information to be transmitted, and the destination of communication (being hidden protocols and the complexities of the communication). The channel itself will decide when is the right time to send the information and what is the best technology.

Apart from the improvements of the communications architecture, the design of a **framework for vertical services deployment** could be very interesting. This framework would offer an easy and seamless integration of new applications with the wireless communication infrastructure and with other future horizontal services. This infrastructure is based on standards and technological paradigms which are highly/ long enough proved in different environments. Furthermore, their interoperability and integration benefits are sufficiently contrasted; one example is the SOA (Service Oriented Architecture) paradigm case. It would be interesting to adopt the Software Engineering best practices and standard of interoperability used in other areas in the railway industry.

Finally, the proposed infrastructure would boost the **development of new vertical services** which can be classified in four categories: (1) driver assistance services, (2) services for passengers, (3) freights tracking services (based on RFID technology) and (4) services for train health monitoring. All these services have in common the need for exchange of digital content (often multimedia) between trains and ground control centres. The ubiquitous nature of connectivity that is provided by the new communications scenario will improve existing railway applications. And furthermore, will facilitate the development of new context-aware and customized services for end users. These advanced features result in fundamental improvements in the field of rail services.

7. Conclusion

In the railway industry, communications were born almost exclusively for the purpose of managing and regulating traffic flow, requiring by the mobile nature of this sector two modes of communication: those that occur between fixed elements of the rail infrastructure, which are based mainly in wired systems; and those which participate in the fixed and mobile elements (communications known as "train-to-earth"), which require a wireless communication channel and traditionally have materialized on the use of analogical communication systems such as traditional phone or radio.

Today, despite the maturity of the railroad industry and the advances in wireless communications technologies, the rail industry continues to base the operation of its priority services in analogical and wired communication technologies, which in fact belong to the past but still are efficient and robust.

New generation of wireless communications technologies, such as those based on conventional mobile technology (GSM, GPRS or UMTS), or broadband solutions (such as WiFi or WiMax), opens countless possibilities of use in the railway industry. As the cost of their deployment is very low, they perfectly complement traditional communication systems, and they have wide bandwidth and wide coverage that enable the deployment of new generation services in this area, some of them directly related to the end user, in order to provide a high quality transport service. On the other hand, the rise of high-speed trains is facilitating EXPANSION GSM-R as a basis for communication between signaling and regulation systems for the railway traffic.

This is precisely the topic on which this chapter is focused. It described a specific architecture of next-generation wireless communications for the rail industry to establish a train-to-earth bidirectional communication channel. This architecture is a single channel of communication between all train applications and those in the control centres. The aim of this communication channel is to standardize the way the data is transmitted between them. Thus, this channel is a resource shared by all the applications that simplifies the complex details related to communications and provides advanced services oriented to communication; services such as the selective treatment of the transmissions based on the nature and volume of the information, the location of the messages destination, the management of priorities and arbitration of communication shifts, attempts management, and so on. Moreover, we illustrate the challenges of bandwidth management in railway wireless broadband communications, and how we have faced to them. We have designed a new system that distributes communication shifts between terrestrial applications and train systems, which require the exchange of large amounts of information.

This chapter summarized the results of the research in train-to-earth wireless communications done during the last five years in collaboration with train manufacturers, railway technology providers and railway operators. Our wireless communications architecture has been incorporated into the manufacturing process of a new series of trains, which is a European-wide revolution since it enables wireless and transparent communication between terrestrial applications and those which are deployed on the trains. Furthermore, this architecture is being the basis for new digital services currently under development which will be in the market in a short time. They have different nature and purpose: from services that control the status of the train, to services for the end-user, and support systems for the train drivers.

8. Acknowledgment

This research was partially supported by Eusko Trenbideak - Ferrocarriles Vascos (a railway company from the north of Spain), Innovate and Transport Engineering (a railway technology provider), the Ministry of Industry, Tourism and Trade of Spain under the Avanza funding program (Grant TSI-020501-2008-148), and the Basque Country Government under the Euskadi+09 funding program (Grant UE09+/70). This support is gratefully acknowledged.

9. References

Aguado, M.; Jacob, E.; Saiz, P.; Unzilla, J.J.; Higuero, M.V. & Matias, J. (2005). Railway signaling systems and new trends in wireless data communication, *Proceedings of IEEE 62nd Vehicular Technology Conference*, pp. 1333-1336, ISBN: 0-7803-9152-7, Sept 2005.

Aguado, M.; Onandi, O.; Agustin, P.S.; Higuero, M. & Jacob Taquet, E. (2008). WiMax on Rails: A Broadband Communication Architecture for CBTC Systems, *IEEE Vehicular Technology Magazine*, Vol. 3, No. 3, pp. 47-56, ISSN: 1556-6072, Nov 2008.

Ambegoda, A.L.A.T.D. ; De Silva, W.T.S.; Hemachandra, K.T.; Samarasinghe, T.N. & Samarasinghe, A.T.L.K. (2008). Centralized Traffic Controlling System for Sri Lanka Railways, in *Proc. of 4th International Conference on Information and Automation for Sustainability (ICIAFS 2008)*, Colombo, Sri Lanka, pp. 145-149, ISBN 978-1-4244-2899-1, Dec 2008.

Berrios, A. (2007). Las comunicaciones ferroviarias: avances y reto. *Anales de mecánica y electricidad*, Vol. 84, No. 1, pp. 64-69.

California Software Labs (2008). *Extending end-to-end QoS to WiFi based WLAN*, CSWL whitepaper, http://www.cswl.com/whitepapers/qos-wireless-lan.html.

Carballedo, R., Perallos, A., Salaberria, I. & Gutiérrez, U. Managing 'train-to-earth' heavy communications: A middleware software to manage broadband wireless communications in the railway scope, in *Proc. of International Conference on Wireless Information Networks and Systems (WINSYS 2010) - 7th IEEE International Joint Conference on e-Business and Telecommunications (ICETE)*, Athens, Greece, pp. 1-6, July 2010.

Carballedo, R., Perallos, A., Salaberria, I., Odriozola, I. & Gutiérrez, U. A backup system based on a decentralized positioning system for managing the railway traffic in emergency situations, in *Proc. of the 13th International IEEE Conference on Intelligent Transportation Systems (ITSC 2010)*, Madeira Island, Portugal, pp. 285-290, Sep. 2010.

European Telecommunications Standards Institute (ETSI) (2008). *Electromagnetic compatibility and Radio spectrum Matters (ERM); Digital Mobile Radio (DMR) General System Design*, technical report: ETSI TR 102 398 V1.1.2, May 2008.

European Telecommunications Standards Institute (ETSI) (2011). *TErrestrial Trunked RAdio (TETRA)*, http://www.etsi.org/website/Technologies/TETRA.aspx

Garstenauer, J. & Pocuca, S. (2011). The future of railway communications, *Proceedings of the 34th International Convention MIPRO*, pp. 421-423, ISBN: 978-1-4577-0996-8, Opatija, Croatia, May 2011.

Gatti, A. (2002). Trains as Mobile devices: the TrainCom Project. *Wireless Design Conference*, London, 2002.

Gutiérrez, U., Salaberria, I., Perallos, A. & Carballedo, R. Towards a Broadband Communications Manager to regulate train-to-earth communications, in *Proc of 15th IEEE Mediterranean Electrotechnical Conference (MELECON 2010)*, La Valletta, Malta, pp. 1600-1605, May 2010.

IEEE 802.11 (2007). *IEEE Standard for Information technology - Telecommunications and information exchange between systems - Local and metropolitan area networks - Specific requirements - Part 11: Wireless LAN Medium Access Control (MAC) and Physical Layer (PHY) Specifications*, ISBN: 0-7381-5655-8, June 2007.

IEEE 802.16.2 (2004). *IEEE Recommended Practice for Local and metropolitan area networks - Coexistence of Fixed Broadband Wireless Access Systems*, ISBN: 0-7381-3986-6, 2004.

International Union of Railways (UIC) (2011). *GSM-R: the railway system for mobile communications*, http://www.uic.asso.fr/uic/spip.php?rubrique851

Laplante, P.A. & Woolsey, F.C. (2003). IEEE 1473: An open source communications protocol for railway vehicles, *IEEE Computer Society*, Vol. 5, Issue 6, pp. 12-16, ISSN: 1520-9202, Nov. 2003.

Marrero, D. ; Macías, E.M. & Suárez, A. (2008). An admission control and traffic regulation mechanism for infrastructure WiFi networks, *IAENG International Journal of Computer Science*, Vol. 35, Issue 1, pp. 154-160, ISSN: 1819-656X, 2008.

Noh-sam P. & Gil-Haeng, L. (2005). A framework for policy-based sla management over wireless lan, in *Proceedings of the Second International Conference on e-Business and Telecommunication Networks (ICETE 2005)*, pp. 173-176, INSTICC Press 2005, ISBN 972-8865-32-5, Reading, UK, October 2005.

Pinto, P.; Bernardo, L. & Sobral, P. (2004). Service integration between wireless systems: A core-level approach to internetworking, in *Proc. of 1st International Conference on E-Business and Telecommunication Networks (ICETE 2004)*, pp. 127-134INSTICC Press 2004, ISBN 972-8865-15-5, Setúbal, Portugal, August 2004.

Salaberria, I.; Carballedo, R.; Gutierrez, U. & Perallos, A. (2009). Wireless Communications Architecture for "Train-to-Earth" Communication in the Railway Industry, *Proceedings of 6th International Symposium on Distributed Computing and Artificial Intelligence 2009 (DCAI2009)*, S. Omatu et al. (Eds.): IWANN 2009, Part II, LNCS 5518, Springer-Verlag, ISBN: 3-642-02480-7, pp. 625-632. June 2009.

Sciutto, G.; Lucchini, M.; Mazzini, D. & Veglia, C. (2007). Technologies to Support the Railway Circulation in Emergency Conditions, in *Proc. of 2nd International Conference Safety and Security Engineering*, Malta, 2007.

Shafiullah, G.; Gyasi-Agyei, A. & Wolfs, P.J. (2007). Survey of Wireless Communications Applications in the Railway Industry, *Proceedings of 2nd International Conference on Wireless Broadband and Ultra-Wideband Communications (AusWireless)*, pp. 27-30, ISBN: 9780769528465, Sidney, Australie, Aug 2007.

Shang-Guan, W.; Cai, B-G.; Wang, J. & Liu, J. (2009). Research of Train Control System Special Database and Position Matching Algorithm," in *Proc. of IEEE Intelligent Vehicles Symposium*, Xian, China, pp. 1039-1044, ISBN 978-1-4244-3503-6, June 2009.

Yaipairoj, S.; Harmantzis, F. & Gunasekaran, V. (2005). A Pricing Model of GPRS Networks with Wi-Fi Integration for "Heavy" Data Users, in *Proceedings of the Second International Conference on e-Business and Telecommunication Networks (ICETE 2005)*, pp. 80-85, INSTICC Press 2005, ISBN 972-8865-32-5, Reading, UK, October 2005.

Part 3

Biological Effects of Wireless Communication Technologies

10

Evaluations of International Expert Group Reports on the Biological Effects of Radiofrequency Fields

Luc Verschaeve

Scientific Institute of Public Health, O.D. Public Health and Surveillance, Brussels and University of Antwerp, Department of Biomedical Sciences Belgium

1. Introduction

Electromagnetic fields, in particular so-called radiofrequencies are used by mobile or wireless communication systems as for example GSM mobile telephones, DECT telephones, wifi etc. Recent years were characterized by a tremendous increase in applications and types of wireless communication systems and this is responsible for an important increase in human exposure to radiofrequency radiation. Discussions on alleged adverse health effects are going on for years and so far no consensus agreement has been reached. These discussions are held amongst scientists as well as amongst laymen from the general public and authorities. Radio, TV, newspapers and magazines often bring erroneous information to the public. But also scientists do not agree. The scientific literature is full of papers showing that these fields can be dangerous and others showing that they are not. This holds true for virtually all possible endpoints and scientific disciplines that were studied, going from *in vitro* studies on cell proliferation, genetic and immunological effects, over animal experimental data on cancer and non cancer issues and human epidemiological investigations. It is not uncommon that controversial results are reported by the same laboratory. This results in claims of 'danger' when reference is made to essentially 'positive' papers (showing adverse biological effects) or claims of innocuity when only papers showing no effects are emphasized. It is clear that all (peer reviewed) scientific data should be considered and carefully analysed in order to come to a best possible 'weight of evidence' evaluation of risk. According to the WHO (World health Organisation) and ICNIRP (International Committee on Non Ionizing Radiation Protection) a single study does not provide the basis for hazard identification. It can at the best form the basis of a hypothesis. Confirmation of the results of any study is needed through replication and/or supportive studies. Only the resulting body of evidence forms the basis for science-based judgments by defining exposure levels for adverse health effects and no observable adverse effects.

This is recognized by most scientists all over the world and this explains why there were and still are many expert groups issued from the scientific community that evaluate(d) the alleged adverse health effects of radiofrequency fields in general, and mobile telephone frequencies in particular. It should be noted that radiofrequencies pose the additional problem (not encountered with other agents) that effects can be thermal or non thermal. At high exposure levels cells or tissues can heat and thermal effects can be observed that are not obtained by normal environmental exposure levels as for example. when exposure is to radiation from a mobile phone base station antenna or when using the handset. Thermal effects are well known but experiments where thermal exposure levels were studied are not relevant in the discussion of "mobile phones and health". Yet, often thermal exposure levels were used, even when the authors of the study claimed that they investigated non thermal exposure levels (wrong experimental set up and dosimetry). It is therefore also important to evaluate not only the biology but also the dosimetric aspects of an investigation.

The purpose of the present chapter is to give an overview of the conclusions of different (inter)national expert groups based on their analyses.

2. Evaluation of different expert group reports (2009-2011)

We found 33 expert group reports that were devoted to health effects of radiofrequency fields and that were published in the period 2009-2011.

2.1 ICNIRP reports (2009)

Statement on the "Guidelines for limiting exposure to time-varying electric, magnetic and electromagnetic fields (up to 300 GHz)".The International Commission on Non-Ionizing Radiation Protection (ICNIRP). Health Physics 97(3):257-259 (2009).
Juutilainen J, Lagroye I, Miyakoshi J, van Rongen E, Saunders R,de Seze R, Tenforde T, Verschaeve L, Veyret B and Xu Z (2009) Exposure to high frequency electromagnetic fields, biological effects and health consequences (100 kHz – 300 GHz). In: Vecchia P., Matthes R., Ziegelberger G., Lin J., Saunders R., Swerdlow A., eds., Review of Experimental Studies of RF Biological Effects (100 kHz – 300 GHz), ICNIRP 16/2009, ISBN 978-3-934994-10-2 pp. 94-319.

The International Committee on Non Ionizing Radiation Protection (ICNIRP) consists of a main commission (12 members) and 4 subcommittee's: epidemiology (5 members), biology (8 members), physics (7 members) and Optics (7 members). Information on ICNIRP can be obtained at http://www.icnirp.de. ICNIRP works in close collaboration with WHO and publishes guidelines and statements (see above) as well as literature reviews that are prepared by their (subcommittee) members. The most recent review on biological effects of radiofrequency radiation is from 2009 (see above). It is a consensus report that was approved by all (sub) committee members and peer reviewed by other experts that do not belong to ICNIRP. The report took all peer-reviewed publications into consideration. It was later on updated and published as single review papers in the scientific literature (van Rongen et al., 2009; Verschaeve et al., 2010; Juutilainen et al., 2011). Recommendations (guidelines) are exclusively based on scientific grounds. Although many countries in the world do adopt the ICNIRP recommendations they are sometimes criticized for insufficient implementation of the precautionary principle. Yet, on pure scientific grounds the ICNIRP papers, recommendations and reviews may be considered of high quality.

Above mentioned ICNIRP documents indicate that it is not possible to deny the existence of non thermal effects following RF-exposure but they consider evidence in favour of such (adverse) effects very weak. Recent *in vitro* and *in vivo* cancer studies show that these effects are unlikely. Also recent epidemiological investigations (e.g., in 2009 already available results from the interphone study) were considered as being indicative for the absence of cancer risk from mobile phones. Other studies that allowed a sufficient 'weight of evidence' evaluation did also not show any indication of health-related biological effects. ICNIRP therefore concluded that there are no indications of non thermal adverse health effects and that their recommendations from 1998 (ICNIRP, 1998) do not need to be adapted.

2.2 Scientific Committee on Emerging and newly Identified Health Risks (SCENIHR), EU, January 2009

Health Effects of Exposure to EMF, Directorate general Health & Consumers, European Commission. January 2009, pp. 83.

SCENIHR produces reports and advises on new technologies which may constitute a health risk for humans. Examples are nanoparticles, but also radiofrequency radiation as those applied in wireless communication systems. A detailed report on health effects of electromagnetic fields was published in 2007 and updated in 2009.

SCENIHR expert group members are selected following a call. Apart from 3 permanent members there were 6 nominated members, all well known in the field and covering different scientific disciplines. They discussed all peer reviewed (English) papers. When other papers were considered the reason for doing so was explained. Evaluation was done according to criteria that were well defined in advance. They included a particular attention to the reported study methods, the number of participants in a study (test and control population), the number of cells or animals that were analysed in the study, possible bias and confounders and dosimetry. Therefore not all papers were given the same weight or importance. Explanations were given when some studies were excluded from the discussion or where given less attention. The focus was on papers that were published after the 2007 report.

The summary and conclusions of the SCENIHR (2009) report were that it is unlikely that radiofrequency radiation is carcinogenic although further studies on long-term cancer effects are needed due to the long latency period for most brain tumours. Some investigations showed non reproducible associations between RF-exposure and self-reported symptoms. Most studies were negative. Overall, recent investigations did not show effects of RF-exposure on reproduction and development, whereas findings of effects on the nervous system (e.g., cognitive effects) were not consistent. Effects on EEG should be further investigated.

SCENIHR concludes that it is still not possible to exclude a small risk from RF-exposure. Therefore uncertainties that were identified in the 2007 report were still present. The weight of evidence analysis is nevertheless rather reassuring. There were no minority opinions.

SCENIHR recommends further research, especially long-term prospective studies, including studies on children.

2.3 Reports from the Dutch health council

Health Council of the Netherlands. Electromagnetic Fields: Annual Update 2008. The Hague: Health Council of the Netherlands, 2008; publication no.2009/02.
http://www.gezondheidsraad.nl/sites/default/files/200902.pdf

The Health Council is an independent scientific advisory body. Its task is to provide the government and parliament with advice in the field of public health and health/healthcare research. The Standing Committee on Radiation and Health deals with questions relating to the health effects of exposure to radiation and questions surrounding the use of medical imaging techniques. Following the rise of technologies such as mobile telephony, attention has in recent years mainly focused on the risks of non-ionizing radiation. Applications, such as high-voltage power lines, also give rise to queries from time to time. The standing committee also monitors scientific developments in the field of ionizing radiation, ultraviolet radiation and ultrasound. Members of the standing committees are carefully selected so as to form a multidisciplinary group of independent experts.

The annual update 2008 (published in 2009) considered two different aspects of RF-bio effects: RF-effects on brain function and 'Electromagnetic Hypersensitivity'. It was prepared by the members of the "electromagnetic field committee" and discussed and approved by the standing committee "Radiation". The report includes a description of the criteria used in the evaluation process. These were inclusion of peer reviewed scientific papers of 'sufficient' quality only, attention for dose-effect relationships and reproducible or consistent results that were supported by quantitative and statistical analyses. Possible working mechanisms were also taken into consideration although absence of such mechanisms did not necessarily exclude plausibility of a causal relationship between exposure and effect. For human studies further attention was paid to 'double blind studies', the constitution of the control populations and other methodological aspects of the study (exposure regimes etc.). Minority opinions were allowed.

The Health Council's conclusion was that effects on brain function were described in some papers but that there were no indications that they might be hazardous. They also concluded that good quality papers do not support the existence of a causal relationship between RF-exposure and symptoms like headache, migraine, fatigue, itching, insomnia etc. But there was a relationship between supposed RF-exposure and subjective symptoms indicating the presence of a nocebo effect. No advises were formulated.

The Dutch health council also published other reports or advises on the subject that we do not consider here (see http://www.gezondheidsraad.nl/en).

2.4 Statens strålskyddsinstitut (SSI = Swedish radiation protection agency)

Recent Research on EMF and Health Risks; Sixth annual report from the Independent Expert Group on Electromagnetic Fields, 2009
http://www.stralsakerhetsmyndigheten.se/Allmanhet/

The Swedish Radiation Protection Agency has appointed an independent international expert group for the evaluation of scientific developments and in order to provide advises on the possible health effects of electromagnetic fields. This working group takes into consideration other expert group reports as a basis for its discussions and reports that should be updated each year. The report from 2009 is the 6th and latest report that was published so far. It concerns *in vitro* and *in vivo* effects of radiofrequencies, in particular genotoxic and non genotoxic endpoints, effects on reproduction, neurodegenerative effects, immunological effects, behavioural effects, cancer etc. Also human studies were evaluated including investigations on brain activity, cognitive functions, sleep disorders, subjective complaints and epidemiological (cancer) studies. The working group consisted of 9 internationally renowned experts.

The report does not give an extensive description of the used methodology but it is clear that peer reviewed scientific papers were carefully evaluated. The conclusion of the report was that "...*there are no new positive findings from cellular studies that have been well established in terms of experimental quality and replication.*" It also stated that "...*recent animal studies have not identified any clear effects on a variety of different biological endpoints following exposure to RF-radiation typical of mobile phone use, generally at levels too low to induce significant heating.*" The SSI furthermore concluded that there are no indications of an increased cancer risk in mobile phone users (up to 10 years of exposure to mobile phone radiation). Absence of cancer risks (as by 2009) is consistent with the results from laboratory investigations in animals as well as with *in vitro* studies that did not identify a possible working mechanism. The working group also considered two studies on children that did not found any effect. In their evaluation of "electromagnetic hypersensitivity" the conclusion was that there were no indications other than the presence of a nocebo effect. The self-declared hypersensitivity is however considered a real health problem (but not caused by the radiation) that should receive sufficient attention.

The SSI did not formulate particular advises but it emphasised the need of further studies, especially on children.

2.5 EFHRAN reports

Report on the analysis of risks associated to exposure to EMF: in vitro and in vivo (animals) studies, July 2010
http://efhran.polimi.it/docs/IMS-EFHRAN_09072010.pdf
Risk analysis of human exposure to electromagnetic fields, July 2010
http://efhran.polimi.it/docs/EFHRAN_D2_final.pdf

Members of the "European Health Risk Assessment Network on Electromagnetic Fields Exposure" (EFHRAN) belong to research institutes from 7 different European countries and are supported by external collaborators from 12 countries. All are international experts in research on non ionizing radiation. Some industrial groups, as for example the European 'consumer voice' in standardisation – ANEC and the GSM Association (GSMA) or the Network Operators' Association AISBL (ETNO) were associated to EFHRAN. The working group evaluated investigations on animals and humans. The role played by EFHRAN members and associated groups in the realisation of the report was not made very clear. The evaluation of effects were done according to a scoring method that is similar to the one used by IARC (International Agency for Research on Cancer). For each endpoint the evidence was evaluated as being "sufficient", "limited", "inadequate" or "inexistent" (= lack of evidence). A critical evaluation was performed of the relevant scientific literature which was based on the data provided by the SCENIHR (2009) report and on data that were published afterwards. The EFHRAN report was devoted to different kinds of non ionising radiation but we will here only consider the evaluation of studies on radiofrequency radiation.

The EFHRAN conclusions were as follows:

Cancer related studies:

-   Limited evidence *in vitro* and lack of evidence with respect to *in vivo* investigations
-   Inadequate evidence for non genotoxic effects
-   Inadequate evidence from cancer studies in humans

Effect on the nervous system:

-   Lack of evidence for effects on the blood brain barrier
-   Limited evidence of effects on stress response genes and gene expression
-   Lack of evidence with respect to behavioural effects
-   Limited evidence from *in vitro* investigations
-   Inadequate evidence in humans related to neurodegenerative diseases and RF-exposure

Effects on reproduction and development:

-   Inadequate evidence concerning development and teratology
-   Inadequate evidence for reproductive effects in animals and *in vitro* studies
-   Inadequate evidence for effects in humans (e.g., behavioural effects in children from RF-exposed mothers)

Other effects:

-   Lack of evidence for auditory effects
-   Inadequate evidence of *in vivo* immunological effects
-   Inadequate evidence for cardiovascular effects in humans
-   No indications of electromagnetic hypersensitivity

2.6 Latin American expert committee on high frequency electromagnetic fields and human health, June 2010

Latin American Expert Committee on High Frequency Electromagnetic Fields and Human Health. Scientific review: Non Ionizing electromagnetic radiation in the radiofrequency spectrum and its effects on human health.
www.wireless-health.org.br/downloads/LatinAmericanScienceReviewReport.pdf

The goal of this study was to comply with the increasing anxiety of the population from Latin American countries with regard to their exposure to non ionizing radiations, especially from wireless communication systems (mobile phones, handset and base station antennas). The report was written by an expert panel which consisted of 5 scientists from different South American countries and a number of renowned international experts. The study was performed on request of the Eduled Institute for Medicine and Health which is a non-profit research- and development institute at Campinas, Sao Paulo (Brazil).

The study reviewed some 350 scientific investigations that were published since February 2010, with emphasize on studies that were performed in South America. Special attention was devoted to Risk Communication and application of the precautionary principle (which are usually not considered in other expert group reports). Attention was also given to regional and international exposure standards and recommendations from international bodies such as ICNIRP (International Committee on Non Ionizing Radiation Protection), IEEE (Institute of Electrical and Electronics Engineers), ITU (International Telecommunication Union) and the FCC (Federal Communication Commission, USA).

This is a well done study but it should be stressed that it is written by a limited number of persons that were assisted by an advisory group with obvious ICNIRP/WHO signature. It is therefore not surprising that the conclusions were similar to those of ICNIRP and WHO. The conclusions were that there is insufficient evidence and lack of consistent data in favour of a causal relationship between low intensity radiofrequency radiation and short term adverse biological effects. The report acknowledge the existence of some alarming studies, e.g., on the blood brain barrier, but they were interpreted as due to thermal effects that are not relevant with respect to public exposures. Provocation studies in humans did not support the presence of health effects below thermal exposure levels. There were no indications of effects from mobile phone radiation on well being and no consistent indications of effects on cognitive functions, neurophysiologic and other physiologic or behavioural disorders. Epidemiological evidence is so far reassuring but it was acknowledged that we should await more studies on long term RF-exposures before any definite conclusion can be reached. The authors also stressed that it is not only important to investigate adverse health effects but that attention should also be paid to the benefits of wireless communication devises. They emphasize the need of correct information of the public via, for example, a central Latin American information centre for the general public and stakeholders. Not only biological effect studies are needed but also studies on socio-economic aspects of the mobile phone technology.

2.7 The Bioinitiative report (2007 – updated 2010)

BioInitiative: A Rationale for a Biologically-based Exposure Standard for Electromagnetic Radiation
www.bioinitiative.org/report/index.htm

This report was written by a number of individual scientists and public health and public policy workers who believe that existing public exposure standards for as well extreme low frequency fields (power lines) as radiofrequency radiation (mobile phones) are inadequate. Notably, not all authors were scientists and not all can be considered 'independent'. Possible conflicts of interest were not assessed. The purpose of this report was to assess scientific evidence on health impacts from electromagnetic radiation below current public exposure limits and to evaluate what changes in these limits are warranted now to reduce possible public health risks in the future. The report is a collection of a number of chapters, called 'sections', written by the individual authors. The sections were not written in a standardised way and there was apparently no consultation or discussion on these sections between the authors. The methods used to collect literature data were not defined. In most cases a selection of the available scientific material has been made in favour of those reporting alarming data (also from the non peer-reviewed literature) whereas negative (reassuring) data were often not reported. The selection criteria for inclusion or rejection of papers were not stated. The report is not a consensus report and the overall summary is often an over exaggeration that does not always comply with the content of the sections.

According to the report it is obvious that exposure to the electromagnetic fields, even at environmental exposure levels, constitute an important health risk for humans and that positive (alarming) data are reported (and considered very likely if not proven) for almost all biological endpoints that were investigated. The report therefore contains recommendations on establishing limits for exposure to electromagnetic fields that are much lower than the limits that are currently applied in many countries all over the world.

The report certainly has some merits but as stated above there are many shortcomings. A detailed evaluation of the Bioinitiative report and its shortcomings is for example given on the website of the Dutch Health Council and will therefore not be further detailed in this paper (http://www.gezondheidsraad.nl/sites/default/files/200817E_0.pdf).

2.8 The AFSSET report (2010)

Agence française de sécurité sanitaire de l'environnement et du travail (Afsset), Comité d'Experts Spécialisés liés à l'évaluation des risques liés aux agents physiques, aux nouvelles technologies et aux grands aménagements, Octobre 2009. Groupe de Travail Radiofréquences, mise à jour de l'expertise relative aux radiofréquences (Saisine n°2007/007) (2009).
www.afsset.fr/index_2010.php

The AFSSET became since 2010 the "French agency for Food, Environment and Occupational Health and Safety (now ANSES)". It was asked by the French government to provide an overview and evaluation of the scientific knowledge on biological effects from mobile phone frequencies. The request was especially focussed on alleged effects on the blood brain barrier and epidemiological investigations on brain cancer in relation with mobile phone and other wireless applications of radiofrequency radiation. A working group was constituted according to strict criteria following a call for experts. Members were experts in the different relevant area of the subject, including medical doctors, biologists, biophysicists, epidemiologists, engineers (dosimetry) and human and social sciences (1 chairman and 12 members). The working group produced a report that was submitted to another expert committee (CES) of 26 members comprising 4 members of the AFSSET working group. There were also approximately 30 external auditors.

The report was written following several (13) meetings that were held between September 2008 and October 2009 comprising 19 auditions. A large database of publications was used essentially including peer reviewed (English) papers. Other reports (SCENIHR, Bioinitiative, etc.) were also consulted in order to identify publications that might have been overlooked.

It may be interesting to know that many of the members were not the usual players involved in research on non ionizing radiation bio effects and no members of other expert groups on the subject. They all possessed of course the necessary expertise to fulfil their tasks.

According to the AFFSET report there are no indications for short or long term adverse health effects as a result of exposure to radiofrequency radiation. Epidemiological investigations were reassuring but nothing can be said about long term effects that were not yet (sufficiently) investigated. Upon receipt of the report from the working group AFFSET concluded a little bit more mitigated. Due to the presence of some studies showing effects and hence remaining uncertainties further research is encouraged.

2.9 IARC (2011)

IARC Monographs on the Evaluation of Carcinogenic Risks to Humans, Volume 102: Non-Ionizing Radiation, Part II: Radiofrequency Electromagnetic Fields [includes mobile telephones, microwaves, and radar] – in press (2011)
www.iarc.fr/en/media-centre/pr/2011/pdfs/pr208_E.pdf

In May 2011, 30 scientists from 14 countries met at the international Agency for Research on Cancer (IARC) in Lyon, France, to assess the carcinogenicity of radiofrequency electromagnetic fields. The results of this meeting will be published in the IARC Monographs (nr. 102; *in press*). This monograph will contain information on (1) exposure data, (2) studies of cancer in humans, (3) studies of cancer in experimental animals, (4) mechanistic and other relevant data, together with a summary and final evaluation and rationale. A summary report is already published (Baan et al., 2011). As for all other evaluations performed by IARC the evaluation of carcinogenic risks to humans of radiofrequency electromagnetic fields resulted from discussions that were held in different working groups (human cancer studies, animal cancer studies and other relevant topics + supporting group related to dosimetry) and in plenary sessions. Working group members were essentially chosen by IARC staff members based on their scientific merits as judged by their peer reviewed publications.

The general principles and procedures as well as the scientific review and evaluation process is well described in the IARC preamble document which can be found on the IARC website (http://monographs.iarc.fr/ENG/Preamble/CurrentPreamble.pdf). All participants have carefully filled in a conflict of interest document well in advance of the meeting as well as at the start of the meeting. Discussions were based on scientific reviews that were written before the meeting by some of the experts on subjects that belong to their field of expertise.

The evaluation of the carcinogenic risks to humans of radiofrequency fields results in a classification in one out of 5 categories (group 1, 2A, 2B, 3 or 4) as indicated in table 1. The decision is based on the human evidence and evidence in experimental animals where the designation "sufficient" evidence, "limited" evidence, "inadequate" evidence or "evidence suggesting lack of carcinogenicity" is given by voting. This results in an overall classification of the carcinogenic risk as indicated in figure 1. The overall evaluation can be changed (e.g., from group 2B to 2A, or 2B to 3) according to the arguments (evaluations) provided by the working group on mechanistic and other relevant data.

Group 1	*Carcinogenic to humans*	107 agents
Group 2A	*Probably carcinogenic to humans*	59
Group 2B	*Possibly carcinogenic to humans*	267
Group 3	*Not classifiable as to its carcinogenicity to humans*	508
Group 4	*Probably not carcinogenic to humans*	1

Table 1. Agents Classified by the *IARC Monographs*, Volumes 1–102 (http://monographs.iarc.fr/ENG/Classification/index.php)

IARC EVALUATION

		EVIDENCE IN EXPERIMENTAL ANIMALS			
		Sufficient	**Limited**	**Inadequate**	**ESLC**
	Sufficient	Group 1			
	Limited	Group 2A		Group 2B	
Evidence In Humans	**Inadequate**	Group 2B		Group 3	
	ESLC	Group 3			Group 4

Mechanistic data can be pivotable when the human data are not conclusive

Fig. 1. IARC evaluation based on evidence from human and animal data (figure provided by IARC).

According to IARC useful information was available regarding associations between the use of wireless phones and glioma, and to a lesser extent acoustic neuroma. The international Interphone study and studies from a Swedish research group (dr. Hardell) were found of most importance in the evaluation process. Both studies were found to be susceptible to bias – due to recall errors and selection for participation- but the working group nevertheless concluded that the findings of an increased risk at the highest exposed groups could not be dismissed as reflecting bias alone. A causal interpretation between exposure to mobile phone radiation and glioma and acoustic neuroma was therefore considered possible. The working group therefore decided that there is limited evidence in humans for the carcinogenicity of radiofrequency radiation. The working group also concluded that there is limited evidence in experimental animals for the carcinogenicity of RF-radiations. Although there was evidence of an effect of RF-radiation on some of the 'other relevant endpoints' the working group reached the overall conclusion that these results provided only weak mechanistic evidence relevant to RF-induced cancer in humans. Therefore, the conclusion is that radiofrequency fields should be classified in group 2B (possible carcinogenic; see Figure 1).

Radiofrequency radiation is thus classified in the same group (2B) than extreme low frequency magnetic fields, coffee and styrene. This raises some questions. Are their effects really comparable? Maybe the classification is not discriminative enough to allow differentiation in the overall EMF frequency range nor does it allow to sufficiently [account] for different qualities of underlying data. According to Leitgeb (2011a,b) other classification systems, e.g., the system developed in 2001 by the German Commission on Radiation Protection (SSK), allows categorization of evidence in other and more classes. Using this system Leitgeb assigned microwave radiation to class E0: "Lack of/or insufficient evidence for causality". This illustrates that a classification in the IARC group 2B should not be interpreted by the public as proof of carcinogenicity at the same level as group 2A and 1. This is of course not correct but very often done.

2.10 French national academy of medicine (2009)

The academy stated that the precautionary principle may not be 'misused' to impose unscientific opinions. Scientific data are needed, not a subjective interpretation of the precautionary principle. According to the Academy " *No mechanism is known through which electromagnetic fields in the range of energies and frequencies used for mobile communication could have a negative effect on health."*

2.11 French academy of medicine, academy of sciences en academy of technologies (2009)

The National Academy of Medicine, the Academy of Science and the Academy of Technologies deplore the conclusions drawn by AFSSET from their experts' report. The three Academies congratulate the experts for their work but roundly criticize the Agency's recommendations. It does not understand why the presentation of the report does not insist on the reassuring aspects that are much more important than the few studies reporting effects. The latter are not to be considered credible alert signals. The academies also do not agree with the AFSSET recommendation to reduce exposure to cellular antennas that they consider scientifically not justified.

2.12 French health ministry (2009)

The website (www.sante.gouv.fr/effets-sur-la-sante.html) of the French Health Ministry was updated in August 2009. It states that the hypothesis that radiation from mobile phone base station antennas can be hazardous to man is no longer valid. It also stated that there are no indications so far that radiation from the handset poses a health risk but did not exclude that this may be the case. The Ministry proposed a number of simple measures to reduce the radiation exposure, especially for children.

2.13 French Parliamentary Office for the Evaluation of Scientific and Technological Choices (OPECST; 2009)

According to the report of this parliamentary organisation one cannot be completely sure that mobile phone radiation is absolutely safe but there are no proven effects so far. For this reason the report states that the ICNIRP guidelines remain valid.

2.14 Report from the Belgian superior health council (2009)

http://www.health.belgium.be/eportal/Aboutus/relatedinstitutions/SuperiorHealthCoun cil/index.htm?fodnlang=en

The Belgian Superior Health Council (SHC) was founded in 1849. It is the scientific advisory body of the Federal Public Service "Health, Food Chain Safety and Environment". In order to guarantee and enhance public health, the council draws up scientific advisory reports that aim at providing guidance to political decision-makers and health professionals. The working group on Non Ionizing radiation of the SHC already made several reports/advises on topics related to wireless communication devices.

This advisory report nr. 8519 on standards for mobile phone masts is one of these. It follows previous advises on this topic and was issued in response to a request from the Minister of Public Health to supply the necessary elements for answering a letter sent by the GSM Operators' Forum (GOF) concerning masts that emit radio waves. In this letter, the GOF claims that the proposed standard of 3 V/m (at 900 MHz) is too rigid.

The SHC stresses that it takes the view that, on account of the scientific uncertainties, the precautionary principle must be applied in this case in order to protect the population and therefore it maintained its proposal of 3V/m. The SHC recommends once again that there should be a policy that favours independent measurements and research (biological effects, epidemiological studies, etc.). This should be done with the assistance of an administration that is competent in this matter and has sufficient staff at its disposal. Advise nr. 8519 (and previous ones) were promulgated before election of the new working group members who do not all agree with the conclusions and advises of the former working group. A revision of the advise in the light of new developments may be envisaged.

2.15 Bundestag (Germany, 2009)

This federal German authority confirmed the validity of the German radiofrequency exposure limits. This is based on the results of German research programmes on mobile phones. According to the Bundestag the exposure limits in force indeed offer sufficient protection against mobile phone radiation.

2.16 The German Mobile Telecommunication Research Programme (DMF, 2009)

The "German Mobile Telecommunication Research Programme" (http://www.emf-forschungsprogramm.de/) started in 2002 and came to an end in 2008. It contained 54 research projects on mobile telecommunication including many different topics (laboratory research, epidemiology, dosimetry) but also aspects of risk communication. The general conclusion was that there is *no reason to question the protective effect of current limit values. Yet, because of the remaining question on health risks from long-term exposure for adults and children and the existence of some studies showing effects one should remain careful with wireless communication technologies.*

2.17 Commission on Radiological Protection (SSK, Germany, August 2009)

The German Commission on Radiological Protection (SSK) has issued a statement in which they reaffirm that there is no scientific evidence of a genotoxic effect (effects on the DNA) of radiofrequency fields or of an influence on gene regulation.

2.18 The Bundesambt für Strahlenschutz (BfS, German, 2009)

According to the German Federal office for radiation protection (BfS) recent studies have failed to demonstrate effects of mobile phone radiation on human fertility. No adverse effects were found on testes and sperm cells. The few papers that showed such effect(s) were considered of low or no scientific value. Experiments on animals have not shown relevant effects whereas *in vitro* studies only showed effects in case of thermal exposure conditions.

2.19 German expert group on children by the Jülich research institute (2009)

http://juwel.fz-juelich.de:8080/dspace/bitstream/2128/3683/1/Gesundheit_16.pdf

This report should be seen as an opinion document written by a limited number of international experts. It gives essentially a summary of different workshops that were held on mobile phones and children. The purpose of the report was to inform the public and authorities about the risks for children from the mobile phone technology. In the report on "Children's Health and RF EMF Exposure" the expert group concluded that *the review of the existing scientific literature does not support the assumption that children's health is affected by RF EMF exposure from mobile phones or base stations.* It is not very clear on what grounds this expert group was constituted. This study was supported by the telecom industry.

2.20 Radiation and Nuclear Safety Authority (STUK, Finland, 2009)

The Finnish Radiation and Nuclear Safety Authority stated in its 2009-report that there are no indications so far for long-term adverse health effects from radiofrequency radiation. However, everybody can reduce its own exposure easily if this is found useful.

2.21 Radiation authority of the five nordic countries (2009)

Five Northern European countries (Denmark, Finland, Iceland, Norway and Sweden) have joined to form the "Radiation Authority of the five Nordic Countries". They have issued a common statement which says that *"the Nordic authorities agree that there is no scientific evidence for adverse health effects caused by radiofrequency field strengths in the normal living environment at present. [...] The Nordic authorities therefore at present see no need for a common recommendation for further actions to reduce these radiofrequency fields." "Furthermore, in terms of overall public exposure, mobile phones are a much more significant source of radiofrequency radiation than fixed antennas. If the number of fixed antennas is reduced, mobile phones will need to use higher power to maintain their connection, thereby the exposure of the general public may increase."*

The authorities emphasize the need of further well conducted research on the alleged effects of radiofrequency fields on health.

2.22 CCARS scientific committee (Spain, 2009)

The "Comité Cientifico Asesor en Radio-frecuencias y Salud" (CCARS) published a literature survey and opinion on mobile phones and health. This was essentially based on the most recent reports and opinions from national and international authorities. They concluded that recent scientific/technical breakthroughs do not justify changes in the present RF benchmark levels and exposure limits for the public and workers.

2.23 Council of ministers of the isle of man (United Kingdom, 2009)

According to a working group report there is no general risk to the health of people living near mobile phone base station antenna. The exposures are limited and well below the guidelines. The group also stated that there is no proven relationship between self reported electromagnetic hypersensitivity and electromagnetic fields. At least some of the symptoms may be related to anxiety about the presence of the new technologies. They finally consider that the precautionary principle can be applied yet, especially with respect to children.

2.24 Institute of Engineering and Technology (IET, 2010)

Position statement on low level electromagnetic fields up to 300 GHz.
www.theiet.org/factfiles/bioeffects/postat02final.clin?type=pdf

This is an update of a previous position statement on *"The possible Harmful effects of low-level electromagnetic fields of frequencies up to 300 GHz"*. It claims that there are still no data in favour of adverse health effects from low level (normal) exposure to the radiofrequency fields. The IET has formulated its statement after consultation of the scientific literature using scientific databases (Medline, biosis, inspec) which provided a total of 813 relevant publications over the period 2008-2009. About half of them were on radiofrequency fields. They included cancer studies (e.g., Interphone study results), laboratory investigations in animals and cells, studies on non thermal working mechanisms and others. The statement also emphasize the need of independent replication studies and asks scientific journals to publish results from well sound scientific research only, whatever the results are. Scientists were encouraged to perform good science and to publish only when their work is of excellent quality.

2.25 Reports from the Health Protection Agency (HPA)

http://www.hpa.org.uk/

The Health Protection Agency (formerly National Radiological Protection Board) issues different reports and information booklets on different aspects of (amongst others) mobile phone effects. According to their 2010 statement "there are thousands of published scientific papers covering research about the effects of various types of radio waves on cells, tissues, animals and people. The scientific consensus is that, apart from the increased risk of a road accident due to mobile phone use when driving, there is no clear evidence of adverse health effects from the use of mobile phones or from phone masts".

2.26 The Austrian ministry of health (2009)

The ministry states in a brochure that there is no scientific evidence that cellular phones are hazardous to man. The brochure yet recommends a reasonable use of a mobile phone and limited use by children.

2.27 Australian Radiation Protection and Nuclear Safety Agency (ARPANSA, 2009)

www.arpansa.gov.au/
www.arpansa.gov.au/pubs/eme/fact1.pdf

In an update of its fact sheet on mobile telephony and health ARPANSA says that "*there is essentially no evidence that microwave exposure from mobile telephones causes cancer, and no clear evidence that such exposure accelerates the growth of an already-existing cancer. More research on this issue has been recommended.* "*Users concerned about the possibility of health effects can minimize their exposure to the microwave emissions by limiting the duration of mobile telephone calls, using a mobile telephone which does not have the antenna in the handset or using a 'hands-free' attachment. "*There is no clear evidence in the existing scientific literature that the use of mobile telephones poses a long-term public health hazard (although the possibility of a small risk cannot be ruled out).*"

2.28 Health Canada, July (2009)

http://www.hc-sc.gc.ca/ewh-semt/radiation/cons/stations/index-eng.php
http://www.hc-sc.gc.ca/ewh-semt/radiation/cons/radiofreq/index-eng.php
http://www.hc-sc.gc.ca/ewh-semt/pubs/radiation/radio_guide-lignes_direct-eng.php

Health Canada is the Federal department responsible for helping Canadians maintain and improve their health, while respecting individual choices and circumstances. It publishes different documents and fact sheets (see for example website addresses given above). According to these the consensus of the scientific community is that RF energy from cell phone towers is too low to cause adverse health effects in humans. In fact, worst-case RF exposure levels emitted from cell phone towers are typically thousands of times below those specified by science-based exposure standards. The RF energy from cell phones also poses no confirmed health risk but it is acknowledged that cell phone use is not entirely risk-free due to distraction, possible interference with some (medical) devices or other sensitive electronic equipment.

2.29 Food and Drug Administration (FDA, USA, 2009 – 2010)

http://www.fda.gov/
http://www.fda.gov/Radiation-
EmittingProducts/RadiationEmittingProductsandProcedures/HomeBusinessandEntertain
ment/CellPhones/ucm116282.htm
http://www.fda.gov/Radiation-
EmittingProducts/RadiationEmittingProductsandProcedures/HomeBusinessandEntertain
ment/CellPhones/ucm116331.htm

The FDA updated its pages on cellular telephones and health. It states that the weight of scientific evidence has not linked cell phones with any health problems. The steps adults can take to reduce RF exposure apply to children and teenagers as well.

2.30 National Cancer Institute (NCI, USA, September 2009)

http://www.cancer.gov/cancertopics/factsheet/Risk/cellphones

A fact sheet from the National Cancer Institute stated that studies thus far have not shown a consistent link between cell phone use and cancers of the brain, nerves, or other tissues of the head or neck. More research is however needed because cell phone technology and how people use cell phones have been changing rapidly.

2.31 US Health Physics Society (2010)

http://hps.org/
http://hps.org/documents/Mobile_Telephone_Fact_Sheet_update_May_2010.pdf

This society also publishes different fact sheets on mobile phones and wireless communication technologies. A recent one on mobile telephones does not deflect from previous ones as it still stated that the available evidence does not show that use of mobile phones or exposure to emissions from their base stations (cell towers) causes brain cancer or any other health effect.

2.32 Committee on Man and Radiation (COMAR, 2009)

http://ewh.ieee.org/soc/embs/comar/

This committee is a technical committee of the "Engineering in Medicine and Biology Society" (EMBS) of the "Institute of Electrical and Electronics Engineers" (IEEE). This committee is particularly interested in the biological effects of non ionizing radiations, including radiofrequency fields. The conclusions from their scientific evaluation stated that the scientific evidence is absolutely not in accordance with what the Bioinitiative project asserted. Indeed the weight of evidence does not support the safety limits recommended by the Bioinitiative group. COMAR recommends on the contrary that the public health officials continue to base their policies on RF safety limits recommended by established and sanctioned international organisations such as ICNIRP, IEEE etc.

2.33 WHO reports

http://www.who.int/en/
http://www.who.int/peh-emf/publications/facts/factsheets/en/

http://www.who.int/mediacentre/factsheets/fs193/en/index.html

WHO published different fact sheets on electromagnetic fields and their effects on human health. An update of the "mobile phone fact sheet 193 (June 2011) is not very different from the previous version(s). It still states that to date, no adverse health effects have been established as being caused by mobile phone use. It also says that it is still too early to fully assess long term effects in humans but that results of animal studies consistently show no increased cancer risk for long-term exposure to radiofrequency fields.

2.34 Council of Europe's Committee on the Environment, Agriculture and Local and Regional Affairs (2011)

Committee on the Environment, Agriculture, and Local and Regional Affairs of the Council of Europe. The potential dangers of electromagnetic fields and their effect on the nvironment. 2011 May 6.
http://assembly.coe.int/main.asp?Link=/documents/workingdocs/doc11/edoc12608.ht
Jowitt T. GSMA slams Euro call for ban on wireless in schools. eWeek Europe. 2011 May 16.
www.eweekeurope.co.uk/news/gsma-slams-euro-call-for-ban-on-wireless-in-schools-29363
http://assembly.coe.int/Mainf.asp?link=/Documents/AdoptedText/ta11/ERES1815.htm

This committee referred to the precautionary principle in order to ask for a reconsideration of the existing guidelines or exposure standards.

This committee consists of 47 members. It can influence decisions of the European Union but is not entitled to adapt existing regulations or to adopt new ones. According to the committee several measures should be taken. These include (1) adoption of reasonable measures to reduce exposure of children to electromagnetic fields, (2) a reconsideration of the ICNIRP guidelines and advises, (3) adoption of campaigns to alert the public, especially concerning health effects on children and adolescents, (4) adoption of measures to protect hypersensitive subjects, (5) encourage new scientific research to develop new less hazardous technologies, (6) A 0.6 V/m exposure limit for radiofrequency technologies such as wifi, WLAN, wiMAX, DECT and mobile phones and indication of SAR-values on the appliances, (7) increasing public information to protect children and a ban on RF-sources in schools (DECT, mobile phones, wifi, WLAN, WiMAX), (8) siting of antenna for wireless communication devises only after a public consultation and all antennas should be at a reasonable distance from dwellings, (9) creation of risk assessment procedures and protection of "early warning scientists", and (10) research in biological effect studies should be encouraged by increasing research funds.

The report does not take into consideration the many other reassuring reports. Its conclusions are not based on a weight of evidence evaluation. The report has the merit that it brings forward the concerns of the public and that it proposes a number of measures that can be taken into consideration. Some of the proposed measures are however not very realistic, especially on the short run.

3. Summary of expert group evaluations

Table 2 gives a summary of the different expert group evaluations together with the main topics to which this evaluation refers and eventually formulated advises. The main result is formulated as "-" when the group concluded that there is no strong or insufficient evidence in favour of adverse health effects, or "+" when in their opinion evidence is sufficient to conclude that there is a real health risk.

EXPERT REPORT	CONCLUSION	ADVISES	+/-
1. ICNIRP (2009) (all topics covered, advises/exposure standards)	No changes needed compared to previous advises	Recommandations (1998) remain valid	-
2. SCENIHR (2009) (all topics covered, *in vitro, in vivo, epidemiological investigations*)	-no cancer risk identified -insufficient evidence for electromagnetic hypersensitivity, cognitive effects and reproductive and developmental disorders -Uncertainties remain	-Need for more long-term investigations - Further research needed on effects on EEG during sleep	-
3. HEALTH COUNCIL OF THE NETHERLANDS (2008-2009) (electromagnetic hypersensitivity and effects on brain activity)	-No indications of effects on brain activity -No causal relationship between RF-exposure and complaints (hypersensitivity)	-	-
4. SSI (2009) (epidemiological investigations, *in vitro, in vivo studies*)	-No strong indications of effects on health	-More research on children needed	-
5. EFHRAN (2010) (human, *in vitro* and *in vivo* studies)	-No strong indications of effects on health. -*In vitro studies show at the most some 'limited evidence'*	.	-
6. LATIN AMERICAN EXPERT GROUP (2010) (all topics covered, includes exposure standards and risk communication)	-Insufficient evidence for adverse health effects from *in vitro* and *in vivo* studies -Epidemiological investigations are reassuring but uncertainty remains regarding long-term effects -Also advantages of mobile phones are highlighted	-Need to continue research -Attention to and funds for socio-economical studies are also needed	-

EXPERT REPORT	CONCLUSION	ADVISES	+/-
7. BIOINITIATIVE REPORT (2007-2010) (all topics covered)	-RF-radiation is hazardous to humans, even at low (daily life) exposure levels (= below the current exposure standards). Hazards were identified for virtually all possible endpoints	-Much stronger exposure standards than the current ones are needed	+
8. BELGIAN SUPERIOR HEALTH COUNCIL (2009-2010) (exposure standards for fixed antennas for mobile communication)	-Previous advises (3V/m at 900 MHz) remain valid	Exposure standards should be 3V/m based on the precautionary principle	(+)
9. AFSSET (2010) (Effects of mobile phones, especially on the blood-brain-barrier and brain cancer)	-So far no indications of short-term and long-term effects -Long-term effects remain uncertain yet	-Further research needed -Exposure levels can be reduced	-
10. FRENCH ACADEMY OF SCIENCES (2009) (all topics covered)	-No risks identified	-	-
11. FRENCH ACADEMY OF SCIENCES AND TECHNOLOGIES (2009) (all topics covered)	-No risks identified	-Reassuring results should also be highlighted	-
12. FRENCH MINISTRY OF HEALTH (2009) (all topics covered)	-No risks from base station antennas -No indications for risks from mobile phones (but still uncertainty)	-	-
13. OPEST (F) (2009) (all topics covered)	-Adverse effects from mobile phone technology are not proven yet	-	-
14. BUNDESTAG (D) (2009) (all topics covered)	-No risks -Adequacy of current German exposure standards is confirmed	-	-
15. SSK (D) (2009) (Genetic effects)	-No scientific evidence in favour of genotoxicity of RF-radiation	-Existing exposure limits should not be adapted	-

EXPERT REPORT	CONCLUSION	ADVISES	+/-
16. BfS (D) (2009) (Fertility)	-No significant effects on testes and sperm	-	-
17. GERMAN EXPERT GROUP ON CHILDREN (Jülich Research Institute) (2009) (risks for children)	-No indications of adverse health effects in children	-	-
18. DMF (D) (2009) (general)	-No reasons to lower current exposure limits	-Further attention needed	-
19. STUK (FIN) (2009) (general)	-No indications of long term effects	-	-
20. RADIATION SAFETY AUTHORITY OF 5 NORDIC COUNTRIES (Scandinavia) (2009) (all topics covered)	-There is no scientific base to conclude that RF-radiation at "normal exposure levels" is hazardous to humans - There is no reason to lower existing exposure standards	-Further research is needed	-
21. SSM (S) (2009) (*in vitro, in vivo*, human studies)	-No significant evolution in research data -No evidence for increased cancer risk	-	-
22. CCARS (E) (2009) (general)	-No increased incidence of brain cancer -Uncertainties remain with respect to long-term effects -No reasons to lower existing exposure limits	-	-
23. COUNCIL OF MINISTERS OF ISLE OF MAN (UK) (2009) (Antennas)	-No health risks for humans -Electromagnetic hypersensitivity related to mobile phones is not proven	-	-
24. INSTITUTE OF ENGINEERING & TECHNOLOGY (IET) (UK) (2010) (all topics covered)	-No indications of health risks	-	-
25. HEALTH PROTECTION AGENCY (HPA) (UK) (2010) (all topics covered)	-No danger from mobile phones (except traffic accidents)	-	-

EXPERT REPORT	CONCLUSION	ADVISES	+/-
26. AUSTRIAN MINISTRY OF HEALTH ((2009) (all topics covered)	- No danger from mobile phones	-Reasonable use of a mobile phone should be recommended, in particular by children	-
27. ARPANSA (AUS) (2009) (all topics covered)	-No evidence for an increased cancer risk from mobile phone radiation	-Advises for reduction of exposure levels for those who wish to do so	-
28. HEALTH CANADA (CAN) (2009) (all topics covered)	-No risks -Current exposure limits remain valid	-	-
29. FDA (USA) (2010) (general)	-No risks from mobile phones (also in children)	-	-
30. NCI (USA) (2009) (all topics covered)	-No adverse effects from a mobile phone -Uncertainty related to long-term effects warrants some care	-	-
31. COMAR (INT) (2009) (all topics covered)	-Scientific data are not at all in accordance with the conclusions and assertions of the Bioinitiative report -Exposure limits (IEEE and other) are certainly adequate	-	-
32. WHO (INT) (2010) (all topics covered)	-Adverse effects from mobile phones are not proven	-	-
33. IARC/WHO (2011) (cancer)	RF-radiation is possibly carcinogenic in humans (group 2B in IARC classification)	-	(+)

Table 2. Summary of the expert group reports (scientific disciplines, conclusions and advises; -/+: overall conclusion in terms of respectively absence of sufficient evidence for adverse health effects (-), or sufficient evidence for adverse health effects (+).

It can be seen from the table that the vast majority of the reports *do not* consider that radiofrequency fields at current exposure levels (especially from mobile phone base-station antennas and handsets) pose a serious health risk to humans. The only exception comes from the Bioinitiative report. All reports, except the Bionitiative report, conclude that there is so far no clear indication of adverse health effects from RF-exposure from applications for wireless communication purposes. They usually remain prudent with regard to long-term bio-effects, not because of strong indications that such effects might occur, but only because there are so far not enough data available to draw a sound conclusion. The same holds true for the IARC evaluation on carcinogenicity where the conclusion "possible carcinogenic" (group 2B) only means that, despite overall reassuring data, there is some limited evidence for carcinogenicity at long term exposures that cannot be ruled out so far. The Belgian Superior Health Council recommended more severe exposure limits (compared to most limits in application) but this recommendation is based on the precautionary principle rather than on solid arguments in favour of hazard or risk.

4. Evaluation of expert group reports based on 10 criteria

An evaluation of the different reports should take into account a great number of aspects. Amongst them the composition of the working group, the topics that were taken into account and the methods that were used are certainly some of the important aspects. We therefore tried to identify the members or participants in the working group activities and tried to see whether they constituted a *multidisciplinary* and *independent* group of experts. Did they evaluate all scientific (peer reviewed) publications, or did they make a selection of papers, and if so, what was the rationale for doing so? Was this satisfactory? Was the report a consensus report? Where minority opinions mentioned?

An evaluation of the reports bases on the answer to these questions can for example be done according to 10 criteria as indicated in Table 3. It is obvious that such an evaluation is always to a certain extent subjective. However, the purpose was not to make a ranking of the expert group reports according to their quality but especially to try to explain why they may (eventually) come to divergent conclusions on radiofrequency induced health effects. Because it is not possible to give in this chapter detailed answers to all the questions for each of the working groups the reports were given a score based on the answers and criteria indicated in table 3 (score of 0 when not a single criterion was met, up to 10 when all criteria were met).

Expert group:
- selection procedure of members and presence or absence of declarations of interest
- composition, complementarity and expertise of expert group members
- possibility to include minority statements

Methods used in the evaluation of the scientific data:
- peer reviewed publications, transparent procedure for selection of data
- method employed

Criteria for evaluation of scientific data:
- transparant and clearly described criteria
- attention to the number of participants/animals/cells considered in the studies
- attention to potential bias and confounding factors
- attention to dosimetry
- evaluation of used study methods and experimental set up in the studies under consideration

---

Table 3. Evaluation of expert group reports based on 10 criteria.

We did not made a full evaluation of all reports because some did not provide sufficient information or were not expert group reports *as such* as they were for example only opinion papers or short evaluations or advises as formulated in leaflets or fact sheets from certain organisations. In such cases a (re)examination of all available scientific data was not necessary and hence not attempted. Here, a "quality comparison" with the "bigger" reports would not be fair. The results of the evaluation are therefore only given as an example for a number of important reports (based on the criteria in Table 3, and summarized in Table 4).

It can be seen that most expert group reports got an excellent score, except the Bioinitiative report. This report certainly has merits and individual sections were often written by well renowned scientists, but overall it was deficient against most of the criteria as indicated before (see also http://www.gezondheidsraad.nl/sites/default/files/200817E_0.pdf). As mentioned before the purpose of the Bioinitiative report was to demonstrate that RF-radiation (at low-exposure levels as from mobile phones and their base station antennas) may be hazardous to humans. The purpose was to indicate that exposure limits should be considerably revised. The report was written in such a way that the outcome was in accordance with these goals.

As indicate above any such evaluation is always subjective to a certain extent, also because it is not always possible to fully appreciate the work that was done. Reports may mention that all peer reviewed papers were consulted but obviously this cannot be verified. They can mention that particular attention was paid to "conflict of interests" of the participating members, or report that literature data was carefully analysed and that particular attention was paid to, for example, aspects of biological dosimetry, but it was also not always possible to understand how this was done. Table 4 nevertheless can be useful as a general appraisal. It shows that most reports got a good to excellent 'score'. Reports from ICNIRP, SCENIHR or the Dutch Health council got a maximal score of '10' as they all fulfilled satisfactorily the 10 criteria of Table 3. All ICNIRP members are experts in non ionizing radiations bio-effects and/or dosimetry. Some questions can be raised on how the members were elected and in how far they constitute a balanced representation of opinions, but the methodology, through literature evaluations by subcommittee members and a careful and strong 'peer reviewed' process of their work can be seen as sufficient guarantee of quality. This justifies a high score although this does not automatically imply that ICNIRP opinions should be accepted without questions. ICNIRP was for example often accused of insufficiently applying the precautionary principle and hence of being not careful enough in its advises. This opinion can be defended. The same holds true for the reports from the Dutch Health Council. All criteria were met (= high score) which does not mean that the council is never criticized or criticisable. It is indeed often criticized, again for not applying the precautionary principle and insisting on absence of proof and lack of convincing data, hence not taking the few alarming data sufficiently into account. The Belgian Superior Health Council is on the contrary often criticized for emphasizing too much on the precautionary principle and providing advises that are scientifically not well sound. We have not extensively described their reports as they were *only* advises from a working group which did not perform a complete literature search and evaluation. The report from the IARC working group on Radiofrequency Electromagnetic Fields (including mobile telephones; Baan et al., 2011, and Monograph Volume 102, *in preparation*) also received a maximum score as it is based on an extensive evaluation of the scientific literature performed by a great number of experts and according to a well described and rigid procedure (see also http://monographs.iarc.fr/ENG/Preamble/index.php). Special attention was also taken to conflict of interests.

We already mentioned that most of the reports express the same opinion. This is not surprising knowing that they are all based on the same scientific data and evidence and usually also similar and well defined criteria. Another reason for fairly concordant conclusions may be yet that different expert groups were often partly composed of the same scientists. The Swedish SSI report was written following constitution of an expert group from which some members were also members from ICNIRP. The same holds true for EFRAN and EDUMED. It is not surprising then that these expert groups expressed the same opinion or did not substantially deviate from the ICNIRP position.

Study	Subject	Expert group	Method	Quality	Score
ICNIRP, 2009	RF– Epidemiology, animals & in vitro studies	+++	++	+++++	10
SCENIHR, 2009	RF-ELF-IF-Static fields ; Epidemiology, *in vitro* & *in vivo* studies	+++	++	+++++	10
Dutch Health Council (2009)	RF– Epidemiology and experimental human studies	+++	++	+++++	10
SSI (IEG), 2009	RF – Epidemiology and *in vitro* & *in vivo* studies	++	+	+++++	8
EFRAN, 2010	RF – Epidemiology and in vitro & in vivo studies	++	++	+++++	9
EDUMED, Latin American Expert Group, 2010	RF- epidemiology, experimental human, *in vitro* & *in vivo* studies	+	++	+++++	8
BIOINITIATIVE, 2007/2010	RF-ELF- epidemiology, experimental human, *in vitro* & *in vivo* studies	+	+	+	3
AFSSET, 2010	RF– especially blood-brain-barrier, epidemiology and psychosocial and cultural aspects	+++	++	++++	9
IARC, 2011	RF– studies on cancer and cancer related aspects (epidemiology, *in vitro* & *in vivo* studies)	+++	++	+++++	10

Table 4. Evaluation of a number of important expert Group reports based on well defined criteria (cf. Table 3).

5. Conclusion

From the more than 30 expert group opinions that were published during the 2009-2011 period the vast majority did not consider that there is a demonstrated health risk from RF-exposure from mobile telephones and other wireless communication devices. Because of remaining uncertainties, especially with respect to long-term exposures, some caution is still expressed. This is the reason why IARC recently classified RF-electromagnetic fields as 2B-carcinogens (= possibly carcinogenic).

6. References

Baan R., Lauby-Secretan B., El Ghissassi F. et al. on behalf of the WHO International Agency for Research on Cancer Monograph Working Group (2011) Carcinogenicity of radiofrequency electromagnetic fields. *Lancet Oncology*, 12, 624-626.

ICNIRP (1998) Guidelines for limiting exposure to time varying electric, magnetic, and electromagnetic fields up to 300 GHz. *Health Phys.* 74, 494-522.

Juutilainen J., Lagroye I., Miyakoshi J., van Rongen E., Saunders R., de Seze R., Tenforde T., Verschaeve L., Veyret B., Xu Z. (2011) Experimental studies on carcinogenicity of radiofrequency radiation. *Crit. Rev. Environ. Sci. Technol.* 41, 1664-1695.

Leitgeb N. (2011a) Editorial. *Wien Med. Wochenschr.* 161,225.

Leitgeb N. (2011b) Comparative health risk assessment of electromagnetic fields. *Wien Med. Wochenschr.*161,251-262.

van Rongen E., Saunders R., Croft R., Juutilainen J., Lagroye I., Miyakoshi J., de Seze R., Tenforde T., Verschaeve L., Veyret B., Xu Z. (2009) Effects of radiofrequency electromagnetic fields on the human nervous system. *J. Toxicol. Environ. Health B Crit. Rev.* 12, 572-597.

Verschaeve L., Juutilainen J., Lagroye I., Miyakoshi J., van Rongen E., Saunders R., de Seze R., Tenforde T., Veyret B., Xu Z. (2010) In vitro and in vivo genotoxicity of radiofrequency fields. *Mutation Res.* 705, 252–268

Part 4

Wireless Sensor Networks and MANETS

11

Multimedia Applications for MANETs over Homogeneous and Heterogeneous Mobile Devices

Saleh Ali Alomari and Putra Sumari
Universiti Sains Malaysia
Malaysia

1. Introduction

Mobile Ad Hoc Networks (MANETs) are considered a vital part in beyond third generation wireless networks (Nicopolitidis et al., 2003). In the matter of fact, they present a new wireless networking paradigm. Any sort of fixed infrastructure is not used by MANETs. They are important sorts of WLANs, therefore, in a distributed and a cooperative environment, MANETs do efficiently function (Murthy and Mano, 2004) (Sarkar et al., 2008). MANETs are networks of self-creating since there is a lack of routers, configuration prior to the network setup, Access Points (APs) and predetermined topology (Wu et al., 2007). MANETs are as well networks of self-administering and self-organizing. This is because in the network creation process, there is no application for central control. On MANETs, it is extremely hard to apply any of the central administration types, for instance, congestion control due to the dynamic nature of the network topology in MANETs, authentication or central routing. In short, several important applications benefited from MANETs, for example, in military, ubiquitous, emergency and collaboration computing.

In this chapter, describe the necessary background for the MANETs over homogeneous and heterogeneous mobile devices. The researcher begin this chapter to introduce the related background and main concepts of the Mobile Ad Hoc Network (MANETs) in Section 1.2, and explained briefly about the existing wireless mobile network approaches, wireless ad hoc networks, wireless mobile approaches in Section 1.2.2. The characteristic of MANETs are in Section 1.2.3. The types of Mobile Ad hoc network in Section 1.2.4. The traffic types in ad hoc networks which include the Infrastructure wireless LAN and ad hoc wireless LAN are presented in Section 1.2.5. In Section 1.2.6 highlight the relevant details about the ad hoc network routing protocol performance issues. The types of ad hoc protocols such as (Table-driven, On-demand and Hybrid) and Compare between Proactive versus Reactive and Clustering versus Hierarchical are in Section 1.2.7. And Section 1.2.8 respectively. The existing ad hoc protocols are presented in Section 1.2.9. The four important issues significant in MANET are Mobility, QoS Provisioning, Multicasting and Security is presented in Section 1.2.10. Furthermore, the practical application and the MANET layers are shown in Section 1.2.11 and Section 1.2.12 respectively . Finally, in Section 1.2.13 the summary of this chapter.

1.1 Overview of MANETs

The main concept of Wireless Local Area Networks (WLANs) refers to MANETs which are also called either infrastructure-based wireless networks or a single hop network

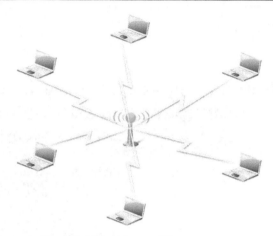

Fig. 1. Illustrates of a single hop WLAN with one AP

(Nicopolitidis et al., 2003) (Murthy and Mano, 2004). Inside a WLAN, the transmission is governed by at least one fixed Access Point (AP) between different mobile nodes. An existing network backbone and the stations contain a bridge as AP functions (Basagni et al., 2004). Both QoS and security issues are efficiently controlled by the AP within a particular network. Inside the network of WLAN, there is no need for different mobile nodes since the AP is the source that does communication through a single hop manner. Wireless network standards are included by the WLAN implementations and developed by Institute of Electrical and Electronics Engineers (IEEE) 802 project (IEEE 802.11, IEEE 802.11b, IEEE 802.11g, IEEE 802.11a, and IEEE 802.11n) and High Performance Radio Local Area Network Type 2 (HiperLAN2). In addition, the European Telecommunications Standardization Institute (ETSI) Broadband Radio Access Networks (BRAN) project (ETSI, 1999) developed the European version of IEEE 802.11a. A frequency of 2.4GHz runs for these standards. However, 5GHz runs for the IEEE 802.11a. For these standards, the transmission rates (bandwidths) are 2 Mbps where as for IEEE 802.11a and IEEE 802.11g, 54 Mbps is run. For IEEE 802.11b, 11 Mbps is run and for IEEE 802.11n, 100 Mbps is run. Note that a single hop WLAN with one AP is shown in Figure 1.

For mobile hosts, a new wireless networking paradigm indicates to a MANET. All sorts of fixed infrastructure are independent to MANET. In order to maintain a connection within the network, nodes (hosts) will rely on each other through a manner that is to be cooperative. Therefore, both computing and ubiquitous communication are considered to be two goals of mobile ad hoc networking. In the matter of fact, both of them are rapidly deployed in such a way they do not rely on a pre-existing infrastructure, for example, Base Station (BS) and Access point (AP) (Perkins et al., 2002). A peer to peer network refers to MANET which has the ability to allow a communication between each wireless client that relies on any infrastructure. MANET can also be defined as a mobile nodes collection of which a highly resource constrained network and a dynamic topology are formed by this collection (Mohapatra and Krishnamurthy, 2005) (Murthy and Mano, 2004). A single hop network refers to WLAN, Major functions within the network are being performed by the cooperation of the nodes. This process represents a mutli-hop network that refers to the MANET. There are such problems entitled in MANETs. These comprise; security, QoS, routing and energy conversation. These problems came due to several reasons: high mobility, resource constrains such as power, storage, and bandwidth (Negi and Rajeswaran, 2004), its cooperative nature and the dynamic topology of nodes operating in MANET's environment. In Defence Advanced Research Projects Agency (DARPA) Packet Radio projects (Jubin et al., 1987), ad hoc networking was initiated for military applications, specifically, for dynamic wireless networks since 1970s. Accordingly, this networking is not considered to be as a new concept. For MANET, a new networking group was formed within the Internet Engineering Task Force (IETF-manet) so that the standard Internet routing support could be developed for mobile IP autonomous segments. In addition, a framework for IP-based protocols in MANET will be developed as well. In the fields of mobile IP-based networks and wireless internet, the increasing improvement in the recent IEEE standards of 802 projects for wireless networks (Broch.J et al., 1998) has raised up. A MANET can be either heterogeneous or homogeneous depending on the type of mobile nodes being involved. When all mobile nodes are of the same type of a MANET, this is called a homogeneous MANET, whereas when different type of mobile nodes are involved, this is otherwise called a heterogeneous MANET. The homogeneous and heterogeneous mobile ad hoc network are shown in In Figure 2 and Figure 3 respectively. The same family of IEEE 802.11 standards is being used by MANETs. More

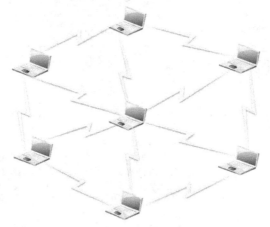

Fig. 2. Illustrates the homogeneous mobile ad hoc network

specifically, in Bluetooth and WLANs, these standards are being used (Morinaga et al., 2002). Table 1 shows a comparison between WLAN and MANET.

1.2 Mobil Ad Hoc Network

With the widespread rapid development of computers and the wireless communication, the mobile computing has already become the field of computer communications in high-profile link. MANET (Sarkar et al., 2008) is a completely wireless connectivity through the nodes constructed by the actions of the network, which usually has a dynamic shape and a limited bandwidth and other features, network members may be inside the laptop, Personal Digital Assistant (PDA), mobile phones and so on. On the Internet, the original mobility is the term used to denote actions hosts roaming in a different domain; they can retain their own fixed IP address, without need to constantly changing, which is Mobile IP technology.

Mobile IP nodes in the main action is to deal with IP address management, by home users and foreign users to the mobile node to packet tunneling, the routing and fixed networks are not different from the original. However, ad hoc network to be provided by mobility is a fully wireless, can be any mobile network infrastructure, without a base station, all

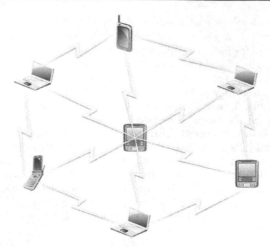

Fig. 3. Illustrates the heterogeneous mobile ad hoc network

Comparison Aspect	(WLAN)	(MANET)
Autonomous terminals	No	Yes
Self-configuration	No	Yes
Mobile host/router	No	Yes
Bandwidth constrained network	No	Yes
Infrastructure-based	Yes (APs/routers/Servers)	No
Power awareness	Does not matter	Yes
Security policy	Centralized	Distributed
Centralized/distributed operation	Centralized	Distributed
Routing	Easy	A bit difficult
Scalability	Easy	A bit difficult
Multicasting	Easy	A bit difficult
Static/ dynamic topology	Static	Dynamic
QoS guarantee	Can be guaranteed easily	A bit difficult
Typical applications	Home, enterprise network	Military/emergency
Single hope / multi hope	Single	multi
Communication mechanism	Base station type access	P2P

Table 1. Illustrates the comparison between WLAN and MANET

the nodes can contact each other at the same time take router work with the Mobile IP completely different levels of mobility. Early use of the military on the Mobile Packet Radio Networked (MPRN)in fact can be considered the predecessor of MANET, when the high-tech communication equipment, the size, weight continuously decreases, power consumption is getting low, Personal Communication System (Personal Communication System, PCs) concept evolved, from the past few years the rapid popularization of mobile phones can be seen to communicate with others at anytime and anywhere, get the latest information, or exchange the required information is no longer a dream. And we have gradually become an integral part of life. Military purposes, as is often considerable danger in field environment, some of the major basic communication facilities, such as base stations, may not be available, in this case, different units, or if they want to communicate between the forces, they must rely on MANET networks infrastructure. In emergency relief, the mountain search and rescue operations at sea, or even have any infrastructure can not be expected to comply with the topographical constraints and the pressure of time under the pressure, ad hoc network completely wireless and can be any mobile feature is especially suited to disaster relief operations when personal communication devices and more powerful, some assembly occasions, if need to exchange large amounts of data, whether the transmission of computer files or applications that display. if can connect with a temporary network structure, then the data transmission will be more efficient without the need for large-scale projection equipment would not have point to point link equipment such as network line or transmission line. The current wireless LAN technology, Bluetooth is has attracted considerable attention as a development plan. Bluetooth's goal is to enable wireless devices to contact with each other, if sentence formation adding the design MANET.

1.2.1 History of Ad Hoc Network

Nowadays, the information technology will be mainly based on wireless technology, the conventional mobile network and cellular are still, in some sense, limited by their need for infrastructure for instance based station, routers and so on. For the MANET, this final limitation is eliminated. The ad hoc network are the key in the evolution of wireless network and the ad hoc network are typically composed of equal node which communication over wireless link without any central control. Although military tactical communication is still considered as the primary application for MANET and commercial interest in this type of networks continues to grow. And all the applications such as rescue mission in time of natural disasters, law enforcement operation, and commercial as rescue and in the sensor network are few commercial examples, but in this time it's become very important in our life and they become use it.

The MANET application is not new one and the original can be traced back to the Defence Advanced Research Projects Agency (DARPA), Packet Radio Networking (PRNET) project in 1972 (Freebersyser and Leiner, 2001, Jubin and Tornow, 1987) which evolved into the survivable adaptive radio networks (SURAN) program. Which was primarily inspired by the efficiency of the packet switching technology for instance the store/forward routing and then bandwidth sharing, it's possible application in the MANET environments. As well commercial rescue in the PRNET devises like repeaters and routers and so on, were all mobile although mobility was so limited in that time, theses advanced protocol was consider good in the 1970s. After few years advance in micro electronics technology and it's was possible to integrate all the nodes and also the network devices into a single unit called ad hoc nodes, and then the advance such as the flexibility, resilience also mobility and independence of fixed infrastructure, and in that time they so interesting to use it immediately among military battlefield, Ad hoc networks have played an important role in military applications and related research efforts. For example, the global mobile information systems (GloMo) simulator (Leiner et al.), the near-term digital radio (NTDR) program and also has been the increase in the police, commercial sector and rescue agencies in use of such networks under disorganized environments. Ad hoc network research stayed long time in the realm of the military. And in the middle of 1990s with advice of commercial radio technology and the wireless became aware of the great advantages of MANET outside the military battlefield domain, and then became so active research work on ad hoc network start in 1995 in the conference session of the Internet Engineering Task Force (IETF) (IETF-MANET). And then in 1996 this works had evolved into MANET, in that time focused to discussion centered in military satellite network, wearable computer network and tactical network with specific concerns begin raised relative to adaptation of existing routing protocols to support IP network in dynamic environments, as well as they make the charter of the MANET Working Group (MANETWG) of the Internet Engineering Task Force (IETF) also the work inside the MANETs relies on other existing IETF standard such as Mobile IP and IP addressing. Most of the currently available solutions are not designed to scale to more than a few hundred nodes. Currently, the research in MANET became so active and vibrant area and the efforts this research community together with the current and future (MANET) enabling radio technology.

Recently, the Ad Hoc Wireless Network and computing consortium was established with the aim to coalescing the interests and efforts to use it anywhere such as academic area and industry and so on. And in order to apply this technology to application ranging for the Home Wireless (HW) to wide area peer to remote networking and communications. And it does will certainly pave the way for commercially viable MANETs and their new and exciting applications, which began to appear in all fields in this life. More recently, the computer has became spread significantly in the all the place and after a pervasive computing environment can be expected based on the recent progresses and advances in computing and communication technologies. Next generation of mobile communications will include both prestigious infrastructure wireless networks and novel infrastructureless MANETs.

1.2.2 Wireless Ad Hoc Networks

MANET is a collection of two or more devices or terminals with wireless communications and networking capability that communicate with each other without the aid of any centralized administrator also the wireless nodes that can dynamically form a network to exchange information without using any existing fixed network infrastructure. And it's an autonomous system in which mobile hosts connected by wireless links are free to be dynamically and some time act as routers at the same time. All nodes in a wireless ad hoc network act as a router and host as well as the network topology is in dynamically, because the connectivity between the nodes may vary with time due to some of the node departures and new node arrivals. The special features of MANET bring this technology great opportunity together with high challenges. All the nodes or devises responsible to organize themselves dynamically to communication between each other and to provide the necessary network functionality in the absence of fixed infrastructure or can call it ventral administration. It implies that maintenance, routing and management, etc, have to be done between all the nodes. This case called peer level multi hopping and that is the main building block for ad hoc network. In the end, conclude that the ad hoc nodes or devices are difficult and more complex than other wireless networks. Therefore, ad hoc networks form sort of clusters to the effective implementation of such a complex process. In Figure 4 shows some nodes forming ad hoc networks, and there are some nodes more randomly in different directions and different speeds.

In the past few years, the people became realized to use all the technology so widely and the people's future living environments are emerging, based on information resource provided by the connections of different communication networks for clients also have seen a rapid expansion in the field of mobile computing because the proliferation not expensive, widely available wireless devices. A new small devices such as personal communication like cell phones, laptops, Personal Digital Assistants (PDAs), handhelds, and there are a lot of traditional home appliances such as a digital cameras, cooking ovens, washing machines, refrigerators and thermostats, with computing and communicating powers attached. Expand this area to became a fully pervasive and so widely. With all of this, the technologies must be

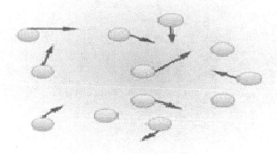

Fig. 4. Illustrates some of the nodes moves randomly in different direction and different speeds

formed the good and new standard of pervasive computing, that including the new standards, new tools, services, devices, protocols and a new architectures.As well as the people in this time, or the users of internet users in ad hoc network through increase in the use of its advantage is that not involve any connection link and the wiring needed to save space, and building low cost, and improve the use, and can be used in mobile phone, because of these advantage local wireless network architecture readily. And beads in these advantages the wireless network can be used in the local area network terminal part of the wireless (Liu and Chang, 2009).

1.2.2.1 Wireless mobile network approaches

The past decade, the mobile network is the only one much important computational techniques to support computing and widespread, also advances in both software techniques and the hardware techniques have resulted in mobile hosts and wireless networking common and miscellaneous.   Now will discuss about to distinct approaches very important to enabling mobile wireless network or IEEE 802.11 to make a communication between each other (part-11, 1997) (part-12, 1999). Firstly infrastructure wireless networks and secondly, infrastructureless wireless networks (ad hoc networks) and will clarify both in bottom.

1.2.2.2 Infrastructure wireless networks

In this architecture that allow the wireless station to make a communication between each other through the Base Station (BS) as shown in Figure 5, and that will handover the offered traffic from the station to another, the same entity will regulate or organize the allocation of radio resources. When a source node likes to communicate with a destination node, the former notifies the BS. At this point, the communicating nodes do not need to know anything about the route from one to another. All that matters is that the both source and destination nodes are within the transmission range for the BS and then if there is any one loses this condition, the communication will frustration or abort.

1.2.2.3 Infrastructureless wireless networks

The mobile wireless network is known as Mobile Ad Hoc Network (MANET). As has been previously defined in the bidder is a collection of two or more devices or nodes or

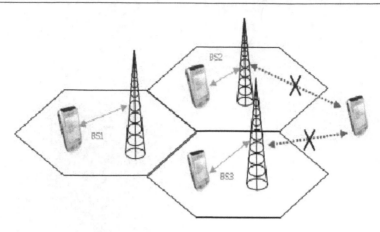

Fig. 5. Illustrates of the infrastructure network

terminals with wireless communications and networking capability that communicate with each other without the aid of any centralized administrator also the wireless nodes that can dynamically form a network to exchange information without using any existing fixed network infrastructure. And it's an autonomous system in which mobile hosts connected by wireless links are free to be dynamically and some time act as routers at the same time (Frodigh et al., 2000). The infrastructureless is important approaches in this technique to communication technology that supports truly pervasive computing widely duo to there is a lot of context information need to exchange between mobile nodes but can not rely on the fixed network infrastructure, but in this time the communication wireless became develops very fast (IETF-manet). In Figure 6 shown a small example for the ad hoc networks, to explain how mobile ad hoc network working.

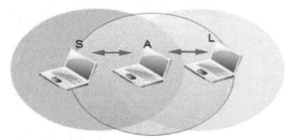

Fig. 6. Illustration of the infrastructureless networks (ad hoc networks)

This figure illustrates the modus operandi of ad hoc networks, there are a three ad hoc network nodes (S, A, L), the source node (S) need to make a communication with the destination node (L) and both of them (S, L) not in the same transmission range of each others, here both they must use the node (A) to send and receive or forewords the packets from source to the destination that means from node to another node.(R) is a node work as host and router in the same time. Additionally, the definition for the router is an entity that determines the path to be used, in order to forward a packet towards the last destination, and then the router chooses the next node to which a packet should be forwarded according to its current understanding of the state of the network.

1.2.3 Characteristics of MANET

Request For Comments (RFC) 2501 document (IETF, 1999) which is published by MANET working group within the IETF describes the main characteristics of MANET which differs from the characteristics of traditional wireless local area networks such as WLANs due to the dynamic and the infrastructureless natures of MANETs (Hekmat, 2006). According to the IETF RFC 2501, MANET has characteristics can be divided into the following:

1. A collection of autonomous terminals means that within a MANET, each mobile node performs its tasks as a router and a host.

2. It contains a dynamic topology which means there are a group of nodes into it that are moving and resulting to a random change rapidly at unpredictable times through the network topology.

3. A distributed operation is contained into it which means that the network's management and control is spread (distributed) in the nodes because of the infrastructure types' absence of which the central control of the network operations is supported. In a MANET, nodes must perform with each other. each node behaves as a router and a host simultaneously in order to have the network functions efficiently implemented, for example, routing and security.

4. It can be deployed as fast as it could be.

5. Pre-existing infrastructure is independent from it.

6. Bandwidth-constrained, variable capacity links compared with the wired network environment, the capacity of the wireless link itself is relatively small, but also susceptible to external noise, interference, and signal attenuation effects.

7. Self-adapts to the propagation patterns and connectivity.

8. Adapts to mobility patterns and traffic.

9. A limited physical security is contained into it, for example, in the absence of any centralized encryption or authentication. In order to reduce security threats, existing techniques of link security are at most applied into the WLANs and the wired networks.

10. It has an energy constrained operation a laptop or handheld computers are often used batteries to provide power, how to save electricity in the context of depletion of system design is also necessary to consider the point.

Mobile networking and MANETs are considered to be of good candidates due to many reasons: its simplicity for usage, robustness, speedy deployment and low cost. Its disadvantages comprise the complexity of routing due to the consistent move of nodes, mobility and dynamic topology, vulnerability of security due to the cooperation principle in MANETs, and the low computing power due to small devices used in MANETs.

1.2.4 Types of mobile ad hoc network

The wireless ad hoc network divided into three main types. Firstly, the quasi-static ad hoc network the nodes may be portable or static, because the power controls and link failures, the resulting network topology may be so active. The sensor network is an example for the quasi-static ad hoc network (Estrin et al., 1999). Secondly, the MANET the entire network may be mobile and the nodes may move fast relative to each other.thirdly, Vehicular Ad Hoc Networks (VANETs) are a kind of network useful for offering traffic information interchange in a collaborative way between vehicles.

1.2.4.1 Mobile Ad Hoc Networking (MANET)

MANET is a group of independent network mobile devices that are connected over various wireless links. It is relatively working on a constrained bandwidth. The network topologies are dynamic and may vary from time to time. Each device must act as a router for transferring any traffic among each other. This network can operate by itself or incorporate into large area network (LAN).

There are three types of MANET. It includes Vehicular Ad hoc Networks (VANETs), Intelligent Vehicular Ad hoc Networks (InVANETs) and Internet Based Mobile Ad hock Networks (iMANET). The set of application for MANETs can be ranged from small, static networks that are limited by power sources, to large-scale, mobile, highly dynamic networks. On top of that, the design of network protocols for these types of networks is face with multifaceted issue. Apart from of the application, MANET need well-organized distributed algorithms to determine network organization, link scheduling, and routing. Conventional routing will not work in this distributed environment because this network topology can change at any point of time. Therefore, we need some sophisticated routing algorithms that take into consideration this important issue (mobile network topology) into account. While the shortest path (based on a given cost function) from a source to a destination in a static network is usually the optimal route, this idea is not easily far-reaching to MANET. Some of the factors that have become the core issues in routing include variable wireless link quality, propagation path loss, fading, interference; power consumed, and network topological changes. This kind of condition is being provoked in a military environment because, beside these issues in routing, we also need to guarantee assets security, latency, reliability, protection against intentional jamming, and recovery from failure. Failing to abide to of any of these requirements may downgrade the performance and the dependability of the network.

1.2.4.2 Mobile ad hoc sensor network

A mobile ad hoc sensor network follows a broader sequence of operational, and needs a less complex setup procedure compared to typical sensor networks, which communicate directly with the centralized controller. A mobile ad hoc sensor or hybrid ad hoc network includes a number of sensor spreads in a large geographical area. Each sensor is proficient in handling mobile communication and has some level of intelligence to process signals and to transmit data. In order to support routed communications between two mobile nodes, the routing protocol determines the node connectivity and routes packets accordingly. This condition has makes a mobile ad hoc sensor network highly flexible so that it can be deployed in almost all environments (Bakht, 2010). The wireless ad hoc sensor networks (Asif, 2009) are now getting in style to researchers. This is due to the new features of these networks were either unknown or at least not systematized in the past. There are many benefits of this network, it includes:

- Use to build a large-scale networks
- Implementing sophisticated protocols
- Reduce the amount of communication (wireless) required to perform tasks by distributed and/or local precipitations.
- Implementation of complex power saving modes of operation depending on the environment and the state of the network.

With the above-mentioned advances in sensor network technology, functional applications of wireless sensor networks increasingly continue to surface. Examples include the replacement of existing detecting scheme for forest fires around the world. Using sensor networks, the detecting time can be reduced significantly. Secondly is the application in the large buildings that at present use various environmental sensors and complex control system to execute the wired sensor networks. In a mobile ad hoc sensor networks, each host may be equipped with a variety of sensors that can be organized to detect different local events. Besides, an ad hoc sensor network requires a low setup and administration costs (Akkaya and Younis, 2005) (Akyildiz et al., 2002).

1.2.4.3 Vehicular Ad Hoc Networks (VANETs)

Vehicular Ad Hoc Networks (VANETs) (Kosch et al., 2006) are a kind of network useful for offering traffic information interchange in a collaborative way between vehicles. They are foreseen to be a great revolution in the driving, providing new services such as Road safety, traffic management, Pollution reduction, Cost reduction in the vehicle security incorporation and public transport.

1.2.5 The traffic types in the ad hoc networks

The traffic types in the ad hoc networks are so differen from the infrastructure wireless network. The traffic types are classified into three types (peer to peer, remote to remote and dynamic traffic) (Mbarushimana and Shahrabi, 2008). Firstly, peer to peer is a communication between two nodes in the same area, that means which are within one hop. Network traffic (in bits per second) is usually fixed. Secondly, remote to remote is a communication between two nodes beyond a single hop, but maintain a stable route between them. This may be the result of a number of nodes, to stay within the range of each other in one area or may move as a group. Movement it's a similar to the standard network traffic. Finally, dynamic traffic it will happen when the nodes are movie dynamically around and then the routers must be reconstructed. This results in a poor connectivity and network activity in short bursts. For example in IEEE 802.11 network and the basic structure divided into two types firstly infrastructures wireless LAN, the second structure ad hoc wireless LAN.

1.2.5.1 Infrastructure wireless LAN

In this kind of network as shown in the Figure 7, the network in any architecture will be an access point; its function is one or more of the wireless local area network and the existing cable network systems to link, so that stations within the wireless local area network and external nodes can connect with each other. It is characterized by a fixed and pre-positioning a good base station location, the static backbone network topology, a good environment and a stable connection, the base station that is doing a good job when you set up detailed plans (Li, 2006).

1.2.5.2 Ad hoc wireless LAN

The ad hoc wireless LAN is an infrastructures relies on infrastructures wireless local area network, which only targeted at local area network within the framework of each machine is able to be linked up into networks, regardless of whether the communication with the outside world, then such a structure, either one or two users can communicate directly with each other, and this structure is composed of at least composed of two or more workstations. Is characterized by no fixed base stations, network will be rapidly changing; dynamic network topology is vulnerable to interference, to automatically form a network without infrastructure

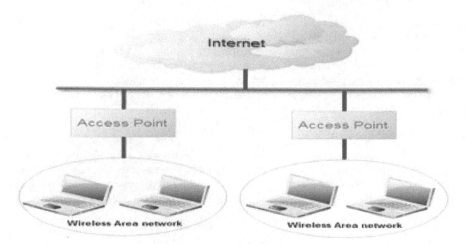

Fig. 7. Illustrates of the infrastructure wireless LAN architecture.

and adapt to topology changes. For more explain shows the Figure 8 for Ad Hoc Wireless network.

Fig. 8. Illustrates of the wireless ad hoc network

1.2.6 Ad hoc network routing protocol performance issues

The MANET with the traditional wired, fixed networks have many different characteristics, so to design a suitable routing protocol for MANET operating environment must also consider the different directions, the following sub-qualitative and quantitative aspects of the discussion:

1. On the qualitative aspects, can be divided into:
    a) Distribution operation: Due to the existence of MANET where there is no prerequisite for the construction of the underlying network, so routing can not rely on a particular node to operate.

b) Loop-freedom: All the routing protocol should be consistent with the characteristics; we must ensure the normal work in order to avoid waste of bandwidth.

c) Demand-based operation: In order to reduce the burden on each node, if the link is not so much the demand should be considered when using On-demand approach to the establishment of the path, and only when the need for a particular path query, the establishment of the path.

d) Proactive operation: With the On-demand concept of the contrary, if the network resources fairly adequate, proactive table-driven approach could speed up the path to the establishment of speed.

e) Security: Because it is the wireless environment, to how to ensure the security of the connection can not be ignored will be part of network security is also a MANET from theory to implementation of the key challenges.

f) Sleep Period operation: As the MANET nodes are generally smaller wireless devices, using the battery as a power supply, how to save power consumption, or for no work, the node goes into sleep mode, can operate more smoothly so that MANET. Also the nodes of a MANETs may stop transmitting or receiving or both, also even receiving requires power for arbitrary time periods and the routing protocol should be able to accommodate such sleep periods without overly adverse consequences. This property may require close coupling with the link-layer protocol through a standardized interface.

2. On the quantity, can be divided into:

a) End-to-end data throughput and delay: Data transmission rate and delay in the case that every routing protocol must take into account the focus should be how to find the best path? Is the maximum bandwidth or minimum latency, or the link to the most stable? Considered more likely to make more complicated routing protocol, but it is possible to significantly improve the transmission quality.

b) Route Acquisition time: While the table-driven generally higher than on-demand performance good, but many of the former to pay the price, which, if properly designed, for example, there is more commonly used in the path cache, or a certain fixed path , can improve the path to the establishment of time.

c) Percentage Out-of-order delivery: Real-time data for this part of the more stringent requirements, and general information will not affect how and upper TCP cooperation is also IP routing work.

d) Efficiency: The simplest method, the smallest control overhead done the most complete, most powerful feature is a common goal for all routing protocol.

1.2.7 Types of ad hoc protocols

Ad hoc network routing protocols is divided to three type of routing protocols, which that depending on a different of routing protocols (Saleh Al-Omari and Putra Sumari, 2010).

1.2.7.1 Oriented routing table (table-driven)

The oriented routing table is an active routing environment in which the intervals between the wireless nodes will send medical information with more paths. Each wireless node is on the basis of information gathered recently to change its route table, when the network topology change makes the original path is invalid, or the establishment of any new path, all nodes will receive updates on the status path. The path will be continuously updated, so that the node in time of peace on its own routing tables is ready, and immediately available when needed. However, such agreements must be periodically to broadcast messages, so a considerable waste of wireless bandwidth and wireless node power, but if you want to reduce the broadcast bandwidth consumption caused by a large number, we should lengthen the interval between each broadcast time, which in turn will result in the path table does not accurately reflect network topology changes.

1.2.7.2 Demand-driven (on-demand)

In the demand driven, When needed to send packets only it began to prepare to send the routing table. When a wireless node needs to send data to another wireless node, the source client node will call a path discovery process, and stored in the registers of this path. The path is not valid until the expiration or the occurrence of conditions of the agreement with the first phase of a ratio of such agreements in each node. A smaller amount of data needed, and do not need to save the entire network environment and the routing information. The main benefit of this agreement is that the use of a lower bandwidth, but the drawback is that not every wireless node that sends packets can always quickly find the path. The path discovery procedure can cause delays and the average delay time is longer (Liu and Chang, 2009).

1.2.7.3 Hybrid

Hybrid is an improvement of the above mentioned two, or the combination of other equipment, such as Global Positioning System (GPS) and other equipment, participate in the study of mechanisms to facilitate the routing of the quick search, and data transmission (Pandey et al., 2005) (Johnson and Maltz, 1999.). However, there are already more than 13 kinds of the above routing protocol have been proposed, following the more representative for several separate presentations, and to compare their individual differences lie.

1.2.8 Compare between proactive versus reactive and clustering versus hierarchical

1.2.8.1 Proactive versus Reactive Approaches

Ad hoc routing protocols can be classified into two types; proactive and On-Demand (reactive) base on each own strategy (Perkins, 2001). Proactive protocols demand nodes in a wireless ad hoc network to keep track of routes to all possible destinations. This is important because, whenever a packet requests to be forwarded, the route is beforehand identified and can be used straight away. Whenever there's modification in the topology, it will be disseminated throughout the entire network. Instances include "destination-sequenced distance-vector" (DSDV) routing (Perkins and Bhagwat, 1994), "wireless routing protocol" (WRP) (Murthy and Garcia-Luna-Aceves, 1996), "global state routing" (GSR) (Chen and Gerla, 1998), and "fisheye state routing" (FSR) (Iwata et al., 2002)and in next section will discuss about everyone.

On-demand (reactive) protocols will build the routes when required by the source node, in order for the network topology to be detected as needed (on-demand). When a node needs to send packets to several destinations but has no routes to the destination, it will start a route detection process within the network. When a route is recognized, it will be sustained by a route maintenance procedure until the destination becomes unreachable or till the route is not wanted anymore. Instances include "ad hoc on-demand distance vector routing" (AODV) (Perkings et al., 2003), "dynamic source routing" (DSR) (J.Broch et al., 2004), and "Cluster Based Routing protocol" (CBRP) (Jiang et al., 1999). Proactive protocols comprise the benefit that new communications with arbitrary destinations experience minimal delay, but experience the disadvantage of the extra control overhead to update routing information at all nodes. To overcome with this limitation, reactive protocols take on the opposite method by tracking down route to a destination only when required. Reactive protocols regularly utilize less bandwidth compared to proactive protocols, however it is a time consuming process for any route tracking activity to a destination proceeding to the authentic communication. Whenever reactive routing protocols must relay route requests,it will create unnecessary traffic if route discovery is required regularly.

1.2.8.2 Clustering versus hierarchical approaches

Scalability is one of the major tribulations in ad hoc networking. The term scalability in ad hoc networks can be defined as the network's capability to provide an acceptable level of service to packets even in the presence of a great number of nodes in the network. If the number of nodes in the network multiply for proactive routing protocols, the number of topology control messages will increases nonlinearly and it will use up a large fraction of the available bandwidth. While in reactive routing protocols, if there are a large numbers of route requests propagated to the entire network, it may eventually become packet broadcast storms. Normally, whenever the network size expands beyond certain thresholds, the computation and storage requirements become infeasible. At a time whenever mobility is being taken into consideration, the regularity of routing information updates may be extensively enhanced, and will deteriorate the scalability issues. In order to overcome these obstacles and to generate scalable and resourceful solutions, the solution is to use hierarchical routing. Wireless hierarchical routing is based on the idea of systematizing nodes in groups and then assigns the nodes with different task within and outside a group. Both the routing table size and update packet size are decreased by comprising only a fraction of the network. For reactive protocols, restricting the scope of route request broadcasts can assists in improving the competency. The best method of building hierarchy is to gather all nodes geographically near to each other into groups. Every cluster has a principal node (cluster head) that corresponds with other nodes. Instances of hierarchical ad hoc routing protocols include "zone routing protocol" (ZRP) (Haas and Pearlman, 2000).

1.2.9 Existing ad hoc protocols

In the ad hoc network there are more than 13 kinds of the above routing protocol have been proposed such as DSDV, GSR, CGSR, WRP, FSR, AODV, DSR, TORA, CBRP, ABR, SSR, CEDAR and ZRP, for more dilates about existing ad hoc network protocols (Saleh Alomari and Putra Sumari, 2010. Further explination for understanding some of the existing mobile ad hoc network are provided in Appendix A figure 10. The comparison between Table Driven, Demand Driven and Hybrid are shown in Table 2,and then show in Table 3 the Table Driven for three kind of protocols such as WRP, CGSR, DSDV and comparison between them, Demand Driven (On-Demand) with six type of protocols such as TORA, DSR, AODV, ABR, CEDAR and SSR and comparison between them shows in Table 4. Finally, shows compare the main characteristics of existing multipath routing protocols in Table 5.

* CEDAR, TORA itself, although it can not also be used in multicasting, but there have been constructed in the two above the multicast routing protocol was proposed.

1.2.10 Challenges and issues of MANETs

For ad hoc networking design and implementation, there lots of factors and challenges which are:

Scalability: in some applications, a MANET can grow to thousands of nodes, such as, battlefield deployments, urban vehicle grids and large environmental sensor fabrics. It is extremely hard to have the scalability handled in a MANET due to the random and unlimited mobility (Perkins et al., 2002).

	Table Driven(Proactive)	Demand Driven(Reactive)	Hybrid
Routing Protocols	DSDV,CGSR,WRP	AODV,DSR,TORA,ABR,SSR	ZRP
Route acquisition delay	Lower	Higher	Lower for Intra-zone; Higher for Inter-zone
Control overhead	High	Low	Medium
Power requirement	High	Low	Medium
Bandwidth requirement	High	Low	Medium

Table 2. Illustrates the comparison between Table Driven, Demand Driven and Hybrid

Table Driven	CGSR	WRP	DSDV
Routing philosophy	Hierarchical	Flat	Flat
Loop-free	Yes	Yes, but not instantaneous	Yes
Number of required tables	2	4	2
Frequency of update transmissions	Periodically	Periodically and as needed	Periodically and as needed
Updates transmitted to	Neighbors and cluster head	Neighbors	Neighbors
Utilize hello messages	No	Yes	Yes
Critical nodes	Cluster head	No	No
Communication complexity	O(x = N)	O(x = N)	O(x = N)

Table 3. Shows the Table-Driven for the three kinds of protocols and comparison between them

Mobility is at most the first designer's enemy of MANET (Murthy and Mano, 2004).

Energy conservation most ad hoc nodes, such as Personal Digital Assistants (PDAs), sensors and Laptops are often power supplied using batteries which have limited power. Therefore, for MANET, energy conservation is considered to be an enormous challenge.

Application/Market penetration: multi-hop technology is not commercial at present. More clearly, the short coverage area's limitation of the wireless products can be justified in its belonging to the standard of IEEE 802.11.

Design/Implementation: manageable, secure, reliable and survivable implementation and design must act for MANET since a bandwidth-constrained operation and a limited physical security are contained in MANETs.

Limited wireless transmission range depends on the wireless technology's capabilities.

Operational/Business-related how to have the network managed and how to bill for services.

The main key issues that affect the design, deployment, and performance of an ad hoc wireless system are summarized as following: scalability, security, energy management, QoS provisioning, deployment considerations, self organization, multicasting, pricing scheme, medium access scheme, routing, transport layer protocols, addressing and service discovery.

On-Demand	TORA	DSR	AODV	ABR	CEDAR	SSR
Overall complexity	High	Medium	Medium	High	High	High
Overhead	Medium	Medium	Low	High	High	High
Routing philosophy	Flat	Flat	Flat	Flat	Core-Extracted	Flat
Loop Free	Yes	Yes	P	Yes	Yes	Yes
Multicast capability	No*	No	Yes	No	No*	No
Beaconing requirements	No	No	No	Yes	Yes	Yes
Multiple route support	Yes	Yes	No	No	No	No
Routes maintained in	Route table	Route cache	Route table	Route table	Route table	Route table
Route reconfiguration methodology	Link reversal	Erase route	Erase route	Localized broadcast query	Dynamic route re-compute	Erase route

Table 4. Shows the Demand Driven (On-Demand) with six types of protocols and comparison between them

	AODV	DSR	CBRP	DSDV	WRP	GSR	FSR
Routing Category	Reactive	Reactive	Reactive	Proactive	Proactive	Proactive	Proactive
TTL Limitation	Yes	Yes	Yes	No	No	No	No
Flood Control	No	No	No	Yes	Yes	Yes	Yes
QoS Support	Yes	Yes	P	Yes	Yes	Yes	Yes
Periodic Update	No	No	No	No	No	No	No
Power Management	No	No	No	No	No	No	No
Multicast Support	Yes	No	No	No	No	No	No
Beaconing	Yes	Yes	Yes	Yes	Yes	Yes	Yes
Security Support	No	No	No	No	No	No	No

Table 5. Shows the comparison of the main characteristics of existing multipath routing protocols

The four important issues significant in MANET are Mobility, QoS Provisioning, Multicasting and Security.

1.2.10.1 Mobility

The mobile user can freely move anywhere and are free to join and move away from the network at anytime. The mobile client can explore the area and can form groups or teams to create a taskforce. In the ad hoc network, the mobile client can have individual random and group mobility and the mobility model can have major impact on the selection of a routing scheme and this directly influences the performance. The mobile clients in MANETs have no physical boundary and their location changes as they move around. This movement of mobile nodes makes the network topology highly dynamic as well as causing the intercommunication patterns between nodes to change frequently in an unpredictable manner (Frodigh et al., 2000), (Satyanarayanan, 2001). Thus, an ongoing communication session suffers frequent path breaks. As a result, broadcasting protocols for MANETs must handle mobility management efficiently (Basagni et al., 1998).

1.2.10.2 QoS provisioning

A network or a service provider offers the QoS to be the performance level of services the user in terms of many performance metrics of QoS such as packet delivery, the average end-to-end delay, and available bandwidth. Between the network and the host, negotiation is mostly needed when providing QoS (i.e. QoS provision). More specifically, this demand is based on the call admission control, resource reservation schemes and priority scheduling. Therefore, when different levels of QoS are provided in a highly changeable environment, an important issue takes place for this provision. (Chakrabarti and Mishra, 2001).

In MANETs, the provision of QoS is made to be more difficult than providing it in fixed wired networks. This difficulty is due to a high change in network topology, the presence of additional bandwidth, and medium and linked constraints. Static constraints such as memory, processing power and bandwidth, will be only taken into account (Basagni et al., 2004). An implementation must be performed for an adaptive QoS within the traditional resource reservation techniques (Ilyas, 2003), in order that multimedia services in MANETs could be efficiently supported.

1.2.10.3 Security

Security attacks consider Ad hoc networks to be highly vulnerable to it. In the matter fact, this is taken into account to be as the main challenges of the developers of MANET. Particular security problems are involved in a MANET. This is referred to several reasons, such as insecure operating environment, shared broadcast radio channel, malicious attacks of a neighbor node, lack of central authority, limited availability of resources, lack of association among nodes, and physical vulnerability. Integrity, availability, confidentiality, non-repudiation and authentication are the most common attributes of MANETs security system (Ilyas, 2003) (Makki et al., 2007).

Survivability of network services despite the denial of service attacks is ensured by the Availability. Certain information is never disclosed to unauthorized entities. This is ensured by confidentiality. A corruption is never happened for a message being sent. This is ensured by Integrity. In order to ensure the identity of the peer node for communications, a node is enabled by authentication. Finally, the message being sent cannot be denied by the origin of a message. This is guaranteed by non-repudiation (Buttyan and Hubaux, 2007). The major security threats that are available in MANETs are denial of service, passive eavesdropping, signaling attacks, resource of service, host impersonation and information disclosure.

1.2.10.4 Multicasting

Multicast is another significant issue of MANETs because the multicast tree is not static in MANETs due to the random movement of nodes in the network. Multiple hops are potentially contained by routes of each pair of nodes. The single hop communication type is less complex than this type of communication. When multicast packets should be sent to groups in several networks, multicast routing becomes essentially. In MANETs, a vital role is played by multicasting through several applications such as in emergency, military operations and rescue operations. Node mobility with the power and bandwidth constraints make multicast routing very challenging in MANETs (Ritvanen, 2004).

1.2.11 Application of MANETs

Mobile ad hoc networks (MANETs) are very flexible networks and suitable for a lot of types of potential applications applied on the Ad hoc networks, as they allow the

Applications	The Possible Service of Ad Hoc Networks
Tactical networks	1)Military communication. 2)Military operations. 3)In the battlefields.
Emergency services	1)Search and rescue operations in the desert and in the mountain and so on. 2)Replacement of fixed infrastructure in case of environmental disasters.3)Policing.4)fire fighting.Supporting doctors and nurses in hospitals.
Coverage extension	1)Extending cellular network access. 2)Linking up with the Internet and so on.
Sensor networks	1)Inside the home: smart sensors and actuators embedded in consumer electronics. 2) Body area networks (BAN). 3) Data tracking of environmental conditions such as animal movements, chemical/biological detection.
Education	1)Classrooms. 2)Ad hoc Network when they make a meetings or lectures. 3) Multi-user games. 4) Wireless P2P networking. 5) Outdoor Internet access Robotic pets. 6)Theme parks.
Home and enterpriser	1)Using the wireless networking in Home or office.2) Conferences, 3)meeting rooms. 4)Personal area networks Personal networks.
Context aware services	1)Follow-on services: call-forwarding, mobile workspace. 2)Information services: location specific services, time dependent services. 3)Infotainment: tourists information.
Commercial and civilian environments	1)E-commerce: electronic payments anytime and anywhere. 2)Business: dynamic database access, mobile offices. 3)Vehicular services: road or accident guidance, transmission of road and weather conditions, taxi cab network, inter-vehicle networks. 4)Sports stadiums, trade fairs, shopping malls and so on. 5)Networks of visitors inside the airports

Table 6. Illustrates some of the application for the ad hoc networks

establishment of temporary communication without any pre-installed infrastructure, the application such as the European telecommunications standard institute (ETSI) also the HIPERLAN/2 standard (Masella, 2001) (Habetha et al., 2001), IEEE 802.11 wireless LAN standard family (Crow. B et al., 1997) and Bluetooth (Bluetooth, 2001) the ad hoc network are very important area in this time and very useful for the military (battlefield) and for the disasters (flood, fire and earthquake and so on),meetings or conventions in which people wish to quickly share information (Chlamtac et al., 2003). And then use it in the emergency search-and-rescue operations, recovery, home networking etc. Nowadays, ad hoc network became so important in our circle life, because can be applied anywhere where there is little or without communication infrastructure or may be the existing infrastructure is expensive to use. The ad hoc networking allows to nodes or devices to keep the connections to the network for as long as it's easy to add and to remove to the end of the network. And there are a lot of varieties of applications for the mobile ad hoc networks, ranging large scale such as dynamic network and mobile and small fixed-constrained energy sources. As well as legacy applications that move from the traditional environment to the Ad Hoc infrastructure environments, a great deal of new services can and will be generated for the new environment, finally as the result the mobile Ad Hoc Network is the important technique for the future and to became for the fourth generation (4G), and the main goals for that to provide propagation the computer environments, that support the users to achieved the tasks to get the information and communicate at anytime, anyplace and from any nodes or devices. And now will present some of these practical applications has been arranged in Table 6.

These are many applications on ad hoc networks as we mentioned above and in Table 6 provide an overview of present and future MANET applications. However, the following is a summary of the major applications in MANETs such as tactical networks (military battlefield), home and enterprise network (personal area network) etc.

- Military battlefield, Military equipment currently is equipped with the state of the art computer equipment. Ad hoc networking help the military with the commonplace network technology to maintain information network between military personnel's, vehicles, and military information head quarters. The basic techniques of ad hoc network originated from this field.

- Commercial sector, ad hoc network can be applied in emergency or rescue operations for disaster relief efforts for example in fire, flood, or earthquake and so on. Emergency rescue operations will go to places where communications are impermissible. Therefore proper infrastructure and rapid deployment of a communication network is badly needed. Information is relayed from one rescue team member to another over a small handheld device. Other commercial application includes for instance ship to ship ad hoc mobile communication and so on.

- Local level, ad hoc networks can autonomously link immediate and temporary multimedia network by using notebook or palmtop computers to distribute and allocate information among conference or classroom participants. Besides, it can also be applied for home networks where devices can be link; other examples include taxicab, sports stadium, boat and small aircraft.

- Personal Area Network (PAN), short-range MANET can simplify the intercommunication between a lot of mobile devices such as a PDA, a laptop, and a cellular phone and there are a lot of new devices in this for MANETs. Wired cables can easily be replaced with wireless connections. Ad hoc network enhances the access to the Internet or other networks by means of Wireless LAN (WLAN), GPRS, and UMTS. The PAN is an upcoming application field of MANET for the future computing technology.

- Personal communications (i.e. cell phones, laptops and ear phone).

- Cooperative environments (i.e. meeting rooms, sports stadiums, boats etc.).

- Conferencing (i.e. using mobile nodes).

- Home Network (almost used for PANs).

- Wireless Mesh Networks (very reliable networks that are closely related to MANETs, the nodes of a mesh network generally are not mobile).

- Hybrid Wireless Networks (the goal is to cost savings, enhanced resilience to failures and performance improvements).

- Wireless Sensor Networks (a very active research area of ad hoc networking which includes fixed networks or mobile sensors (Sarkar et al., 2008).

1.2.12 MANET layers

The network architecture can be described using a reference the model. More obviously, the layers of software and hardware are described by this model so that data could be sent among two points, besides, to make it capable for interpellating of multiple devices/applications in a network. In order to increase compatibility in the network between different components from different manufacturers, reference models are required for so (White, 2002). Seven layers are contained in the International Organization for Standardization (ISO/IEC, 2003) which proposed the Open Systems Interconnection (OSI) reference model. In the matter of fact, these layers are ordered from the lowest to the highest layer. The lowest layer represents layer one whereas the highest layer represents layer seven as shown in Figure 9. In other words, these layers are respectively ordered as: application layer, presentation layer, session layer, transport layer, network layer, data link layer and physical layer (from the highest to the

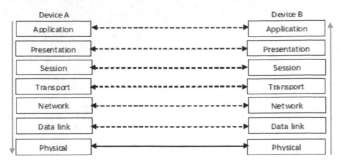

Fig. 9. Illustrates of the original International Organizations for Standardization (ISO) and Open Systems Interconnection (OSI) reference model.

lowest). The transmission of bits is handled by the physical layer through a communications channel. In addition, other physical specifications are taken into account.Such specifications comprise; modulation techniques, connectors and media choice. The access of multiple nodes is coordinated by the functions of data link layer along to a shared medium, control and address information, error detection code, flow control, Medium Access Control (MAC) addressing and so on. Network layer is responsible for creating, maintaining and ending network connection. It transfers a data packet from node to node within the network. In other words, it is responsible for congestion control, IP addressing, and internet working. The transport layer provides an end to end error-free network connection, and makes sure the data arrives at the destination exactly as it left the source. In order to establish sessions between users, the session layer is the layer that controls such a process. At the same time, a series of functions necessary for presenting the data package properly to the sender or receiver are performed by the presentation layer, for example, such as compression and encryption. The application layer is considered to be as the highest layer that provides the user the ability to efficiently access the network. Frequent reconnection and disconnection with peer applications are handled by this layer as a main role of it. Another role of it is to have services and data transmission among users supported, such as, electronic mail and remote file transfer.

1.2.13 Summary

In this chapter, described the necessary an overview for the current literature of Mobile Ad Hoc Network (MANET), covering the main concepts of MANET and the existing wireless mobile network approaches, wireless ad hoc networks, wireless mobile approaches, characteristic, applications, challenges, MANET layers and MANET issues. In particular, mobile ad hoc networks have been classified into two types, MANET and mobile ad hoc sensor network. The traffic types in ad hoc networks which include the Infrastructure wireless LAN and ad hoc wireless LAN are presented in Section 1.2.5. In Section 1.2.6 highlight the relevant details about the ad hoc network routing protocol performance issues. The types of ad hoc protocols such as (Table-driven, On-demand and Hybrid) and Compare between Proactive versus Reactive and Clustering versus Hierarchical are in Section 1.2.7. And Section 1.2.8 respectively. The existing ad hoc protocols are presented in Section 1.2.9. The four important issues significant in MANET are Mobility, QoS Provisioning, Multicasting and Security is presented in Section 1.2.10. Furthermore, in Section 1.2.11 and Section 1.2.12 shows the practical application and the layers of the MANET.

2. Appendix A

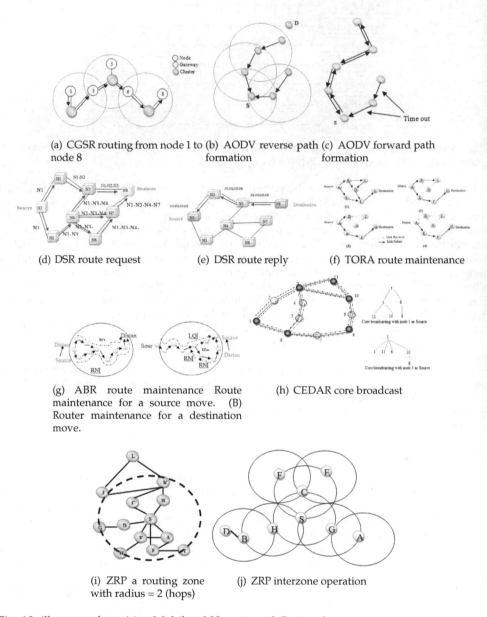

(a) CGSR routing from node 1 to node 8

(b) AODV reverse path formation

(c) AODV forward path formation

(d) DSR route request

(e) DSR route reply

(f) TORA route maintenance

(g) ABR route maintenance Route maintenance for a source move. (B) Router maintenance for a destination move.

(h) CEDAR core broadcast

(i) ZRP a routing zone with radius = 2 (hops)

(j) ZRP interzone operation

Fig. 10. illustrates the exiting Mobile ad Hoc network Protocols

3. References

[1] Akkaya, K. and Younis, M. (2005) A survey on routing protocols for wireless sensor networks. In Ad-hoc Networks (2005). Vol.3, N0.2,pp. 325-349.

[2] Akyildiz, I. F., Su, W., Sankarasubrmanina, Y. and Cayirci, E. (2002) A Survey on Sensor Networks. IEEE Communications Magazine (August 2002), pp.102-114.

[3] Asif, H. M. (2009) http://mobius.cs.uiuc.edu/publications/SECON04.pdf. Computer Engineering Department Saudi Arabia King Fahd University of Petroleum and Minerals.

[4] Crow. B, Widjaja, I.and Sakai, P. (1997) Investigation of the IEEE 802.11 medium access control (MAC) sublayer functions. In Proceedings of the INFOCOM '97,Sixteenth Annual Joint Conference of the IEEE Computer and Communications Societies., pp. 66-43.

[5] Bakht, H. (2010) WIRELESS INFRASTRUCTURE, Sensor networks and ad-hoc networking. http://www.computingunplugged.com/issues/issue200410/0000139 8001.html, (accessed on 4 Feb 2011).

[6] Basagni, S., Conti, M., Giordano, S.and Stojmenovic, I "" (2004) Mobile Ad Hoc Networking, A John Wiley and Sons, Inc., Publication.

[7] Basagni, S., Chlamtac, I., Syrotiuk, V. R. and Woodward, B. (1998) A distance routing effect algorithm for mobility (DREAM) for wireless networks. Proceedings 4th Annual ACM/IEEE International Conference on Mobile Computing Networking (MobiCom), pp. 76-84.

[8] Bluetooth (2001) Specification of the Bluetooth system, Core, v1.1. Bluetooth SIG. (http://www.bluetooth.com)

[9] Broch. J, David .A and David. B. (1998) "A Performance comparison of multi-hop wireless ad hoc network routing protocols". Proc. IEEE/ACM MOBICOM'98, pp.85-97.

[10] Buttyan, L. and Hubaux, J.-P. (2007) "Security and Cooperation in Wireless Networks, Thwarting Malicious and Selfish Behavior in the Age of Ubiquitous Computing", A Graduate Textbo ok, Available on http : //secowinet.epf l.ch under a permission from Cambridge University Press, Draft Version 1.3, Feb. 2007.

[11] Charabarti, S. and Mishra, A. (2001) QoS Issues in Ad Hoc Wireless Networks. IEEE Communications Magazine, February 2001.

[12] Chen, T. and Gerla, M. (1998) Global state routing: A new routing scheme for ad hoc wireless networks. in Proceedings of IEEE ICC'98, Vol. 1, No. 7-11, pp.171 - 175.

[13] Chlamtac, I., Conti, M. and Liu, J. J (2003) Mobile ad hoc networking: imperatives and challenges. Ad Hoc Networks, Vol. 1, No. 1, pp. 13-64.

[14] Estrin, D., Govindan, R. and Hedemann, J. (1999) New Century Challenges: Scalable Coordination in Sensor Networks. ACM, Mobicom, 1999.

[15] Etsi, E. T.(1999) High Performance Radio Lo cal Area Network Type 2 (Hiperlan2), Broadband Radio Access Networks (BRAN) project, 1999. On the URL:http : //portal.etsi.org/archiv ed/radio/hiperlan/hiperlan.asp.

[16] Freebersyser, J. A. and Leiner, B. (2001) A DoD perspective on mobile ad hoc networks. In: Perkins, C.(Ed.) Ad Hoc Networking, Addison Wesley, Reading, MA, 2001, pp. 29-51.

[17] Frodigh, M., Johansson, P. and Larsson., P. (2000) Wireless ad hoc networking: the art of networking without a network Ericsson Review. Vol.5, No.4, pp. 248-263.

[18] Haas, Z. J. and Pearlman, M. R. (2000) The Zone Routing Protocol (ZRP) for Ad Hoc Networks. Internet draft, http://www.ics.uci.edu/ atm/adhoc/papercollection/ haas-draft-ietf-manet-zone-zrp-00.txt.

[19] Habetha, J., Mangold, S. and Wiegert, J. (2001) 802.11 versus HiperLAN/2 - a comparison of decentralized and centralized MAC protocols for multihop ad hoc radio

networks. in Proceedings of 5th World Multiconference on Systemics, Cybernetics and Informatics, Orlando, USA, pp. 33-40.

[20] Hekmat, R. (2006) Ad-hoc Networks: Fundamental Properties and NetworkTopologies, A book published by Springer.

[21] IEEE (2004) Std 802.16-2004 TM, IEEE Standard for Local and Metropolitan Area Networks - Part 16: Air Interface for Fixed Broadband Wireless Access Systems.

[22] IEEE (2005a) Part 16: Air Interface for Fixed and Mobile Broadband Wireless Access Systems. IEEE 806.16e, IEEE P802.16e/D12.

[23] IEEE (2005b) Std 802.16e-2005TM, IEEE Standard for Local and Metropolitan Area Networks, Part 16: Air Interface for Fixed and Mobile Broadband Wireless Access Systems, Feb 2006.

[24] IEEE (2011) 802 Working group. [Online] http://www.ieee802.org/IEEE 802 LAN/MAN Standards Committee.

[25] IETF-MANET IETF MANET Working Group. http:// www.ietf.org/ html.charters/ manetcharter.html.

[26] IETF-MANET IETF Working Group: Mobile Adhoc Networks (manet). http://www. ietf.org/html. charters /manetcharter.html.

[27] IETF (1999) RFC 2501 - Mobile Ad hoc Networking (MANET): Routing Protocol Performance Issues and Evaluation Considerations. http://www.faqs.org /rfcs /rfc-sidx26.html.

[28] Ilyas, M. (2003) The Handbook of Ad Hoc Wireless Networks (Electrical Engineering Handbook) [Hardcover], CRC Press 2003.

[29] ISO/IEC (2003) Draft ITU-T Recommendation and Final Draft International Standardof Joint Video Specification (ITU-T Rec. H.264 | ISO/IEC 14496-10 AVC). Joint Video Team of ITU-T and ISO/IEC JTC 1, JVT G050r1.

[30] Iwata, A., Chiang, C.-C., PEI, G., Gerla, M. and Chen, T.-W. (2002) Scalable Routing Strategies for Ad Hoc Wireless Networks. IEEE Journal on Selected Areas in Communications, Special Issue on Ad-Hoc Networks, Vol. 17, Issue. 8, pp.1369 - 1379.

[31] Broch. J, Johnson. D and D. Maltz, T. (2004) he Dynamic Source Protocol for Mobile Ad hoc Networks. http://www.ietf.org/internet-drafts/draft-ietf-manet-dsr- 10.txt, IETF Internet draft , 19 July 2004.

[32] Jiang, M., Li, J. and Tay, Y. C. (1999) Cluster Based Routing Protocol. August 1999 IETF Draft. http://www.ietf.org/internet-drafts/draft-ietf-manetcbrp-spec-01.txt.

[33] Johnson, J. B. D. B. and Maltz, D. A. (1999.) The dynamic source routing protocol for mobile ad hoc networks. IETF MANET Working Group, Internet-Draft, October 1999.

[34] Jubin, J. and Tornow, J. D. (1987) The DARPA Packet Radio Network Protocols. proceedings of the IEEE, January, 1987, vol. 75, no. 1, pp.21-32.

[35] Saleh Ali AL-OMARI, and Putra Sumari. (2010) AN OVERVIEW OF MOBILE AD HOC NETWORKS FOR THE EXISTING PROTOCOLS AND APPLICATIONS International journal on applications of graph theory in wireless ad hoc networks and sensor networks, Vol. 2, No.1, pp. 87-110.

[36] LEHR, W. and MCKNIGHT, L. W. (2003) Wireless Internet Access: 3G vs. WiFi? Telecommunication Policy, Vol. 27,No.5, pp. 351-370.

[37] Leiner, B., Ruth, R. and Sastry. A. R (1996)" Goals and challenges of the DARPA GloMo program. IEEE Personal Communications, December 1996., Vol. 3, No. 6, pp. 34-43.

[38] Li, X. (2006) Multipath Routing and QoS Provisioning in Mobile Ad hoc Networks. Department of Electronic Engineering. Queen Mary, University of London, PhD thesis.

[39] Liu, C.-H. and Chang, S.-S. (2009) The study of effectiveness for ad-hoc wireless network. Proceedings of the 2nd International Conference on Interaction Sciences: Information Technology, Culture and Human, Vol. 403, No.6, pp.412-417.

[40] Makki, S. K., Reiher, P. and Makki, K. (2007) Mobile and Wireless Network Security and Privacy, Springer Science and Business Media, LLC, 2007.

[41] Masella, A. K. A. A. (2001) Serving IP quality of service with Hiper- LAN/2. Computer Networks: The International Journal of Computer and Telecommunications Networking - Wireless networking, Vol. 37, issue. 1, pp. 17-24.

[42] Mbarushimana, C. and Shahrabi, A. (2008) Type of service aware routing protocol in mixed traffic Mobile Ad Hoc Networks. IEEE International Symposium On Wireless Communication Systems (ISWCS '08),Reykjavik pp.677-681.

[43] Mohapatra, P. and Krishnamurthy, S. V. (2005) Ad Hoc Networks Technologies and Protocols", Springer, 2005.

[44] Morinaga, N., Kohno, R. and Sampei, S. (2002) Wireless Communication Technologies New Multimedia Systems", Kluwer Academic Publishers, 2002.

[45] Murthy, C. S. R. and Mano, B. (2004) Ad Hoc Wireless Networks: Architectures and Protocols, Prentice Hall PTR.

[46] Murthy, S. and Garcia-luna-aceves, J. J. (1996) An efficient routing protocol for wireless networks. ACM Mobile Networks and Applications Journal, pp.183-197.

[47] Maltz, J.B. and D. Johunson, (2005). "Lessons from a full-Scale multi-hop wireless ad hoc network test bed". IEEE Personal communications magazine.

[48] Nicopolitidis, P., Obaidat, M. S., Papadimitriou, G. I. and Pomportsis, A. S. (2003) Wireless Networks. John Wiley and Sons, Ltd.

[49] Pandey, A. K.,and Fujinoki, H(2005)." Study of MANET routing protocols by GlomoSim simulator. International Journal of Network Management", November 2005, Vol 15, pp.393 -410.

[50] Part-11 (1997) IEEE Computer Society LAN MAN Standards Committee, Wireless LAN medium access control(MAC) and physical layer (PHY) specifications, IEEE standard 802.11, 1997. The Institute of Electrical and Electronics Engineers, New York, NY.

[51] Part-12 (1999) IEEE Computer Society. IEEE standard for information technology telecommunications and information exchange between systems - local and metropolitan networks - specific requirements.

[52] Part-16 (2004) IEEE Standard for Local and metropolitan area networks Part 16: Air Interface for Fixed Broadband Wireless Access Systems, IEEE Std 802.16-2004, 2004.

[53] Perkings, C. E., M, E., belding-royer and Das. R. S. (2003) Ad Hoc On-Demand Distance Vector (AODV) Routing. http://www.ietf.org/internetdrafts/draft-ietf-manet-aodv-13.txt, IETF Internet draft,RFC.

[54] Perkins, C. E. (2001) Ad hoc networking: an introduction, Addison-Wesley Longman Publishing Co., Inc. Boston, MA, USA I'2001,pp. 1 - 28, ISBN: 0-201-30976-9.

[55] Perkins, C. E. and Bhagwat, P. (1994) Highly Dynamic Destination-Sequenced Distance-Vector Routing (DSDV) for Mobile Computers. n Proceedings of ACM SIGCOMM, pp.234-244.

[56] Perkins, D. D., Hughes, H. D. and Owen, C. B. (2002) Factors A ecting the Performance of Ad Hoc Networks. IEEE internet computing, East Lansing, 2002, MI 48824-1226.

[57] Park,V. and S. Corson. (2001)." Temporally-ordered Routing algorithm (TROA)". Internet Draft, draftietf-manet-tora-spec-04-txt. July,2001.

[58] Ritvanen, K. (2004) Multicast Routing and Addressing. Helsinki University of Technology Department of Computer Science and Engineering, A Seminar on Internetworking.

[59] Sarkar, S. K., Basavaraju, T. G. and Puttamadappa, C. (2008) Ad Hoc Mobile Wireless Networks: Principles, Protocols, and Applications, Auerbach Publications Taylor and Francis Group.

[60] Sinha .P , R.Sivakumar, V. Bharghavan,"CEDAR: a Core-Extraction Distributed Ad hoc Routing algorithm". IEEE INFOCOM'99, Vol 4, No.2, pp. 120-127.

[61] Satyanarayanan, M. (2001) IEEE Pervasive Computing:Vision and Challenges. Personal Communication, Vol. 8, No. 2, pp. 10 -17.

[62] White, C. (2002) Data Communications and Computer Networks, Published by Thomson, Third Edition, 2002.

[63] Wu, S.-L., Yu-chee and TSENG (2007) Wireless Ad Hoc Networking. Auerbach Publications -Taylor and Francis Group.

12

Power Management in Sensing Subsystem of Wireless Multimedia Sensor Networks

Mohammad Alaei and Jose Maria Barcelo-Ordinas
Computer Architecture Department, Universitat Politecnica de Catalunya, Barcelona
Spain

1. Introduction

A wireless sensor network consists of sensor nodes deployed over a geographical area for monitoring physical phenomena like temperature, humidity, vibrations, seismic events, and so on. Typically, a sensor node is a tiny device that includes three basic components: a sensing subsystem for data acquisition from the physical surrounding environment, a processing subsystem for local data processing and storage, and a wireless communication subsystem for data transmission. In addition, a power source supplies the energy needed by the device to perform the programmed task. This power source often consists of a battery with a limited energy budget. In addition, it is usually impossible or inconvenient to recharge the battery, because nodes are deployed in a hostile or unpractical environment. On the other hand, the sensor network should have a lifetime long enough to fulfill the application requirements. Accordingly, energy conservation in nodes and maximization of network lifetime are commonly recognized as a key challenge in the design and implementation of WSNs.

Experimental measurements have shown that generally data transmission is very expensive in terms of energy consumption, while data processing consumes significantly less (Raghunathan et al., 2002). The energy cost of transmitting a single bit of information is approximately the same as that needed for processing a thousand operations in a typical sensor node (Pottie & Kaiser, 2000). The energy consumption of the sensing subsystem depends on the specific sensor type. In some cases of scalar sensors, it is negligible with respect to the energy consumed by the processing and, above all, the communication subsystems. In other cases, the energy expenditure for data sensing may be comparable to, or even greater (in the case of multimedia sensing) than the energy needed for data transmission. In general, energy-saving techniques focus on two subsystems: the communication subsystem (i.e., energy management is taken into account in the operations of each single node, as well as in the design of networking protocols), and the sensing subsystem (i.e., techniques are used to reduce the amount or frequency of energy-expensive samples).

1.1 Power consumption in sensing subsystem

In fact, the energy consumption of the sensing subsystem not only may be relevant, but it can also be greater than the energy consumption of the radio or even greater than the energy consumption of the rest of the sensor node (Alippi et al., 2007). This can be due to many different factors (Raghunathan et al., 2006):

- Power hungry transducers. Some sensors intrinsically require high power resources to perform their sampling task. For example, sensing arrays such as CCDs or multimedia sensors (Akyildiz et al., 2007) such as CMOS image sensors generally require a lot of power. Also chemical or biological sensors (Diamond, 2006) can be power hungry as well.
- Long acquisition time. The acquisition time may be in the order of hundreds of milliseconds or even seconds, especially in the case of multimedia sensors. Hence the energy consumed by the sensing subsystem may be high, even if the sensor power consumption is moderate. In this case reducing communications may be not enough, but energy conservation schemes have to actually reduce the number of acquisitions (i.e. data samples). It should also be pointed out that energy-efficient data acquisition techniques are not exclusively aimed at reducing the energy consumption of the sensing subsystem. By reducing the data sampled by source nodes, they decrease the number of communications as well. Actually, many energy-efficient data-acquisition techniques have been conceived for minimizing the radio energy consumption, under the assumption that the sensor consumption is negligible.
- Power hungry A/D converters. Sensors like acoustic and seismic transducers generally require high-rate and high-resolution A/D converters. The power consumption of the converters can account for the most significant power consumption of the sensing subsystem, as in (Schott et al., 2005).

1.2 Multimedia sensing subsystem

One of the main differences between multimedia sensor networks and other types of sensor networks lies in the nature of how the image sensors perceive information from the environment. Most scalar sensors provide measurements as 1-dimensional data signals. However, image sensors are composed of a large number of photosensitive cells. One measurement of the image sensor provides a 2-dimensional set of data points, which we see as an image. The additional dimensionality of the data set results in richer information content as well as in a higher complexity of data processing and analysis. In addition, a camera's sensing model is inherently different from the sensing model of any other type of sensor. Typically, a scalar sensor collects data from its vicinity, as determined by its sensing range. Multimedia nodes are characterized by a directional sensing model, called Field of View (FoV, see Figure 1), and can capture images of distant/vicinal objects/scenes within its FoV from a certain direction. The object covered by the camera can be distant from the camera and the captured images will depend on the relative positions and orientation of the cameras towards the observed object (Soro & Heinzelman, 2005; Tezcan & Wang, 2008; Adriaens et al., 2006). Because of non-coincidence between neighborhood and sensed region by multimedia nodes, coverage-based techniques in WSN do not satisfy WMSN requirements.

Accordingly, the amount of power consumed in the sensing subsystem of a multimedia sensor node is considerably more than of a scalar ordinary sensor. For example, a temperature sensor (texas instrument, 2011) as a scalar sensor consumes 6μW for sensing the environment. To have a view of multimedia sensors power consumption, table 1 shows the power consumed by four classes of cameras that are available today either as prototypes or as commercial products. At the lowest end of the spectrum is tiny Cyclops (Rahimi et al., 2005) that consumes a mere 46mW and can capture low resolution video. CMU-Cams (Rowe et al., 2002) are cell-phone class cameras with on-board processing for motion detection, histogram computation, etc. At the high-end, web-cams can capture high-resolution video at full frame rate while consuming 200mW, whereas Pan-Tilt-Zoom cameras are re-targetable sensors that produce high quality video while consuming 1W. It is noticeable that the mentioned power amounts are the power consumed by the camera sensors without considering the power consumed by the host motes, see (Tavli et al, 2011) for a survey of visual network platforms.

Fig. 1. The Field of View (FoV) of a multimedia sensor node.

Multimedia Sensor	Power of image capturing	Capability in image capturing
Cyclops	42 mW	Fixed angle lens, 352×288 at 10 fps
CMU-Cam	200 mW	Fixed angle lens, 352×288 up to 60 fps
Web-Cam	200 mW	Auto focus lens, 640×480 at 30 fps
High-end PTZ Camera	1 W	Pan-tilt-zoom lens, 1024×768 up to 30fps

Table 1. Power consumption and capabilities of four classes of camera sensors.

On the other hand, given the large amount of data generated by the multimedia nodes, both processing and transmitting image data are quite costly in terms of energy, much more so than for other types of sensor networks. Furthermore, visual sensor networks require large bandwidth for transmitting image data. Thus both energy and bandwidth are even more constrained than in other types of wireless sensor networks.

In this chapter, we describe a power efficient mechanism for managing the sensing subsystem of multimedia sensor nodes for surveillance in WMSNs. For this purpose, the deployed multimedia nodes are clustered according to their common covering regions and the clusters are managed to schedule the members to collaboratively survey the sensing area in a duty-cycled manner. With avoiding acquisition of redundant and correlated data, not only the sensing subsystem of nodes save its energy, but also the transmission and processing subsystems meet an optimized amount of data to be transmitted/processed and thus can conserve their residual energy. Therefore, the network lifetime is considerably prolonged.

The chapter is organized as follows. In section 2 we present an overview of work related to sensor management and scheduling policies. A surveillance mechanism with its details in grouping, management and scheduling multimedia nodes to be energy efficient is explained in section 3. Finally, the future work and conclusions are derived.

2. Sensor management and scheduling policies

In redundantly deployed multimedia sensor networks a subset of cameras can perform continuous monitoring and provide information with a desired quality. This subset of active cameras can be changed over time, which enables balancing of the cameras energy consumption, while spreading the monitoring task among the cameras. In such a scenario the decision about the camera nodes activity and the duration of their activity is based on sensor management policies. *Sensor management policies* define the selection and scheduling (that determines the activity duration) of the camera nodes activity in such a way that the visual information from selected cameras satisfies the application specified requirements while the use of camera resources is minimized. Various quality metrics are used in the evaluation of sensor management policies, such as the energy-efficiency of the selection method or the quality of the gathered image data from the selected cameras. In addition, camera management policies are directed by the application; for example, target tracking usually requires selection of cameras that cover only a part of the scene that contains the non-occluded object, while monitoring of large areas requires the selection of cameras with the largest combined FoV. While energy-efficient organization of camera nodes is oftentimes addressed by camera management policies, the quality of the data produced by the network is the main concern of the application.

The problem of finding the best camera candidates is investigated in (Soro & Heinzelman, 2007). In this work, the authors propose several cost metrics for the selection of a set of camera nodes that provide images used for reconstructing a view from a user-specified view point. Two types of metrics are considered: coverage aware cost metrics and quality-aware cost metrics. The *coverage-aware cost metrics* consider the remaining energy of the camera nodes and the coverage of the indoor space, and favor the selection of the cameras with higher remaining energy and more redundant coverage. The *quality-aware cost metrics* favor the selection of the cameras that provide images from a similar view point as the user's view point. Thus, these camera selection methods provide a trade-off between network lifetime and the quality of the reconstructed images.

Monitoring of large areas (such as parking lots, public areas, large stores, etc.) requires complete coverage of the area at every point in time. Such an application is analyzed in (Dagher et al., 2006), where the authors provide an optimal strategy for allocating parts of the monitored region to the cameras while maximizing the lifetime of the camera nodes. The optimal fractions of regions covered by every camera are found in a centralized way at the base station. The cameras use JPEG2000 to encode the allocated region such that the cost per bit transmission is reduced according to the fraction received from the base station.

Oftentimes, the quality of a reconstructed view from a set of selected cameras is used as a criterion for the evaluation of camera selection policies. In the work (Park et al., 2006) distributed look-up tables are used to rank the cameras according to how well they image a specific location, and based on this, they choose the best candidates that provide images of the desired location. Their selection criterion is based on the fact that the error in the captured image increases as the object gets further away from the center of the viewing frustum. Thus, they divide the frustum of each camera into smaller unit volumes (subfrustums). Then, based on the Euclidian distance of each 3D point to the centers of subfrustums that contain this 3D point, they sort the cameras and find the most favorable camera that contains this point in its field of view. The look-up table entries for each 3D location are propagated through the network in order to build a sorted list of favorable cameras. Thus, camera selection is based exclusively on the quality of the image data provided by the selected cameras, while the resource constraints are not considered.

In order to reduce the energy consumption of cameras, the work (Zamora & Marculescu, 2007) explores distributed power management of camera nodes based on coordinated node wake-ups. The proposed policy assumes that each camera node is awake for a certain period of time, after which the camera node decides whether it should enter the low-power state based on the timeout statuses of its neighboring nodes. Alternatively, camera nodes can decide whether to enter the low-power state based on voting from other neighboring cameras.

Selection of the best cameras for target tracking has been discussed often (Pahalawatta et al., 2004; Ercan et al., 2006). Pahalawatta et al. present a camera selection method for target tracking applications used in energy-constrained visual sensor networks. The camera nodes are selected by minimizing an information utility function (obtained as the uncertainty of the estimated posterior distribution of a target) subject to energy constraints. However, the information obtained from the selected cameras can be lost in the case of object occlusions. This occlusion problem is further discussed by Ercan et al. where they propose a method for camera selection in the case when the tracked object becomes occluded by static or moving occluders. Finding the best camera set for object tracking involves minimizing the MSE of the object position's estimates. Such a greedy heuristic for camera selection shows results close to optimal and outperforms naive heuristics, such as selection of the closest set of cameras to the target, or uniformly spaced cameras. The authors here assume that some information about the scene is known in advance, such as the positions of static occluders, and the object and dynamic occluders prior probabilities for location estimates.

As a conclusion, in multimedia sensor networks, sensor management policies are needed to assure balance between the opposite requirements imposed by the wireless networking and vision processing tasks. While reducing energy consumption by limiting data transmissions is the primary challenge of energy-constrained visual sensor networks, the quality of the image data and application, QoS, improve as the network provides more data. In such an environment, the optimization methods for sensor management developed for wireless sensor networks are hard to directly apply to multimedia sensor networks. Such sensor management policies usually do not consider the event-driven nature of multimedia sensor networks, nor do they consider the unpredictability of data traffic caused by a monitoring procedure. Thus, more research is needed to further explore sensor management for multimedia sensor networks. Since sensor management policies depend on the underlying networking policies and vision processing, future research lies in the intersection of finding the best trade-offs between these two aspects of visual sensor networks. Additional work is needed to compare the performance of different camera node scheduling sensor policies, including asynchronous (where every camera follows its own on-off schedule) and synchronous (where cameras are divided into different sets, so that in each moment one set of cameras is active) policies. From an application perspective, it would be interesting to explore sensor management policies for supporting multiple applications utilizing a single visual sensor network.

The presented mechanism in the following section groups multimedia nodes in clusters based on their common sensing region of the whole deployment region. The clusters monitor the environment independently but in each cluster the members collaborate in data acquisition in an intermittent manner. The scheduling and activity times in each cluster are determined based on the cluster population and the scale of overlapping between FoV of cluster members. So, the data transmissions are not limited in this kind of sensor management but the volume of sensed data is reduced by management in only sensing subsystem and applying coordination among cluster members to optimize capturing image times and to avoid redundant sensing of the same data in the overlapped FoVs. On the other hand, the sensing region is divided between clusters and each cluster monitors its domain with its exclusive frequency and member scheduling. Thus, clusters are not synchronized for sensing the region whiles each point of the sensing region is monitored frequently according to the number of nodes that cover that point by their sensing subsystem.

3. The surveillance mechanism

3.1 Preliminary

We assume wireless sensor nodes with fixed lenses providing a θ angle FoV, densely deployed in a random manner. The assumption of fixed lenses is based on the current WMSN platforms (Tavli et al, 2011). Almost all of them (SensEye, MicrelEye, CITRIC, Panoptes, Meerkats) (Kulkarni et al., 2005; Kerhat et al., 2007; Chen et al., 2008; Feng et al., 2005; Margi et al., 2006) have fixed lenses and only high powered PTZ cameras have movement capabilities. We consider a monitor area with N wireless multimedia sensors, represented by the set $S = \{S_1, S_2, ..., S_N\}$ randomly deployed. Each sensor node is equipped to learn its location coordinates and orientation information via any lightweight localization technique for wireless sensor networks. It is not the purpose of this chapter to define mechanisms to find this location. Without loss of generality, let us assume that nodes in the set S belong to a single-tier network or the same tier of a multitier architecture.

Our policy in order to applying collaboration among multimedia sensor nodes in the surveillance mechanism is clustering the network nodes based on their similarity in sensing the environment. The criterion applied in this purpose is the clustering scale of FoVs of nodes. The nodes having a large region of common area in their FoV, have a similar view of the sensing area then can cooperate in a established group, (Alaei & Barcelo, 2010).

3.2 Cluster formation and cluster membership

Now, let us consider the set $S = \{S_1, S_2, ..., S_N\}$ of wireless multimedia nodes belonging to the same tier of a network randomly deployed. The cluster formation algorithm is executed in a centralized manner by the sink after deploying the network. The main reasons in choosing a central architecture are the following: (i) for a distributed architecture, each node should notify to the rest of the nodes about its location A_i and its orientation α_i (i = 1,...,N). In a centralized architecture the nodes should notify to the sink their location and orientation. Note that this notification can be done using any energy efficient sensor routing protocol and only is necessary at bootstrap phase. All phases of the clustering algorithm are executed only one time, right after node deployment. (ii) In many WSN applications, the sink has ample resources (storage, power supply, communication and computation) availability and capacity which make it suitable to play such a role. (iii) Collecting information by a sink node is more power efficient compared to spreading this information to each and every other node within the network. (iv) Having the global view of the network at the sink node facilitates provision algorithms for closer-to-optimal cluster determination; the global knowledge can be updated at the sink when new nodes are added or some nodes die. Such maintenance tasks can be regarded as a normal routine for the sink. (v) Finally, using a centralized scheme can relieve processing load from the sensors in the field and help in extending the overall network lifetime by reducing energy consumption at individual nodes. The following phases are performed to establish and form clusters, (Figure 2):

- *Bootstrap*: At node bootstrap, each sensor {S_i, i = 1,...,N} transmits its position (x_i,y_i) and orientation α_i to the sink. To accomplish this step any efficient sensor routing algorithm can be used. Thus, the clustering algorithm is not bound to how the sink receives this information. If there is an un-connected node in the network, it cannot announce itself and thus will not be considered in the algorithm.

- *Cluster Formation*: (i) Initially, the sink creates an empty cluster associated with an un-clustered multimedia node of S. Thus, that node will be clustered as the first member (*i.e.*, Cluster-Head (CH)) of the established cluster. (ii) Then, the sink finds the qualified un-clustered nodes for joining to the CH by computing the area of overlapped polygons of their FoV. From position and orientation of nodes, the sink computes the overlapped region between each un-clustered multimedia node and the CH of the established cluster. For calculating the FoV overlapping area of two nodes, we first survey the intersection of their FoVs. Second, if they intersect each other, we find the intersection region and at last, compute the area of the polygon. For this purpose, in the first step, we define the equations of the sides of FoVs using the vertex coordinates. Then, the intersection of each side of each FoV to all sides of the other is calculated. A decomposition approach is used for calculating the area of the overlapping region of FoVs. If the computed overlapped region is equal or greater than the threshold considered as the *Clustering Scale* (γ) -the minimum region that has to be overlapped between two node FoVs to be grouped in a cluster-, the un-clustered node will be clustered as a member of the established cluster. (iii) When no more nodes can be added to the cluster, the sink takes a new un-clustered node, begins a new cluster and goes to step (ii).

- *Membership notification*: we assume that the sink uses any energy-efficient sensor routing algorithm to notify to each first-member of every cluster about its cluster-ID and what are the members of the cluster. Then, each first-member sends a packet to the members of his cluster notifying them about the cluster which they belong to.

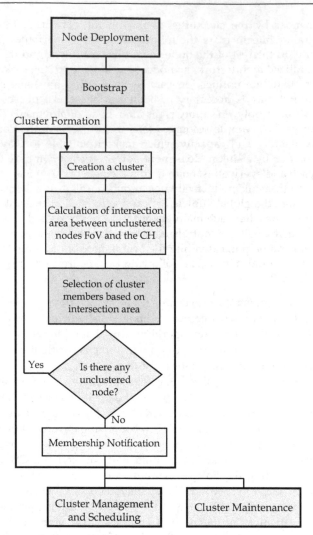

Fig. 2. Clustering Procedure.

The algorithm is executed by the sink once upon deployment and thus all nodes will become clustered. If a node joins to the network hereinafter, it has to send its position and orientation to the sink for announcing itself as a new node. The sink computes the FoV of the new node and finds the first cluster that can accept it as a new member. For this purpose, the sink computes the overlapping regions between FoV of the new node and the CH of each cluster and checks whether he is satisfying the cluster membership test. Then, the sink sends a message to the CH in order that this node re-organizes the cluster with the new member. Depending on the application, this notification may suppose a new reconfiguration in the monitoring task (*i.e.*, a new duty-cycle period). On the other hand, each node periodically sends a Hello message to the CH notifying its current residual energy. When a node dies, the CH will notify the rest of the members about the new cluster set and will reconfigure any parameter related to the cluster. The CH also periodically compares the residual energy of cluster members and its residual energy to select the new CH with the maximum residual energy in the cluster. If the CH decides to entrust CH role to another cluster member, notifies to the cluster members about the new CH. Note that the beaconing among cluster members implies low overhead since clusters have few nodes and hello periods can be on the order of duty-cycle sensing periods.

3.2.1 Intra-cluster collaboration

Let us see the potential of cooperative node monitoring in clusters in terms of sensor area coverage. We define the Maximum Cluster Coverage Domain (MCCD) parameter for a cluster as the maximum monitoring area which is covered by that cluster. Since each cluster is established considering the clustering scale equal to γ, the MCCD can be computed as follows (C_{size} is the size of the cluster):

$$MCCD = \gamma \cdot A_{FoV} + (1-\gamma) \cdot A_{FoV} \cdot C_{size} = (C_{size} - \gamma \cdot (C_{size} - 1)) \cdot A_{FoV} = \beta \cdot A_{FoV} \qquad (1)$$

where:

$$1 \le \beta = C_{size} - \gamma \cdot (C_{size} - 1) \qquad (2)$$

The effective cluster covering domain can be inferior to the MCCD calculated by Equation (1) since some nodes can overlap more than the region determined by γ. Since MCCD gives us an upper bound on the area covered by the cluster, using MCCD will allow us worst-case dimensioning. Factor β represents the increment of area that the cluster senses with respect to an individual sensor. When each node of a cluster obtains an image from its FoV, a part of the related MCCD with a ratio at least equal to $1/\beta$ respect to the MCCD is captured whereas this part includes overlapped areas of other nodes in the cluster. Sensing the environment by each member delivers information not only from the FoV of the active node but also from some overlapped parts of other nodes in the same cluster: at least $\gamma \cdot A_{FoV}$ of the area is common to the first-member and more than $1/\beta$ of the MCCD is monitored. For example, in a cluster consisting of just 2 members, assuming a clustering scale of $\gamma = 0.5$, the MCCD is $1.5 \cdot A_{FoV}$. Thus, when each of the two members of the cluster is activated and monitors the environment, an area of one FoV is captured that is at least 2/3 of the whole MCCD of the cluster. Consequently, scheduling and coordination among members in order to sense the field in a collaborative manner may yield a gain in energy saving and performance efficiency even with a low number of members in the cluster.

3.2.2 Cluster formation evaluation

All sensor nodes have been configured with a FoV vertex angle of $\theta = 60°$ and R_S of 20 m. A sensing field spanning an area of 120m × 120m has been used. Sensor densities were varied to study the cluster formation from sparse to dense random deployments. Figures illustrate the average results of 50 independent running tests whereas each test corresponds to a different random deployment. Once a random deployment is defined, cluster formation is obtained from node location, angle of orientation and FoVs of nodes, using the described method whose complexity is O(N.logN). Furthermore, as it was mentioned before, each node sends a packet to the sink in the bootstrap phase, then the sink notifies each CH via one packet his membership set for that cluster (phase 3) and then the CHs notify cluster nodes about their cluster membership and any related parameter. Thus, the average overhead of the algorithm is forwarding N packets from the nodes to the sink and forwarding N_C packets from the sink to first-members and forwarding $N_C \cdot (\mu_{Csize}-1)$ packets from CHs to cluster nodes; where N is the number of nodes, N_C is the average number of clusters and μ_{Csize} is the average cluster size. So the total overhead will be: $N + N_C + N_C \cdot (\mu_{Csize}-1)$ packets. The maintenance overhead is $N_C \cdot (\mu_{Csize}-1)$ beacons every keep-alive period, where the keep-alive period can be a multiple of the sensing duty-cycle period.

3.2.2.1 Number of clusters and cluster-size

The average number of clusters, μ_{NC}, and the average cluster-size (μ_{Csize}) in a tier/network for different node densities with several clustering scales are shown in Figures 3 and 4. Increasing the node density does not only cause an increment in the number of clusters but also yields more overlapping areas among FoVs and thus raises the cluster-size. However, the clustering scale (γ) also impacts in the cluster membership selection process. The clustering scale determines the minimum region that is required to be overlapped between the FoV of each node belonging to a given cluster and the FoV of the CH of that cluster. So, γ determines the minimum intersection part of FoV of each member with the CH of an established cluster. Lower clustering scales obligate less overlapping areas for cluster membership and increase the domain covered by a given cluster since more nodes will be conforming to the membership rule. Increasing the clustering scale restricts node membership because of higher required overlapping areas between FoVs of nodes. Thus, higher clustering scales result in lower cluster-sizes, less MCCD and thus higher number of clusters.

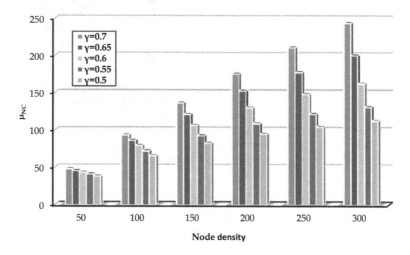

Fig. 3. Average number of established clusters.

Sparse networks have low average cluster-size, μ_{Csize}, because sparse deployments result in low overlapping areas. Moreover, high values of γ also will produce low μ_{Csize}. The result will be lower potential for node coordination. On the other hand, dense wireless multimedia sensor networks can particularly benefit from higher cluster sizes and thus more potential for node coordination.

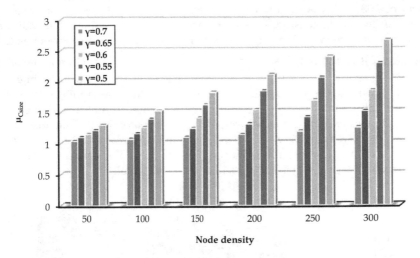

Fig. 4. Average size of established clusters.

Finally, Figure 5 shows the cumulative probability function for the cluster-size in the network for different node densities assuming a clustering scale of $\gamma = 0.5$. For example, in a network consisting of 250 nodes, 28% of clusters have a single member which does not have enough overlapping with others to satisfy the clustering scale, 32% of clusters have a cluster size of 2, 21% of 3, 12% of 4 and 7% of them consisting of more than four members.

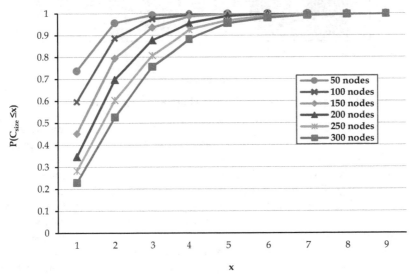

Fig. 5. The cluster size cumulative distribution function ($\gamma = 0.5$).

3.2.2.2 Coverage

Figure 6 illustrates the percentage of area that is covered by the random deployment in terms of node density. As it is shown in the figure, for covering 95% of the area, a dense deployment of 300 nodes is required. As the figure shows, the rate of increment of the covered area for low node densities is faster than for high node densities. This indicates that after a new node is added in a dense deployment, low new coverage area is obtained.

For example, the first 100 nodes cover 75% of the field, but the next 100 nodes will only cover 15% of new area. The conclusion is that dense networks are able to cover high areas at the cost of high overlapping and sensing redundancy, but this overlapping can be used for improving reliability if nodes belonging to the same cluster work in a coordinated manner. Furthermore, the existence of obstacles produces a reduction of the sensing area because of FoV occlusion effect, (Tezcan & Wang, 2008). So, employing dense networks of low-cost, low-resolution and low-power multimedia sensor nodes instead of sparse networks of high-power, high-resolution sensors (*e.g.*, PTZ) will be more beneficial.

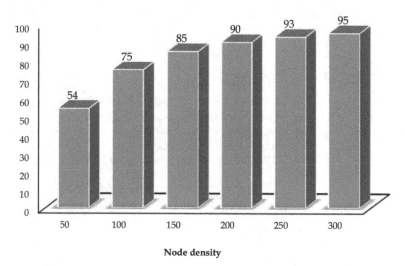

Node density

Fig. 6. Percentage of the covered area with respect to the whole deployment area.

Applications that are interested in multiple views will also benefit from this situation, since there will be several nodes monitoring the same area from several perspectives. Applications that are interested in detecting objects and are not interested in having an instantaneous multiple-view of the object may benefit from collaborative node processing in terms of energy savings. For the first set of applications clustering of nodes may serves as an indicator of triggering simultaneous multi-perspective pictures. For the second set of applications, clustering may serve as a baseline framework for collaborative node scheduling avoiding redundant sensing and processing and thus increasing network lifetime. Other applications that are interested in correlated data (*e.g.*, Distributed Video Coding, DVC) may use clustering in order to exploit multi-view correlations to build joint encoders (Pereira et al., 2008).

3.3 Cooperative node selection and scheduling

In monitoring mechanisms, usually cameras should perform duty-cycled monitoring over the area that they sense. That means that every T (Figure 7.a) seconds the sensors in the monitored area will awake and monitor the area. This is the situation for a planned network in which every sensor is placed in such a position that there is no overlapping among sensors. Nevertheless, this duty-cycle scheduling will produce high power consumption in those situations in which there are overlapping sensors, since camera nodes with overlapping areas do not cooperate to sense the area and thus they redundantly monitor the area.

In this section, we explain a cooperative mechanism based on the clustering method that coordinates nodes belonging to the same cluster to work in a collaborative manner to monitor the sensing area. The objective of this mechanism is to increase power conservation by avoiding similar sensing and redundant processing at the same time. Also, collaborative sensing by nodes that have FoVs intersecting each other yields to more reliability: cluster members will monitor the region sequentially and if a moving object is not detected in one image capturing, it will be in the vicinal FoVs at the next capturing times. Thus, the other members in the same cluster may detect the object.

Let us divide the environment in domains covered by clusters of nodes (MCCD, Section 3.2). All clusters concurrently sense their domains. In each cluster, members are awakened sequentially in an intermittent manner by the CH with a time interval related to the cluster-size and the scale of clustering (see Figure 7.b); (*i.e.*, $T_{interval}$ is the time between awakening two consecutive members of a cluster). In this way, each node of a given cluster periodically participates in capturing an image from its unique perspective and surveillance the environment and finally sleeps again with a cluster-based period called T_P. Formulas for these periods are derived in Section 3.3.1.

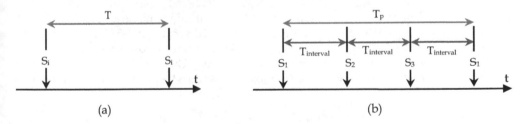

(a)   (b)

Fig. 7. (a) Period of awakening a given node in the un-cooperative scheduling. (b) Scheduling for a cluster consisting of three members (S_1, S_2, S_3).

3.3.1 Cluster-based T_P and $T_{interval}$ computation

Let us consider as baseline mechanism a non-collaborative duty-cycled scheme in which every node awakes with an interval period of time T and monitor the area (i.e., takes a picture and performs object detection) as tier 1 in (Kulkarni et al., 2005). The objective of the collaborative mechanism is to produce a cluster-based duty-cycled scheduler in which: (i) Each node is awakened and senses the area with a reliable period of $T_P > T$ taking advantage of the overlapping among nodes in the cluster, thus, saving energy and increasing network lifetime. Each cluster will have its own T_P interval, determined according to the cluster-size and the clustering scale. (ii) During the sleeping period of each member of a given cluster, other nodes belonging to the cluster are awakened with intervals of $T_{interval} < T$ (that is equal to: T_p / C_{size}) in a sequential manner.

The area sensed by each cluster is related to the MCCD area. In order to compute T_p we will consider the MCCD area. By awaking each member of a given cluster, in average, a part of the related MCCD with a ratio equal to $1/\beta$ is captured (Equation (2)). Note that the MCCD is an area of $\beta.A_{FoV}$ and is sensed by C_{size} overlapping members, thus sensing the environment by each node delivers information not only from the FoV of the awakened node but also from some overlapped parts of the FoV of other nodes in the same cluster. Then, we may define the node interval duty-cycle period as:

$$T_P = T \cdot \frac{C_{size}}{\beta} = T \cdot \frac{C_{size}}{C_{size} - \gamma \cdot (C_{size} - 1)} \tag{3}$$

Note that the T_P is proprietary for each cluster in terms of its cluster-size and clustering scale. As it was mentioned before, the MCCD calculated by Equation (1) is the maximum covering domain of a cluster while the effective cluster covering domain may be less than MCCD since some members of a given cluster may overlap more than the region determined by γ. Consequently, a given cluster can cover an area less than $\beta \cdot A_{FoV}$. Thus, using β gives us the lowest interval T_P and thus the most reliable one since lower values of β would increase the interval T_P. On the other hand, members of a cluster are awakened sequentially to sense their environment in an intermittent way with time intervals equal to $T_{interval}$:

$$T_{interval} = \frac{T_P}{C_{size}} = \frac{T}{C_{size} - \gamma \cdot (C_{size} - 1)} \leq T \tag{4}$$

Let us consider Figure 6.b and for example a cluster with three members, $C = \{S_1, S_2, S_3\}$, cluster-head S_1 and $\gamma = 0.5$. Every node will be awakened every $T_P = 1.5 \cdot T$ seconds and the area will be monitored every $T_{interval} = 0.5 \cdot T$ seconds. As can be observed, every sensor is awakened with a period higher than the non-collaborative scheme but the area is monitored more times. Then, the area duty-cycled frequency is increased while the sensor duty-cycled frequency is reduced.

Table 2 shows the evolution respects of T_p and $T_{interval}$ to T as a function of γ for several C_{size}. We first have to notice that for a clustering scale factor $\gamma = 1$, $T_p = T$, while for $\gamma < 1$, $T \leq T_p \leq T/(1-\gamma)$. Then, the duty-cycle frequency at which a specific node is awakened is decreased by a factor that at least is $(1-\gamma)$ times the frequency of the non-collaborative scheme. On the other hand some sensor of the cluster will be on duty every $T_{interval}$ seconds. Note that $T_{interval}$ will be lower than T and will be smaller as C_{size} increases. This means that the area is monitored more frequently although every specific sensor monitors with less frequency. The reason is justified in how clusters are formed. Any sensor of the cluster overlaps with the first-member by at least an area of $\gamma \cdot A_{FoV}$. Thus, when a sensor enters in duty, he will monitor an area equal to $\gamma \cdot A_{FoV}$ overlapped with the first-member and an area equal to $(1-\gamma) \cdot A_{FoV}$ that in the worst case does not overlap with any other member of the cluster. Sensing the whole cluster area with $T_{interval}$ equal to T would result in that an area equivalent to $(1-\gamma) \cdot A_{FoV}$ would be monitored every $C_{size} \cdot T$, a value that can be very high. However, using Equation (3), monitoring of the area equivalent to $(1-\gamma) \cdot A_{FoV}$ is guaranteed by a monitoring interval that is not superior to $T/(1-\gamma)$, that is much lower than $C_{size} \cdot T$.

C_{size} / γ	5	4	3	2
0.5	1.67	1.60	1.5	1.33
0.55	1.79	1.70	1.58	1.38
0.6	1.92	1.82	1.67	1.43
0.65	2.08	1.95	1.77	1.48
0.7	2.27	2.11	1.88	1.54

(a)

C_{size} / γ	5	4	3	2
0.5	0.334	0.4	0.5	0.665
0.55	0.358	0.425	0.527	0.690
0.6	0.384	0.455	0.557	0.715
0.65	0.416	0.488	0.590	0.740
0.7	0.454	0.528	0.627	0.770

(b)

Table 2. (a) T_p/T , (b) $T_{interval}/T$ for different cluster sizes and clustering scales.

Sleep/wake up protocols has extensively been studied in the area of wireless sensor networks, mainly for the radio subsystem, (Anastasi et al., 2009). Our clustering algorithm works on the sensing subsystem. It is important to notice that executing object detection does not imply sending packets to the sink. Thus, the sleep/wake up algorithm can be decoupled with the radio subsystem. Sleep/wake up can be based on periodic duty-cycle synchronized by the first-member: every T_p period, the sensing subsystem wakes up and performs object detection. However, clock drifts can cause cluster de-synchronization. To handle resynchronization, the system makes use of the beaconing scheme for cluster maintenance: nodes receive periodical beacons from the first-member and vice versa in order to detect new members or to detect members that have died. Beaconing duty-cycling belongs to the radio subsystem and it is independent of the sensing subsystem. That means that waking up the sensor to send a beacon is independent of waking up the sensor to take a picture and perform object detection. Thus, the cluster-head may resynchronize cluster members without need of waking up the sensing subsystem.

3.4 Lifetime prolongation evaluation

To evaluate the scheduling scheme in terms of power conservation, we compare the cooperative scheduled scheme with a single-tier network or a tier of a multi-tier architecture consisting of N nodes monitoring without coordination among them as (Rahimi et al., 2005; Kulkarni et al., 2005; Feng et al., 2005), in which, nodes are awakened with a time period of T. We note that the evaluation is over the sensing subsystem and that the radio subsystem (*i.e.*; transmission and reception of packets) is not taken into account.

The energy consumed in the network for object detection by N nodes during a duty-cycle interval of T in the non-collaborative scheduling is:

$$E = N \cdot (T_{sleep} \cdot P_{sleep} + E_{w\_up} + E_{cap} + E_{detect})$$

(5)

where T_{sleep} and P_{sleep} are the period and power consumption for a node in sleep mode. E_{w_up}, E_{cap} and E_{detect} respectively are the energies consumed in waking up a node, capturing a picture and performing object detection.

Let us now consider the cooperative scheduling algorithm in a clustered tier/network. Both, the interval between waking up consecutive nodes in the same cluster and the period of waking up a given node are functions of the cluster-size of the cluster which the nodes belong to. In one hand, in clusters with high cluster-size, $T_{interval}$ is small and thus cluster duty-cycle frequency is increased. On the other hand, higher number of nodes in the cluster causes longer periods T_P for awaking a given node of the cluster and thus yields an enhancement for power conservation in cluster's members. Assuming average cluster-size for all clusters in the tier/network, T_P will be:

$$T_P = \frac{T \cdot \mu_{C_{size}}}{\mu_{C_{size}} - \gamma \cdot (\mu_{C_{size}} - 1)}$$

(6)

where T is the base period for waking nodes in the base un-coordinated tier. Figure 8 shows the evolution of T_p normalized by T (i.e.; μ_{Csize}/β) for several node densities and clustering scales, γ. We may observe that the node average duty-cycle frequency is reduced by factors that are, for example, on the order of 0.78 for a 200 node network and a scale factor of $\gamma = 0.6$.

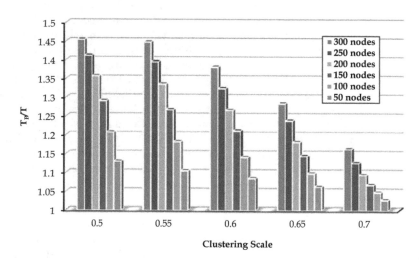

Fig. 8. T_P/T for several node densities and clustering scales.

Consequently, the total amount of averaged consumed energy by nodes for object detection in the coordinated tier during T_P will be:

$$E_P = E + N \cdot P_{sleep} \cdot (T_P - T) \tag{7}$$

From (6) and (7):

$$E_P = E + \frac{\gamma \cdot T \cdot (\mu_{C_{size}} - 1)}{\mu_{C_{size}} - \gamma \cdot (\mu_{C_{size}} - 1)} \cdot N \cdot P_{sleep} \tag{8}$$

So:

$$\frac{E_P}{T_P} = \frac{E \cdot (\mu_{C_{size}} - \gamma \cdot (\mu_{C_{size}} - 1))}{T \cdot \mu_{C_{size}}} + \frac{\gamma \cdot (\mu_{C_{size}} - 1) \cdot N \cdot P_{sleep}}{\mu_{C_{size}}}$$

$$\frac{E_P}{T_P} = (1 - \frac{\mu_{C_{size}} - 1}{\mu_{C_{size}}} \cdot \gamma) \cdot \frac{E}{T} + \frac{N \cdot \gamma \cdot (\mu_{C_{size}} - 1)}{\mu_{C_{size}}} \cdot P_{sleep} \quad where \ (0 < \gamma < 1) \ and \ (\mu_{C_{size}} > 1)$$

Therefore, the consumed power is:

$$P_P = \lambda \cdot P + \sigma \cdot P_{sleep} \tag{9}$$

where:

$$\lambda = (1 - \frac{\mu_{C_{size}} - 1}{\mu_{C_{size}}} \cdot \gamma) \quad , \quad 0 < \lambda < 1$$

$$\sigma = \frac{N \cdot \gamma \cdot (\mu_{C_{size}} - 1)}{\mu_{C_{size}}} \quad , \quad 0 < \sigma < \gamma \cdot N$$

Parameter P in Equation (9) is the power consumed in the network with the base un-coordinated mechanism. The consumed power in our scheme (P_P) is reduced by a factor λ with respect to P. The λ factor depends on the average cluster-size and the clustering scale factor. As can be observed from Equation (9) increasing μ_{Csize} produces lower values of λ, and thus a saving in energy with respect the uncoordinated system. For example a $\mu_{Csize} = 1.5$ (100 nodes with $\gamma = 0.5$) produces a $\lambda = 1 - \gamma/3 = 0.83$ while a $\mu_{Csize} = 2.15$ (200 nodes with $\gamma = 0.5$) produces a $\lambda = 1 - 0.53 \gamma = 0.73$. The other term ($\sigma \cdot P_{sleep}$) in Equation (9) is due to the fact of taking nodes to sleep mode in intervals of duration ($T_P > T$) and then nodes sleep $T_P - T$ more time than in the un-clustered scheme.

Figure 9 illustrates the impact of factor λ in Equation (9) in terms of node densities for several clustering scales. From this figure we can see that in high node density tiers, the factor λ is more beneficial since μ_{Csize} is higher and thus there is more potential of cooperation among nodes.

Figure 10 shows the consumed power (P) in the base un-coordinated tier for object detection in four cases of period of duty-cycle for different node densities. The consumed power has been computed for nodes consisting of Cyclops as camera sensor embedded in the host MICA II, similar to the tier 1 in (Kulkarni et al., 2005).

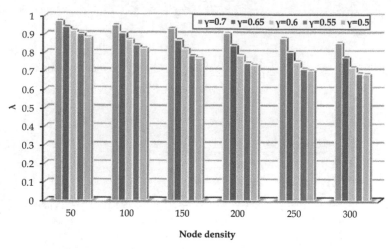

Fig. 9. Factor λ in cooperative scheduling for several clustering scales.

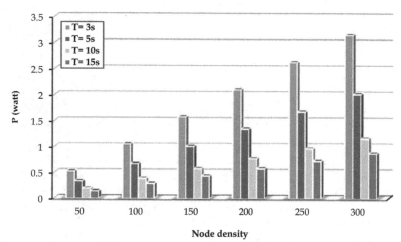

Fig. 10. Consumed power (P) for a non-cooperative tier/network of nodes consisting of Cyclops.

For instance, in the case without coordination, the power consumed in a tier consisting of 200 nodes that performs monitoring with a duty cycle of T=5 second, is 1.344 watts. In the coordinated network with the same number of nodes and a clustering scale of 0.5, the power consumed by the network would be reduced by a factor λ of 0.737 (see Figure 9) at the cost of increasing 52.60 mW, ($\sigma \cdot P_{sleep}$). This means a tier power consumption of 1.344·0.737 + 0.0526 = 1.043 Watts implying a reduction of 22.39%. Thus, in this case, the Prolongation Lifetime Ratio (PLR) would be of 1.344/1.043 = 1.289. Figure 11.a,b shows the prolongation lifetime ratio assuming a clustering scale of 0.5 and 0.6 for different node densities in four cases of duty-cycle (T). Tiers with high number of nodes have higher capability for cooperation and thus their nodes can conserve considerable amount of energy comparing to sparse networks and consequently, have longer prolonged lifetime. The figure indicates the more prolongation lifetime for dense tiers.

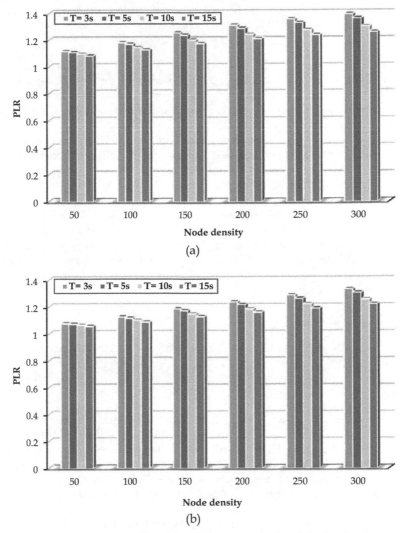

Fig. 11. Prolongation Lifetime Ratio (PLR) for different node densities in the clustered tier with a clustering scale equal to (a) 0.5. (b) 0.6, in four states of base awakening period.

4. Future work

In the clusters established by the depicted mechanism, each cluster member has a common sensing region with the CH. The clusters do not have any intersection and each cluster monitors its covering domain with only intra-cluster collaboration. Clustering with the capability of intersection and cooperation among clusters can increase the scale of efficiency of monitoring performance and power conservation of cluster members. In a monitoring mechanism utilizing intra and inter cluster cooperation, sensing regions are allocated to intersected clusters thus can be monitored with a higher frequency and/or consuming less amount of energy although the node selection and scheduling procedure will be more complicated. Some initial work has been done in (Alaei & Barcelo, 2010).

5. Conclusion

In this chapter a mechanism for management the wireless multimedia sensor nodes, was described. The mechanism, first, clusters nodes according to their scale of similarity in covering the environment; second, selects and schedules members of established clusters to monitor the sensing region which is divided among clusters. The members of each cluster are scheduled with an exclusive frequency based on the number of members in the cluster and the scale of overlapping among fields of view of the cluster members and thus the monitoring efficiency is increased. Moreover, because of the established intra cluster coordination and collaboration, sensing subsystem of multimedia nodes are optimized to avoid redundant and overlapped sensing. Thus, the capability of energy saving is considerably enhanced with respect to ordinary duty-cycling manners of environment monitoring by WMSNs. On the other hand, optimizing the data sensed by sensing subsystem results in conservation of energy in the transmission and processing subsystems since they meet less amounts of multimedia data to be transmitted and/or processed by the network nodes. Results show how this mechanism prolongs the network lifetime along with a better monitoring performance.

6. Acknowledgment

This work is partially supported by the EuroNF NoE and grants TIN2010-21378-C02-01 and SGR2009-1167.

7. References

Adriaens, J.; Megerian, S. & Potkonjak, M. (2006). Optimal worst-case coverage of directional field-of-view sensor networks, *Proceedings of the 3rd IEEE Communication Society Conference on Sensor, Mesh and Ad Hoc Communications and Networks (IEEE SECON)*, pp. 336–345, ISBN 1-4244-0626-9, Reston, VA USA, September 25–28, 2006

Akyildiz, I. F.; Melodia, T. & Chowdhury, K. R. (2007). A Survey on Wireless Multimedia Sensor Networks. *Computer Networks*, Vol. 51, Issue 4, (March 2007), pp. 921-960, ISSN 1389-1286

Alaei, M. & Barcelo-Ordinas, J.M. (2010). A method for clustering and cooperation in Wireless Multimedia Sensor Networks. *Sensors*, Vol. 10, No. 4, (March 2010), pp. 3145-3169, ISSN 1424-8220

Alaei, M. & Barcelo-Ordinas, J.M. (2010). MCM: multi-cluster-membership approach for FoV-based cluster formation in wireless multimedia sensor networks, *Proceedings of The 6th International Wireless Communications and Mobile Computing Conference (IWCMC 2010)*, pp. 1161-1165, ISBN 978-1-4503-0062-9, Caen, France, June 28-July 2, 2010

Alippi, C.; Anastasi, G.; Galperti, C.; Mancini, F. & Roveri, M. (2007). Adaptive Sampling for Energy Conservation in Wireless Sensor Networks for Snow Monitoring Applications, *Proceedings of IEEE International Workshop on Mobile Ad-hoc and Sensor Systems for Global and Homeland Security (MASS-GHS07)*, Pisa, Italy, October 8, 2007

Anastasi, G.; Conti, M.; Francesco, M. & Passarella, A. (2009). Energy conservation in wireless sensor networks: A survey. *Ad Hoc Networks*, Vol. 7, Issue 3, (May 2009), pp. 537–568, ISSN 1570-8705

Chen, P.; Ahammed, P.; Boyer, C.; Huang, S.; Lin L.; Lobaton, E.; Meingast, M.; Oh, S.; Wang, S.; Yan, P.; Yang, A.Y.; Yeo, C.; Chang, L.C.; Tygar, D. & Sastry, S.S. (2008). CITRIC: A low-bandwidth wireless camera network platform. *Proceedings of the 2nd ACM/IEEE International Conference on Distributed Smart Cameras (ICDSC 2008)*, pp. 1–10, ISBN 978-1-4244-2665-2, Palo Alto, CA, USA, September 7–11, 2008

Dagher, J. C.; Marcellin, M. W. & Neifeld, M. A. (2006). A method for coordinating the distributed transmission of imagery, *IEEE Transactions on Image Processing*, Vol. 15, No. 7, (July 2006), pp. 1705–1717, ISSN 1057-7149

Diamond, D. (2006). Energy Consumption Issues in Chemo/Biosensing using WSNs, *Energy and Materials: Critical Issues for Wireless Sensor Networks Workshop*, June 30, 2006.

Ercan, A.; Gamal, A. E. & Guibas, L. (2006). Camera network node selection for target localization in the presence of occlusions, *Proceedings of the ACM SenSys Workshop on Distributed Smart Cameras*, 2006.

Feng, W.C.; Kaiser, E.; Shea, M.; Feng, W.C & Baillif, L. (2005). Panoptes: scalable low-power video sensor networking technologies. *ACM Transactions on Multimedia Computing, Communications, and Applications*, Vol. 1, Issue 2, (May 2005), pp. 151–167, ISSN 1551-6857

Kerhet, A.; Magno, M.; Leonardi, F.; Boni, A. & Benini, L. (2007). A low-power wireless video sensor node for distributed object detection. *Journal of Real-Time Image Processing*, Vol. 2, No. 4, October 2007, pp. 331–342, ISSN 1861-8219

Kulkarni, P.; Ganesan, D.; Shenoy, P. & Lu, Q. (2005). SensEye: A multi tier camera sensor network, *Proceedings of the 13th ACM International Conference on Multimedia (ACM MM 2005)*, pp. 229–238, ISBN 1-59593-044-2, Singapore, November 6–11, 2005

Margi, C.B.; Lu, X.; Zhang, G.; Stanek, G.; Manduchi, R. & Obraczka, K. (2006). Meerkats: A power-aware, self-managing wireless camera network for wide area monitoring, *Proceedings of International Workshop on Distributed Smart Cameras (DSC 06) in conjunction with SenSys06*, ISBN 1-59593-343-3, Boulder, CO, USA, October 31, 2006

Pahalawatta, P. V.; Pappas, T. N. & Katsaggelos, A. K. (2004). Optimal sensor selection for video-based target tracking in a wireless sensor network, *Proceedings of the International Conference on Image Processing (ICIP '04)*, pp. 3073– 3076, ISBN 0-7803-8554-3, Singapore, October 24-27, 2004

Park, J.; Bhat, P. & Kak, A. (2006). A look-up table based approach for solving the camera selection problem in large camera networks, *Proceedings of the International Workshop on Distributed Smart Cameras (DCS '06) in conjunction with SenSys06*, ISBN 1-59593-343-3, Boulder, CO, USA, October 31, 2006

Pereira, F.; Torres, L.; Guillemot, C.; Ebrahimi, T.; Leonardi, R. & Klomp, S. (2008). Distributed video coding: Selecting the most promising application scenarios. *Signal Processing: Image Communication*, Vol 23, Issue 5, (June 2008), pp. 339–352, ISSN 0923-5965

Pottie, G. & Kaiser, W. (2000). Wireless Integrated Network Sensors. *Communication of the ACM*, Vol. 43, N. 5, (May 2000), pp. 51-58

Raghunathan, V.; Ganeriwal, S. & Srivastava, M. (2006). Emerging techniques for long lived wireless sensor networks. *IEEE Communications Magazine*, Vol. 44, Issue 4, (April 2006), pp. 108- 114, ISSN 0163-6804

Raghunathan, V.; Schurghers, C.; Park, S. & Srivastava, M. (2002). Energy-aware Wireless Microsensor Networks. *IEEE Signal Processing Magazine*, Vol. 19, Issue 2, (March 2002), pp. 40-50, ISSN 1053-5888

Rahimi, M.; Baer, R.; Iroezi, O.I.; Garcia, J.C.; Warrior, J.; Estrin, D. & Srivastava, M. (2005). Cyclops: in situ image sensing and interpretation in wireless sensor networks, *Proceeding of the 3rd ACM Conference on Embedded Networked Sensor Systems (SenSys 05)*, pp.192–204, ISBN 1-59593-054-X ,San Diego, CA, USA, November 2–4, 2005

Rowe, A.; Rosenberg, C. & Nourbakhsh. I. (2002). A Low Cost Embedded Color Vision System. *Proceedings of the international IEEE/RSJ Conference on Intelligent Robots and Systems (IROS 2002)*, pp. 208-213, ISBN 0-7803-7398-7, Lausanne, Switzerland, Sep.30-Oct.4, 2002

Schott, B.; Bajura, M.; Czarnaski, J.; Flidr, J.; Tho, T. & Wang, L. (2005). A modular power-aware microsensor with >1000X dynamic power range, *Proceedings of the Fourth International Symposium on Information Processing in Sensor Networks (IPSN 2005)*, pp.469-474, ISBN 0-7803-9201-9, UCLA, Los Angeles, California, USA ,April 25-27 2005

Soro, S. & Heinzelman, W. (2007). Camera selection in visual sensor networks, *Proceedings of the IEEE Conference on Advanced Video and Signal Based Surveillance (AVSS '07)*, pp. 81–86, ISBN 978-1-4244-1696-7, London, UK, September 5- 7, 2007.

Soro, S. & Heinzelman, W. (2005). On the coverage problem in video-based wireless sensor networks. *Proceedings of the 2nd IEEE International Conference on Broadband Communications and Systems (BroadNets)*, pp. 932–939, ISBN 0-7803-9276-0, Boston, MA, USA, October 3–7, 2005

Tavli, B.; Bicakci, K.; Zilan, R. & Barcelo-Ordinas, J.M. (2011). A Survey of Visual Sensor Platforms. *Journal on Multimedia Tools and Applications*, ISSN 1573-7721, June 2011

Tezcan, N. & Wang, W. (2008). Self-orienting wireless multimedia sensor networks for occlusion-free viewpoints. *Computer Networks*, vol. 52, issue 13, (September 2008), pp. 2558–2567, ISSN 1389-1286

Zamora, N. H. & Marculescu, R. (2007). Coordinated distributed power management with video sensor networks: analysis, simulation, and prototyping, *Proceedings of the 1st ACM/IEEE International Conference on Distributed Smart Cameras (ICDSC '07)*, pp. 4–11, ISBN 978-1-4244-1354-6, Vienna, Austria, September 26-28, 2007.

http://www.ti.com/product/tmp103?DCMP=analog_signalchain_mr&HQS=Other%252bPR%252btmp103-pr

Permissions

The contributors of this book come from diverse backgrounds, making this book a truly international effort. This book will bring forth new frontiers with its revolutionizing research information and detailed analysis of the nascent developments around the world.

We would like to thank Dr. Ali Ekşim, for lending his expertise to make the book truly unique. He has played a crucial role in the development of this book. Without his invaluable contribution this book wouldn't have been possible. He has made vital efforts to compile up to date information on the varied aspects of this subject to make this book a valuable addition to the collection of many professionals and students.

This book was conceptualized with the vision of imparting up-to-date information and advanced data in this field. To ensure the same, a matchless editorial board was set up. Every individual on the board went through rigorous rounds of assessment to prove their worth. After which they invested a large part of their time researching and compiling the most relevant data for our readers. Conferences and sessions were held from time to time between the editorial board and the contributing authors to present the data in the most comprehensible form. The editorial team has worked tirelessly to provide valuable and valid information to help people across the globe.

Every chapter published in this book has been scrutinized by our experts. Their significance has been extensively debated. The topics covered herein carry significant findings which will fuel the growth of the discipline. They may even be implemented as practical applications or may be referred to as a beginning point for another development. Chapters in this book were first published by InTech; hereby published with permission under the Creative Commons Attribution License or equivalent.

The editorial board has been involved in producing this book since its inception. They have spent rigorous hours researching and exploring the diverse topics which have resulted in the successful publishing of this book. They have passed on their knowledge of decades through this book. To expedite this challenging task, the publisher supported the team at every step. A small team of assistant editors was also appointed to further simplify the editing procedure and attain best results for the readers.

Our editorial team has been hand-picked from every corner of the world. Their multi-ethnicity adds dynamic inputs to the discussions which result in innovative outcomes. These outcomes are then further discussed with the researchers and contributors who give their valuable feedback and opinion regarding the same. The feedback is then collaborated with the researches and they are edited in a comprehensive manner to aid the understanding of the subject.

Apart from the editorial board, the designing team has also invested a significant amount of their time in understanding the subject and creating the most relevant covers. They scrutinized every image to scout for the most suitable representation of the subject and create an appropriate cover for the book.

The publishing team has been involved in this book since its early stages. They were actively engaged in every process, be it collecting the data, connecting with the contributors or procuring relevant information. The team has been an ardent support to the editorial, designing and production team. Their endless efforts to recruit the best for this project, has resulted in the accomplishment of this book. They are a veteran in the field of academics and their pool of knowledge is as vast as their experience in printing. Their expertise and guidance has proved useful at every step. Their uncompromising quality standards have made this book an exceptional effort. Their encouragement from time to time has been an inspiration for everyone.

The publisher and the editorial board hope that this book will prove to be a valuable piece of knowledge for researchers, students, practitioners and scholars across the globe.

List of Contributors

Pero Latkoski and Borislav Popovski
Faculty of Electrical Engineering and Information Technologies / Ss Cyril and Methodius University – Skopje, Macedonia

Anum L. Enlil Corral-Ruiz and Felipe A. Cruz-Pérez
Electrical Engineering Department, CINVESTAV-IPN, Mexico

Genaro Hernández-Valdez
Electronics Department, UAM-A, Mexico

Vyacheslav Tuzlukov
Kyungpook National University, South Korea

Yao-Liang Chung and Zsehong Tsai
Graduate Institute of Communication Engineering, National Taiwan UniversityTaipei, Taiwan, R.O.C.

Panagiotis Lytrivis and Angelos Amditis
Institute of Communication and Computer Systems (ICCS), Greece

Niharika Kumar, Siddu P. Algur and Amitkeerti M. Lagare
RNSIT, BVB College of Engineering, Motorola Mobility, India

Lars Häring
Department of Communication Systems, University of Duisburg-Essen, Germany

Seyed Reza Abdollahi, H.S. Al-Raweshidy and T.J. Owens
WNCC, School of Eng. and Design, Brunel University, Uxbridge, London, UK

Itziar Salaberria, Roberto Carballedo and Asier Perallos
Deusto Institute of Technology (DeustoTech), University of Deusto, Spain

Luc Verschaeve
Scientific Institute of Public Health, O.D. Public Health and Surveillance, Brussels and University of Antwerp, Department of Biomedical Sciences, Belgium

Saleh Ali Alomari and Putra Sumari
Universiti Sains Malaysia, Malaysia

Mohammad Alaei and Jose Maria Barcelo-Ordinas
Computer Architecture Department, Universitat Politecnica de Catalunya, Barcelona, Spain

Printed in the USA
CPSIA information can be obtained
at www.ICGtesting.com
JSHW011503221024
72173JS00005B/1181